histology and embryology

The National Medical Series for Independent Study

histology and embryology

Kurt E. Johnson, Ph.D.
Associate Professor of Anatomy
George Washington University
Medical Center
Washington, D.C.

A WILEY MEDICAL PUBLICATION
JOHN WILEY & SONS
New York • Chichester • Brisbane • Toronto • Singapore

Harwal Publishing Company, Media, Pa.

Library of Congress Cataloging in Publication Data

Johnson, Kurt E.
Histology and embryology.

(The National medical series for independent study)
(A Wiley Medical publication)
Includes index.
1. Histology. 2. Embryology. I. Title. [DNLM:
Histology. 2. Histology—Examination questions.
3. Embryology. 4. Embryology—Examination questions.
QS 504 J67h]
QM553.J628 1984 611'.018 83-12717
ISBN 0-471-86826-4

2 3 4 5 6 7 8 9 10

This book is dedicated respectfully to Dr. Sherman J. Silber, FACS. He knows that the epididymis is a long coiled tube and not an anastomosing network of tubes—an impressive bit of anatomic knowledge.

Contents

	Preface	ix
	Acknowledgments	xi
	Introduction	xiii
	Pretest	1
1	Techniques Used in Microscopic and Developmental Anatomy	11
2	Cell Biology	19
3	Epithelium	29
4	Connective Tissue	39
5	Cartilage and Bone	51
6	Muscular Tissue	61
7	Nervous Tissue	71
8	Peripheral Blood	83
9	Bone Marrow and Hematopoiesis	89
10	Immune System	97
11	Cardiovascular System	109
12	Respiratory System	119
13	Upper Gastrointestinal Tract and Development of the Face	129
14	Esophagus and Stomach	139
15	Intestines	149
16	Liver, Gallbladder, and Pancreas	159

17	Skin	169
18	Thyroid and Parathyroid	181
19	Adrenal Glands	189
20	Pituitary Gland	197
21	Female Reproductive System	207
22	Male Reproductive System	223
23	Urinary System	235
24	The Eye	247
25	The Ear	257
	Post-test	265
	Index	279

Preface

Histology and Embryology evolved from class outlines and handouts that were distributed to first- and second-year medical students at Duke University and George Washington University. Each chapter covers the essential features of the microscopic anatomy of an organ or organ system and includes some discussion of developmental anatomy. Approximately 100 figures appear in *Histology and Embryology*, most of which are light and electron micrographs. The micrographs that are incorporated in the study questions give medical students valuable practice with the picture-type questions that appear in National Board examinations.

The author hopes that students find *Histology and Embryology* a useful and comprehensive study guide.

Kurt E. Johnson

Acknowledgments

I received a good deal of help in this project from my colleagues both past and present. Dr. Michael K. Reedy of Duke University provided me with a truly outstanding micrograph illustrating muscle tissue. I am especially grateful to my fellow anatomists at George Washington University for the micrographs they provided. These contributors include Dr. Ernest N. Albert, Dr. Frank Allan, Dr. Daniel P. DeSimone, Dr. Marilyn J. Koering, Dr. Jeffrey M. Rosenstein, Dr. Frank J. Slaby, and Dr. Raymond J. Walsh.

Dr. Mark R. Adelman of the Uniformed Services University of the Health Sciences read the manuscript. Dr. Helen A. Padykula of the University of Massachusetts donated a classic electron micrograph of muscular tissue. Dr. John A. Long of UCLA gave a fine electron micrograph of the ultrastructure of human adrenal cortex. Dr. Bela Gulyas of NICHD donated micrographs of corpus luteum.

I wish to acknowledge Lois Gottlieb for her help in locating specimens for me to photograph and Judy Gunther for her artwork. I also want to thank the secretarial staff in the Anatomy Department for their skillful typing. The editorial assistance of Debra L. Dreger also is gratefully acknowledged.

I appreciate all of these contributions and acknowledge that any errors contained herein are my own.

Introduction

Histology and Embryology is one of seven basic science review books in a series entitled *The National Medical Series for Independent Study*. This series has been designed to provide students and house officers, as well as physicians, with a concise but comprehensive instrument for self-evaluation and review within the basic sciences. Although *Histology and Embryology* would be most useful for students preparing for the National Board of Medical Examiners examinations (Part I, FLEX, and FMGEMS), it should also be useful for students studying for course examinations. These books are not intended to replace the standard basic science texts, but, rather, to complement them.

The books in this series present the core content of each basic science area using an outline format and featuring a total of 300 study questions. The questions are distributed throughout the book at the end of each chapter and in a pretest and post-test. In addition, each question is accompanied by the correct answer, a paragraph-length explanation of the correct answer, and specific reference to the outline points under which the information necessary to answer the question can be found.

We have chosen an outline format to allow maximum ease in retrieving information, assuming that the time available to the reader is limited. Considerable editorial time has been spent to ensure that the information required by all medical school curricula has been included and that each question parallels the format of the questions on the National Board examinations. We feel that the combination of the outline format and board-type study questions provides a unique teaching device.

We hope you will find this series interesting, relevant, and challenging. The authors, as well as the John Wiley and Harwal staffs, welcome your comments and suggestions.

Introduction

Pretest

QUESTIONS

Directions: Each question below contains five suggested answers. Choose the **one best** response to each question.

1. The tongue has filiform, fungiform, and circumvallate papillae. Which statement best describes these papillae?

(A) The predominant papillae are fungiform
(B) Fungiform papillae are located at the tongue root
(C) Circumvallate papillae are located all over the dorsal surface
(D) Circumvallate papillae contain taste buds
(E) Filiform papillae contain taste buds

2. The adrenal cortex shows a striking zonation. One of these zones, the zona glomerulosa, has cells distinguished by which of the following activities?

(A) Cortisol secretion
(B) Mineralocorticoid secretion
(C) Pregnenolone synthetase
(D) 17-α-Hydroxylase activity
(E) Response to adrenocorticotropic hormone

3. Which of the following statements is true concerning the presence of fixation artifacts in histologic specimens?

(A) Most preparative techniques do not introduce artifacts
(B) Live cells can be examined microscopically without fixation and thus without artifacts
(C) Glutaraldehyde fixation does not introduce artifacts
(D) Staining introduces artifacts under most circumstances
(E) The cell nucleus can be interpreted as a fixation artifact

Questions 4 and 5

The epithelium pictured below was taken from a 24-year old patient.

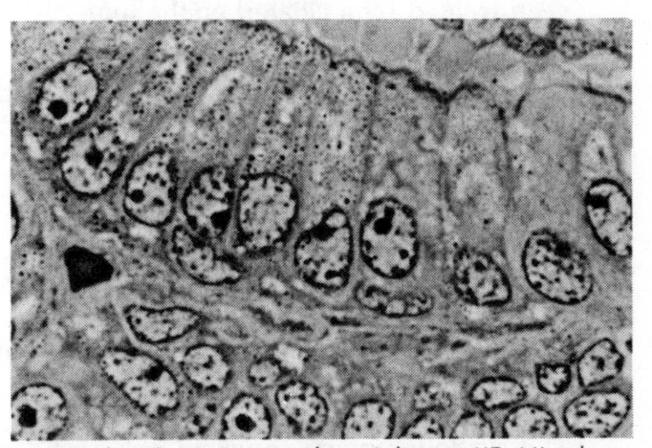

Reprinted with permission from Johnson KE: *Histology: Microscopic Anatomy and Embryology.* New York, John Wiley, 1981, p 219.

4. What type of epithelium is pictured above?

(A) Columnar
(B) Cuboidal
(C) Pseudostratified
(D) Transitional
(E) Stratified columnar

5. In which of the following sites in the body can this type of epithelium be found?

(A) Respiratory system
(B) Vagina
(C) Oviduct
(D) Seminal vesicle
(E) Duodenum

6. All of the following statements concerning granulopoiesis are true EXCEPT

(A) heterochromatinization of the nucleus occurs in association with specific granule accumulation
(B) metamyelocytes contain both specific and azurophilic granules
(C) specific granules invariably outnumber azurophilic granules
(D) promyelocytes lack specific granules
(E) myeloblasts lack specific granules

7. Which of the following statements best describes the prostate gland?

(A) It functions normally without androgenic stimulation
(B) Its secretions are rich in proteolytic enzymes
(C) It is lined by a ciliated epithelium
(D) It stores spermatozoa
(E) It contributes little to the volume of the ejaculate

8. All of the following statements describing parathormone are true EXCEPT

(A) it is a low molecular weight polypeptide hormone
(B) it is antagonistic to the effects of calcitonin
(C) it inhibits osteocytic osteolysis
(D) it promotes renal tubular resorption of calcium
(E) it stimulates renal tubular excretion of phosphate

9. The ciliary body, located between the edge of the retina and the edge of the lens, is characterized by all of the following EXCEPT

(A) an epithelium that contains cones
(B) an epithelium that secretes aqueous humor
(C) an epithelium that is continuous with the retina
(D) ciliary muscles
(E) ciliary processes that suspend the lens

10. Which of the following components of the respiratory system does not have ciliated cells in its epithelium?

(A) Olfactory mucosa
(B) Nasal cavity
(C) Trachea
(D) Bronchiole
(E) Alveolus

11. An acidophilic structure is best identified by

(A) a net negative charge
(B) staining with toluidine blue
(C) staining with eosin or orange G
(D) a positive Feulgen reaction
(E) a positive periodic acid-Schiff (PAS) reaction

12. Connective tissue has all of the following components EXCEPT

(A) cells derived from mesoderm
(B) cells that commonly secrete collagen
(C) extracellular fibers
(D) amorphous ground substance
(E) a basal lamina

13. A macrophage has all of the following features EXCEPT

(A) an irregularly shaped nucleus
(B) many lysosomes
(C) many surface projections
(D) few microfilaments
(E) the ability to engage in active pinocytosis

14. All of the following statements concerning T cells are true EXCEPT

(A) they are similar to B cells when seen with the light microscope
(B) they are derived from bone marrow stem cells and require thymosin for their differentiation
(C) they are abundant in the thymus and in the spleen
(D) they have fewer surface immunoglobulins than B cells
(E) they undergo functional maturation in the spleen

Directions: Each question below contains four suggested answers of which **one or more** is correct. Choose the answer

A if **1, 2, and 3** are correct
B if **1 and 3** are correct
C if **2 and 4** are correct
D if **4** is correct
E if **1, 2, 3, and 4** are correct

15. The central veins of the liver drain into which of the following vessels or structures?

(1) Biliary apparatus
(2) Hepatic connective tissue
(3) Hepatic sinusoids
(4) Sublobular veins

16. Components of adipose tissue include

(1) collagen fibers
(2) fibroblasts
(3) amorphous ground substance
(4) elastic fibers

17. A proximal convoluted tubule is characterized by

(1) an epithelial lining that has many microvilli
(2) cytoplasm that is strikingly eosinophilic and packed with mitochondria
(3) the ability to actively resorb protein from the glomerular filtrate by pinocytosis
(4) increased permeability to water when stimulated by antidiuretic hormone (ADH)

18. The dermis is characterized by

(1) a lack of chondroitin sulfate in the extracellular matrix
(2) avascularity
(3) greater thinness in thin skin than in thick skin
(4) numerous collagen fibers and elastic fibers

19. Features of endochondral bone formation include

(1) cartilage degeneration that is directly proportional to the increase in bone
(2) frequent cell division
(3) bones that grow in length and girth
(4) osteoprogenitor cells that differentiate into osteoblasts

20. Steps taken with freeze-fracture etching include

(1) glutaraldehyde fixation
(2) glycerol treatment to prevent the formation of ice crystals
(3) heavy metal coating to emphasize relief of specimens
(4) slow freezing (over 1 to 3 hours)

21. Hyaline cartilage matrix contains

(1) large amounts of collagen
(2) chondroitin sulfate
(3) glycoproteins
(4) many elastic fibers

22. The auditory tube of the ear is characterized by

(1) direct communication with the cavity containing the ossicles
(2) parts that are lined by pseudostratified epithelium
(3) mucous glands in the mucosa and lymphoid nodules in the lamina propria
(4) ceruminous glands

23. Testosterone is characterized by which of the following statements?

(1) It is made from cholesterol in the Leydig cells
(2) It is controlled by the hypothalamus
(3) It is required for spermatogenesis and prostatic secretion
(4) It stimulates luteinizing hormone releasing hormone (LH-RH) production

24. Chief cells are abundant in the fundic gastric glands and contain

(1) prominent nucleoli
(2) strongly basophilic cytoplasm
(3) well-developed Golgi apparatus
(4) abundant smooth endoplasmic reticulum

SUMMARY OF DIRECTIONS

A	B	C	D	E
1,2,3 only	1,3 only	2,4 only	4 only	All are correct

25. Components of the tracheal microanatomy include

(1) smooth muscle
(2) scattered mucous cells
(3) short cells
(4) adipose tissue

26. Blood thyroxine levels are regulated in a classic feedback-loop system. Components of this system include

(1) thyroid-stimulating hormone (TSH)
(2) thyrotropin releasing hormone (TRH)
(3) thyroxine
(4) thyroglobulin

27. Tears are characterized by which of the following statements?

(1) They are produced mainly by the lacrimal glands
(2) They are important for cleansing the eyes
(3) Secretions from the conjunctiva help prevent their evaporation
(4) Secretions from the meibomian glands help prevent their evaporation

28. Secretions of the enterochromaffin system include

(1) secretin
(2) polypeptide hormones
(3) serotonin
(4) glucagon

29. There are two types of bone marrow—red marrow and yellow marrow. How do these marrow types differ?

(1) Yellow marrow is more hematopoietically active than red marrow
(2) Yellow marrow contains more fat cells than red marrow
(3) Only yellow marrow is found in newborns
(4) Only red marrow is found in adult skull bones

30. Schwann cells are characterized by which of the following statements?

(1) They are neural crest derivatives
(2) They are responsible for myelination of peripheral neurons
(3) They are separated from one another by nodes of Ranvier
(4) They sometimes exhibit defects called Schmidt-Lantermann clefts

31. Components of mature enamel include

(1) glycoproteins
(2) collagen
(3) hydroxyapatite
(4) cellular processes

32. Eccrine sweat glands are found over most of the body. Features of these glands include

(1) abundance in the palm of the hand
(2) connection to the body surface by a coiled duct
(3) the ability to produce a copious secretion to cool the body
(4) the ability to produce a hypertonic solution of sodium, chloride, and urea

Directions: The groups of questions below consist of lettered choices followed by several numbered items. For each numbered item select the **one** lettered choice with which it is **most** closely associated. Each lettered choice may be used once, more than once, or not at all.

Questions 33–36

For each description of cardiovascular layers, select the appropriate layer or layers.

(A) Tunica intima
(B) Tunica adventitia
(C) Both
(D) Neither

33. Homologous to the endocardium
34. Homologous to the epicardium
35. Intrinsic blood vessels in the aorta found here
36. Absent in muscular arteries

Questions 37–40

For each description of cell characteristics, select the appropriate leukocytes.

(A) Granulocytes
(B) Agranulocytes
(C) Both
(D) Neither

37. Can be phagocytic
38. Lack specific granules
39. Lack lysosomes
40. Have lobulated nuclei with two to four lobes

Questions 41–43

For each description of cells involved in ovarian follicular development, select the appropriate cell type.

(A) Granulosa cells
(B) Granulosa-lutein cells
(C) Theca interna cells
(D) Theca-lutein cells
(E) None of the above

41. Form after ovulation and secrete progesterone
42. Secrete glycoproteins
43. Secrete androgens

Questions 44–48

For each of the following descriptions of sites or structures of the pancreatic acinar cell, select the appropriate lettered component shown in the micrograph below.

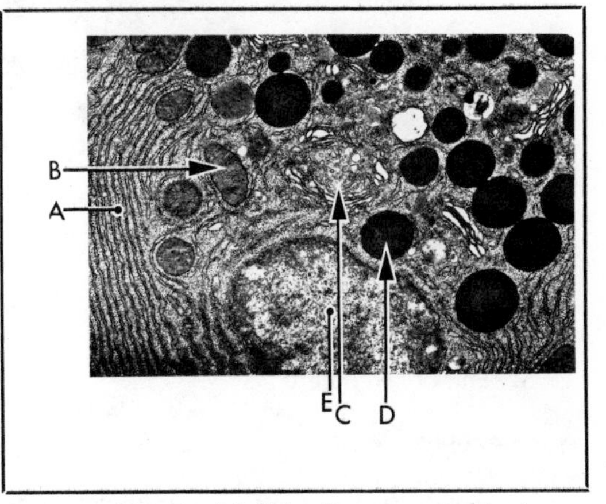

Courtesy of Dr. Frank J. Slaby, Department of Anatomy, George Washington University.

44. Granules of secretion product
45. Membranous structure involved in processing secretion products for transport
46. Site of oxidative phosphorylation
47. Site of polypeptide chain synthesis
48. Site of localization of glycosyltransferase enzymes

Questions 49–52

For each of the following characteristics of types of cells found in the hypophysis, select the appropriate cell type.

(A) Lactotrops
(B) Thyrotrops
(C) Both
(D) Neither

49. Present in the pars distalis
50. Present in the median eminence in the neurohypophysis
51. Produces a low molecular weight polypeptide hormone
52. Produces a high molecular weight glycoprotein hormone

Questions 53–57

For each description of different regions within a sarcomere, select the lettered area shown in the micrograph below with which it is most likely to be associated.

Courtesy of Helen A. Padykula, Department of Anatomy, University of Massachusetts.

53. Z line
54. I band
55. H band
56. Middle of sarcomere
57. End of sarcomere

Questions 58–60

For each description of components of an ileal villus, select the lettered structure in the micrograph below with which it is most likely to be associated.

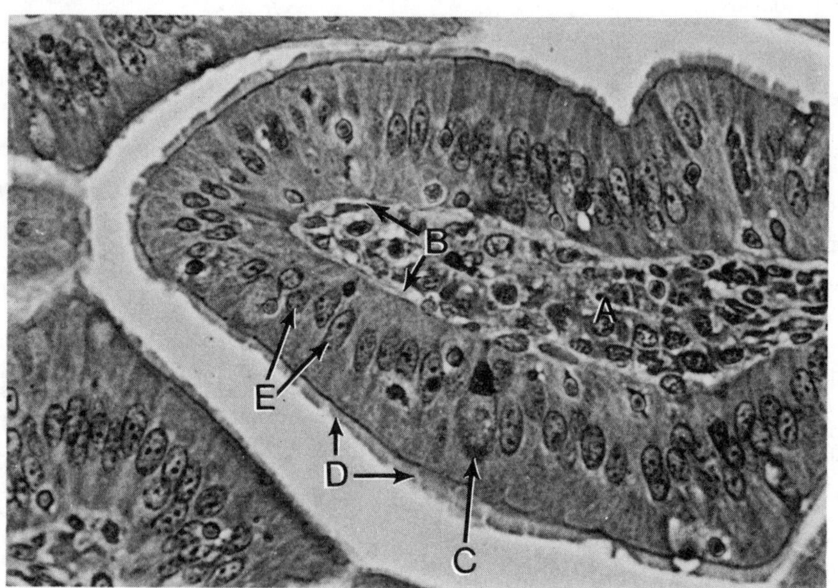

Reprinted with permission from Johnson KE: *Histology: Microscopic Anatomy and Embryology.* New York, John Wiley, 1981, p 219.

58. Protective mucus is secreted here
59. Microvilli here increase the surface area available for absorption
60. Polymorphonuclear leukocytes, plasma cells, and lymphocytes occur here

ANSWERS AND EXPLANATIONS

1. The answer is D. (*Chapter 13 IV B*) The tongue has three types of papillae. Filiform papillae, the most numerous, have no taste buds associated with them. Fungiform papillae are scattered among the filiform papillae on the dorsal surface of the tongue. Circumvallate papillae are restricted to the root of the tongue. Both fungiform and circumvallate papillae have taste buds.

2. The answer is B. (*Chapter 19 IV A 4*) The zona reticularis and zona fasciculata both secrete cortisol, under the influence of adrenocorticotropic hormone (ACTH). A key enzyme in cortisol synthesis, 17-α-hydroxylase, would be expected to occur in both zones. Pregnenolone synthetase is an enzyme active in an early stage of steroid biosynthesis, and its product, pregnenolone, is a precursor for the synthesis of many steroids. Thus, pregnenolone synthetase is widely distributed in the adrenal cortex. Aldosterone, a mineralocorticoid, is synthesized in the zona glomerulosa.

3. The answer is B. (*Chapter 1 IV D 1,4*) Almost all forms of preparative techniques for histologic examination of specimens produce some sort of artifact. Phase-contrast microscopy, on the other hand, allows scientists to examine live cells without fixation and therefore without the introduction of fixation artifacts. While staining increases the contrast of structures, in most instances it does not introduce artifacts. The cell nucleus is not, of course, an artifact.

4 and 5. The answers are: 4-C, 5-D. (*Chapter 3 III B*) This is an example of pseudostratified epithelium. It might be mistaken for columnar epithelium, but close inspection reveals the curved basal nuclei typical of pseudostratified epithelium. Notice the prominent clear area in the apices of the tall cells. This represents the dilated cisternae of the Golgi apparatus. This secretory epithelium is found in the seminal vesicle; it produces some of the material in the ejaculate. The pseudostratified epithelium of the respiratory system is ciliated.

6. The answer is C. (*Chapter 8 III B; Chapter 9 V B*) Azurophilic granules outnumber specific granules in the early stages of granulopoiesis but not in the late stages. Metamyelocytes have granules that are typical of a mature granulocyte but have an immature nuclear morphology. Myeloblasts and promyelocytes both lack specific granules.

7. The answer is B. (*Chapter 22 VIII B*) The prostate requires androgenic stimulation but does not actually store spermatozoa. The prostate has glands lined by a pseudostratified epithelium without cilia. Its secretions contribute a considerable amount to the volume of the ejaculate.

8. The answer is C. (*Chapter 18 IV E*) Parathormone is a low molecular weight hormone whose antagonistic effect on calcitonin causes a systemic increase in serum calcium. Parathormone promotes osteolysis and renal resorption of calcium, and it stimulates phosphate excretion.

9. The answer is A. (*Chapter 24 III B*) The ciliary body is covered by a double-layered epithelium. This epithelium is continuous with the retina but lacks photoreceptors characteristic of the more posterior part of the retina. Also, the ciliary epithelium secretes the aqueous humor. Many long ciliary processes project from the ciliary body toward the lens. The major components of the ciliary body, however, are the ciliary muscles. Cones are found in the retina, not in the ciliary body.

10. The answer is E. (*Chapter 12 II A, B; III A; IV B; V B*) The nasal cavity has a pseudostratified ciliated columnar (PCC) epithelium. The olfactory mucosa also has a PCC epithelium; here the epithelium is tall and modified for olfaction. The PCC epithelium of the trachea contains six types of cells, all of which help to perform the diverse functions of this tubular structure. Ciliated cells are commonly found in bronchioles. Alveolar epithelium, however, lacks ciliated cells and contains type I and type II cells.

11. The answer is C. (*Chapter 1 III B 1,3; C 2; E 1,5*) Acidophilic structures have a net positive charge and typically will stain with eosin or orange G, two negatively charged acidic dyes. Acidophilic structures will not stain with toluidine blue because toluidine blue is positively charged and so stains basophilic structures. There is no direct connection between acidophilia and the periodic acid-Schiff (PAS) status of a specimen, nor is there a connection between acidophilia and the Feulgen reaction, which is a test for DNA.

12. The answer is E. (*Chapter 4 VI C; VIII; XI A*) Practically all of the body's connective tissue is derived from mesoderm. Also, connective tissue has extracellular fibers, collagen fibers, and amorphous ground substance. A basal lamina is an epithelial, not connective tissue, characteristic.

13. The answer is D. (*Chapter 10 III A, B*) As part of its motility system, a macrophage has numerous microfilaments. Since it is an active phagocyte, this cell also has many lysosomes and surface projections and is active in pinocytosis. A macrophage has an irregularly shaped nucleus.

14. The answer is E. (*Chapter 10 IV A, C*) T cells undergo functional maturation in the thymus, where they are stimulated to differentiate by the thymic hormone, thymosin. T cells do not undergo functional maturation in the spleen, but they are abundant there and in the thymus.

15. The answer is D (4). (*Chapter 16 III A, B*) Blood coming to the liver flows from the branches of the portal veins, through the hepatic sinusoids, and into the central veins. The central veins empty into sublobular veins. The central veins do not nourish the connective tissue and biliary apparatus of the liver. This function is performed by branches of the hepatic artery.

16. The answer is E (all). (*Chapter 4 VI C; VII C; VIII; IX A*) Almost all types of connective tissues have at least some elastic fibers, and adipose tissue is no exception. Likewise, amorphous ground substance is found in varying amounts in all connective tissue types. In addition to these components, adipose tissue contains scattered collagen fibers and fibroblasts.

17. The answer is A (1, 2, 3). (*Chapter 23 III A 3, 4, 6*) The proximal convoluted tubule (PCT), like the distal convoluted tubule (DCT), has numerous apical microvilli. The PCT is very active in both ion and macromolecule transport (by pinocytosis) and has many mitochondria in its cells. These organelles produce adenosine triphosphate (ATP) for transport. Antidiuretic hormone (ADH) makes the collecting tubules more permeable to water, but it has no effect on either the PCT or the DCT.

18. The answer is D (4). (*Chapter 17 VI A–C*) The dermis of thin skin is thick, and the dermis of thick skin is thin. The dermis is highly vascular; its blood supply is important for thermoregulation. It contains collagen, hyaluronate, and chondroitin sulfate in its extracellular matrix. The dermis also has many elastic fibers.

19. The answer is E (all). (*Chapter 5 IX C*) During endochondral bone formation, cartilage is replaced by bone. Bones increase in length and girth partly as a result of cell division. Osteoprogenitor cells are the main source of osteoblasts.

20. The answer is A (1, 2, 3). (*Chapter 1 VI B*) In freeze-fracture etching, specimens that are unfixed or lightly fixed with glutaraldehyde are infiltrated first with glycerol. Rapid freezing of the water-glycerol solution in cells prevents ice crystal formation and reduces the destructive effects of freezing. After the frozen specimen has been fractured with a razor blade, it is coated with heavy metal to bring out details of its relief.

21. The answer is A (1, 2, 3). (*Chapter 5 III E*) Glycoproteins and proteoglycans are contained in the amorphous ground substance of cartilage. The matrix of hyaline cartilage contains one particularly prominent proteoglycan. It is a copolymer of protein with three glycosaminoglycans—chondroitin sulfate, hyaluronate, and keratan sulfate. This matrix also contains large amounts of type I collagen but only a few elastic fibers.

22. The answer is A (1, 2, 3). (*Chapter 25 II B; III D*) The auditory tube connects the middle ear to the pharynx and is lined, in places, by a pseudostratified epithelium. The auditory tube can have mucous glands and sometimes exhibits lymphoid nodules in close association with it; however, there are no ceruminous glands in the auditory tube.The external auditory meatus, an indentation in the side of the head, leads to the tympanic membrane and is lined by modified skin with ceruminous glands.

23. The answer is A (1, 2, 3). (*Chapter 22 V A 6*) Testosterone is made from cholesterol in the Leydig cells and is under the control of the hypothalamus. Spermatogenesis and prostatic secretion are testosterone-dependent. Testosterone inhibits luteinizing hormone releasing hormone (LH-RH) production.

24. The answer is A (1, 2, 3). (*Chapter 14 IV B 3 b*) Chief cells are abundant in the glands of the fundic portion of the stomach. These cells secrete pepsinogen, a protein, into the stomach. Like all cells specialized for the synthesis of protein for export, chief cells have prominent nucleoli, well-developed Golgi apparatus, and cytoplasmic basophilia due to an abundance of rough endoplasmic reticulum. Abundant smooth endoplasmic reticulum is a characteristic feature of steroid-secreting cells.

25. The answer is E (all). (*Chapter 12 III A, B*) The trachea is lined by a pseudostratified ciliated columnar epithelium that contains six types of cells. Two of these are mucous cells and short cells. Secretions from mucous cells form a continuous layer on the tracheal epithelium. Short cells are abundant and rest on the basement membrane of the epithelium. C-shaped rings of cartilage keep the trachea open; where the cartilage is absent there is a band of smooth muscle. The tracheal adventitia contains blood vessels, nerves, and adipose tissue.

26. The answer is E (all). (*Chapter 18 II D 3*) Blood thyroxine levels are regulated in a classic feedback-loop system. Low thyroxine levels indirectly stimulate the secretion of thyrotropin releasing hormone (TRH); the secretion of TRH then directly stimulates the secretion of thyroid-stimulating hormone (TSH). TSH stimulates thyroid follicular cells to engulf stored thyroglobulin, which is degraded into active thyroxine. The thyroxine then is secreted from the thyroid. When blood thyroxine levels are high again, TSH secretion is inhibited.

27. The answer is E (all). (*Chapter 24 IV A–C*) Tears are secreted by lacrimal glands and help to cleanse the cornea. Oily secretions from the meibomian glands and mucus from the conjunctival glands form a thin film on the tears, helping to prevent evaporation of tears.

28. The answer is E (all). (*Chapter 14 V B*) The enterochromaffin system is a group of cells scattered throughout the epithelium of the gastrointestinal tract. These cells secrete a collection of different polypeptide hormones including secretin, cholecystokinin (a polypeptide hormone), serotonin, and glucagon. The enterochromaffin cells function in the regulation of gastrointestinal tract motility.

29. The answer is C (2, 4). (*Chapter 9 II A, B 4*) Red marrow is characterized by active hematopoiesis. Yellow marrow is hematopoietically inactive; lipid-laden fat cells far outnumber hematopoietic cells in yellow marrow. Practically all marrow in newborns is red; however, yellow marrow replaces most red marrow by the time an individual reaches puberty. In an adult, red marrow is found only in the skull bones, clavicle, vertebrae, sternum, and pelvic bones.

30. The answer is E (all). (*Chapter 7 I E; VI D*) Schwann cells are derived from neural crest and form myelin sheaths around the axons of peripheral neurons. Gaps between Schwann cells are called nodes of Ranvier. Defects in the myelin sheaths are called Schmidt-Lantermann clefts.

31. The answer is B (1, 3). (*Chapter 13 III B, C*) Enamel covers the crown of every tooth and is a secretion product of ameloblasts. Its peculiar hardness is a result of its lack of cellular processes. Enamel is composed of proteins, glycoproteins, and hydroxyapatite, and it is chemically distinct from dentin. Dentin, not enamel, contains collagen.

32. The answer is A (1, 2, 3). (*Chapter 17 III A 2*) Eccrine sweat glands are found over most of the body, including the palms of the hands. These glands have coiled ducts and produce a large volume of sweat when the body becomes overheated. Sweat is a hypotonic, not hypertonic, solution of water, sodium chloride, urea, and other components.

33–36. The answers are: 33-A, 34-B, 35-B, 36-D. (*Chapter 12 II C; IV; V*) The tunica adventitia and the epicardium both are located at the abluminal portion of the cardiovascular system. They are similar histologically. The tunica intima and the endocardium both are located at the luminal portion of the cardiovascular system. They are similar histologically. The tunica adventitia contains the vasa vasorum.

The vasa vasorum comprise the intrinsic blood supply of the large vessels such as the aorta. Muscular arteries have both an intima and an adventitia.

37–40. The answers are: 37-C, 38-B, 39-D, 40-A. (*Chapter 8 III B, C*) Neutrophils are examples of granulocytes and monocytes are examples of agranulocytes; both are phagocytes. All leukocytes have a nuclear envelope and mitochondria. Granulocytes contain specific granules and, like agranulocytes, have many lysosomes. Unlike agranulocytes, granulocytes have several prominent nuclear lobes.

41–43. The answers are: 41-B, 42-A, 43-C. (*Chapter 21 III B 1–4, II; III C–E; III F 2 a, b*) Granulosa cells secrete glycoproteins into the spaces surrounding them in primary ovarian follicles. The glycoproteins coalesce into Call-Exner bodies, which in turn are thought to fuse to form the antrum of the secondary follicle. In the preovulatory follicle, the granulosa cells secrete liquor folliculi, which accumulates in the antrum and causes its volume to increase. A mound of granulosa cells, called the cumulus oophorus, surrounds the oocyte in the preovulatory follicle. Theca interna cells secrete androgens, which are converted to estrogens in the granulosa cells. At ovulation, the mature follicle ejects the oocyte together with an attached layer of granulosa cells called the corona radiata. The follicle then becomes a corpus luteum, formed from both granulosa and theca interna cells. The granulosa cells become granulosa-lutein cells, cease to secrete glycoproteins, and begin to secrete progesterone. The theca interna cells become theca-lutein cells, which are thought to secrete steroids other than progesterone.

44–48. The answers are: 44-D, 45-C, 46-B, 47-A, 48-C. (*Chapter 2 III B; V B; VIII B*) This electron micrograph is of a pancreatic acinar cell, a classic example of a cell specialized for protein synthesis. It has ribosome-studded endoplasmic reticulum for protein synthesis and mitochondria for adenosine triphosphate (ATP) production. It also has granules of secretion product and a Golgi apparatus with glycosyltransferases for adding carbohydrate moieties to products destined for secretion.

49–52. The answers are: 49-C, 50-D, 51-A, 52-B. (*Chapter 20 V B 3 c, d*) Lactotrops are acidophils that secrete a low molecular weight polypeptide hormone called prolactin. Thyrotrops are basophils that secrete a high molecular weight glycoprotein hormone called thyroid-stimulating hormone (TSH). Both lactotrops and thyrotrops are found in the pars distalis of the adenohypophysis.

53–57. The answers are: 53-A, 54-B, 55-D, 56-E, 57-A. (*Chapter 6 IV B 1*) Thin filaments insert at the Z line (A) and are the sole component of the I band (B). There is no overlap between thick and thin filaments in the H band (D). The M line (E) is the middle of the sarcomere, and the Z line is the end of the sarcomere. As the extent of thick and thin filament overlap changes during the contraction and relaxation cycle, the I band changes length.

58–60. The answers are: 58-C, 59-D, 60-A. (*Chapter 15 II C 3*) The ileal villus is a connective tissue core of lamina propria (A), where formed elements of the blood can be found. The epithelium covering the villus is composed of tall columnar cells (E), with many associated microvilli in a brush border (D) and mucus-secreting goblet cells (C) resting on a basement membrane (B).

1 Techniques Used in Microscopic and Developmental Anatomy

I. INTRODUCTION

A. SPECIMEN PREPARATION

B. STAINING METHODS

C. METHODS USED TO EXTEND HUMAN PERCEPTION.

C. METHODS USED TO EXTEND HUMAN PERCEPTION. The following methods are used to diagnose disease, to learn more about structure-function relationships, and to elucidate causes of disease of unknown etiology in order to cure disease.

1. **Microscopy.**
   a. Light Microscopy.
      (1) Brightfield
      (2) Phase contrast
      (3) Differential interference contrast
   b. Electron Microscopy.
      (1) Transmission electron microscopy (TEM)
      (2) Scanning electron microscopy (SEM)
   c. Freeze-Fracture Etching (FFE).
2. **Differential Centrifugation.**

II. SPECIMEN PREPARATION

A. FIXATION

1. For light microscopy, specimens usually are immersed in a solution of an appropriate buffer and formaldehyde.
2. For electron microscopy, specimens usually are immersed in a solution of buffer and glutaraldehyde followed by several rinsing steps and postfixation in buffered osmium tetroxide.

B. DEHYDRATION

1. Dehydration usually is done with increasing concentrations of **ethanol** known as a graded ethanol series.
2. Dehydration can be done with other organic solvents such as **acetone** or dioxane.
3. Dehydration introduces some artifacts (e.g., lipid extraction).

C. EMBEDDING

1. **Paraffin** is used for light microscopy.
2. **Epoxy** resins are used for TEM.
3. For SEM specimens, embedding is not performed.

D. SEM PREPARATION

1. Following dehydration in a graded series of ethanol, specimens are **critical point dried**, which serves to dehydrate the specimens without subjecting them to the surface tension

forces found at liquid-gas interfaces. Specimens are infiltrated with carbon dioxide under a high pressure which keeps the carbon dioxide in a liquid state. The dehydrating agent, namely ethanol, is freely miscible with the liquid carbon dioxide. Following complete exchange of the acetone with carbon dioxide, the pressure on the specimen is gradually released, allowing the liquid carbon dioxide to evaporate and leaving a completely dehydrated specimen.

2. Following critical point drying, specimens are coated with a thin layer of metal in a vacuum apparatus. Typically, the metals used are mixtures of gold and palladium.

E. SECTIONING

1. Specimens for light microscopy, after being embedded in paraffin, are cut into thin sections, typically **5–10 μm** thick, with a sharp metal knife or razor blade.

2. Specimens for TEM are cut into extremely thin sections, typically **100 nm** or less in thickness, with sharp glass or diamond knives.

III. STAINING METHODS AND TERMINOLOGY

A. GENERAL INFORMATION

1. The function of staining is to increase the inherent contrast of a specimen.

2. Specimens for **light microscopy** are stained with various combinations of dyes which bind to proteins, nucleic acids, and other macromolecules in the section.

3. Specimens for **electron microscopy** are stained with solutions of heavy metals such as lead and uranium. These heavy metals increase the electron density of materials in the section and thus increase the electron scattering that occurs when the specimen is irradiated with a beam of electrons in the microscope.

4. **Histochemical staining** often is used to localize certain substances, enzymatic activities, or antigens in a section.

B. ACIDOPHILIA

1. Some tissue components are said to be acidophilic, exhibiting acidophilia, because they **bind acidic dyes** such as **eosin** or **orange G**.

2. Acidophilia and eosinophilia often are used interchangeably although they are not exactly the same, as not all acidophilic substances are eosinophilic.

3. Acidophilic substances bear a net **positive charge** and bind negatively charged (i.e., acidic, anionic) dyes.

C. BASOPHILIA

1. Some tissue components are basophilic, exhibiting basophilia.

2. These components **bind basic dyes** such as **hematoxylin, methylene blue**, and **toluidine blue**.

3. Basophilic substances bear a net **negative charge** and bind positively charged (i.e., basic, cationic) dyes.

4. RNA has a net negative charge because of its phosphate groups and so binds a dye like toluidine blue. Thus, the cytoplasm of cells with large amounts of endoplasmic reticulum, studded with ribosomes containing RNA, is described by histologists as basophilic.

D. METACHROMASIA

1. Some substances, such as cytoplasmic granules in mast cells, are metachromatic (or said to exhibit metachromasia) when stained.

2. Metachromasia, which literally means a change in color, is a property of certain dyes, usually basic ones.

3. For example, when toluidine blue is bound in tissue to a substance that occurs at a low concentration, it stains that substance blue, but when it is bound to a substance that occurs at a high concentration, it stains that substance purple (i.e., metachromatically).

4. The molecular basis of this phenomenon can be understood by considering the staining of cartilage.

a. The extracellular matrix of cartilage contains a high concentration of a compound called chondroitin sulfate, which bears a highly negative net charge because of a large number of sulfate groups. The nuclei of cells in cartilage, however, contain a low concentration of negatively charged molecules, such as DNA. The nuclei of chondrocytes stain blue; the extracellular matrix surrounding them stains purple.
b. This change in color occurs because at high concentrations there are interactions between the electrons of closely packed dye molecules which cause their absorption maxima to shift to different wavelengths. Dilute solutions of toluidine blue actually are blue, that is, orthochromatic or of normal color; concentrated solutions are purple, that is, metachromatic or of changed color.

E. PERIODIC ACID-SCHIFF (PAS) REACTION

1. This test is used to identify carbohydrates by exposing a tissue section to periodic acid oxidation followed by staining with Schiff's reagent.

2. The reaction can be used to identify **glycogen**, a glucose polymer stored in the cytoplasm of many different cells.
   a. The hydroxyl groups of the glucose in glycogen are oxidized to aldehydes by the periodate oxidation.
   b. The free aldehydes react strongly with bisulfite groups on colorless leukofuchsin in Schiff's reagent to yield a magenta condensation product. (Schiff's reagent is basically just leukofuchsin in solution.)

3. In addition to glycogen, several other structures within cells are PAS-positive.
   a. Many cells contain a carbohydrate-rich coat called the **glycocalyx**, which is strongly PAS-positive because it contains many glycoproteins.
   b. Nearly all epithelia rest on a **basement membrane**, which is strongly PAS-positive because it contains numerous glycoconjugate-rich macromolecules.

4. Glycogen can be distinguished from other PAS-positive substances by means of a histochemical reaction in which slides are preincubated with the enzyme α-amylase. Because this enzyme selectively destroys glycogen, all PAS-positive, α-amylase-sensitive structures in cells can be identified as glycogen.

5. Mild acid hydrolysis of DNA creates Schiff's reagent-reactive groups (or aldehydes) in the deoxyribose of DNA and provides the chemical basis of a specific and sensitive test for DNA called the **Feulgen reaction**.

F. IMMUNOHISTOCHEMISTRY. Specific antibodies to certain tissue antigens can be tagged with probes. Various antigens then can be localized within tissue sections by incubating the sections in solutions that contain the tagged antibodies.

1. For light microscopy, antibodies commonly are tagged with **fluorescent probes** (or dye molecules). Specimens then are illuminated with a high intensity light source that has the appropriate filters to create the correct excitatory wavelengths and a barrier filter to filter out all the light emitted from the specimen except for the emission wavelengths of the fluorescent tags.

2. For electron microscopy, electron-dense probes such as **ferritin** can be attached to specific antibodies.

G. ENZYME HISTOCHEMISTRY can be used to localize enzymes within tissues.

1. Frozen sections are incubated in reaction mixtures that create reaction products from enzymatic activities in the specimens. Precipitation of the reaction products then is created in such a way as to minimize diffusion of the reaction products away from the site of production.

2. For example, alkaline phosphatase activity in the renal brush border can be demonstrated by incubating frozen sections with phosphorylated substrates in alkaline buffers and then precipitating the phosphates with lead salts which show black or brown in the sections.

IV. LIGHT MICROSCOPY

A. COMPONENTS OF A LIGHT MICROSCOPE

1. Light source for specimen

2. Substage condenser for gathering light into objectives

3. Stage for holding and moving specimen

4. Objectives for image formation and magnification
5. Oculars for image formation and magnification

B. NUMERICAL APERTURE

1. Numerical aperture (NA) = n sin θ, where **n** is the refractive index of the medium between a slide and objective, and θ is one half of the angle of the cone of light gathered by the objective.
2. Low-power objectives have a relatively low NA, whereas high-power objectives have a relatively high NA.
3. The reason it is desirable to use objectives with a high NA lies in the theory of image formation.

C. THEORY OF IMAGE FORMATION FOR BRIGHTFIELD MICROSCOPY

1. Each specimen produces a complex diffraction pattern in the back focal plane of the microscope.
2. This diffraction pattern is resynthesized into a recognizable magnified image by the oculars.
3. Electromagnetic radiation interacts with all matter by diffraction. Regular objects, such as diffraction gratings, produce a simple diffraction pattern.
4. The spacings between periodic structures in the diffraction grating are inversely proportional to the angle of diffracted rays.
5. Closely spaced diffraction gratings produce a diffraction pattern where the zero-order, first-order, and all subsequent diffraction fringes are widely spaced.
6. Widely spaced diffraction gratings produce a diffraction pattern where the different orders of diffraction fringes are closely spaced.
7. High-magnification lenses with a high NA make small structures larger and also make fine details resolvable.
8. Oil-immersion objectives employ a medium called immersion oil between a specimen and the objective. Immersion oil has a refractive index of about 1.5 (i.e., higher than air and matched to optical glass).
9. The use of oil-immersion objectives allows high-NA lenses to gather diffracted informational rays from the finer details of a specimen, thus improving resolution.

D. PHASE-CONTRAST MICROSCOPY

1. This technique allows direct examination of living cells without fixation or staining.
2. Different organelles and domains within a cell have chemical differences that are reflected in slight differences in refractive indices.
3. Phase-contrast microscopy employs specially designed substage condensers and objectives that convert slight differences in the refractive index into noticeable differences in the relative light intensity passing through a specimen.
4. This technique is completely harmless to living cells and allows direct observation of cell structure such as nuclei and mitochondria.

E. DIFFERENTIAL INTERFERENCE CONTRAST MICROSCOPY

1. This technique converts differences in the refractive index in a specimen into an appearance of low relief. Thus, the nucleus and various particulate cytoplasmic inclusions appear to have depth to them.
2. This optical trickery is quite useful for making detailed observations on living cells.

V. ELECTRON MICROSCOPY

A. GENERAL COMMENTS

1. The purpose of the electron microscope is to view fixed and sectioned or metal-coated specimens under high magnification with resolution of fine detail.
2. The electron microscope illuminates a specimen with a stream of electrons with short

wavelengths rather than with photons, and the lenses used for image formation are magnetic rather than glass.

3. The short wavelengths of electrons give a high theoretical resolving power to the electron microscope. Nevertheless, the actual resolving power in a modern electron microscope is typically less than 0.5 nm; this resolution, however, is easily high enough to examine all cellular organelles and some macromolecules in detail.

B. TRANSMISSION ELECTRON MICROSCOPY (TEM)

1. Thin plastic sections are stained with heavy metal salts such as lead citrate, uranyl acetate, or both after being suspended between the wires of a small circular metal screen.
2. The specimen scatters electrons and causes the formation of an image.
3. The image then is recorded on photographic plates which later can be printed to make electron micrographs.

C. SCANNING ELECTRON MICROSCOPY (SEM)

1. A solid specimen is processed and mounted on a metallic specimen holder.
2. After being placed in the microscope and bombarded with electrons, an image is formed on a screen and usually photographed to generate a permanent record of the specimen's appearance.
3. Electron micrographs show fine details of a specimen, down to the molecular level of resolution in some cases.
4. Because of the differences in the way specimens are irradiated and the way in which images are formed, TEM gives higher resolution information than SEM on fine details of the structure of cell slices, revealing the morphology of the surfaces (i.e., in cross section) and the internal elements of cells as well.
5. SEM reveals the three-dimensional character of the surface features of cells and organisms at lower resolution than TEM.

VI. FREEZE-FRACTURE ETCHING (FFE)

A. RATIONALE OF TECHNIQUE

1. FFE is a relatively new technique but one that has become increasingly important to investigators in cell biology, especially because of its ability to reveal details about membrane structure and intercellular junctions.
2. With FFE, the disruption of specimens by fixatives is reduced to a minimum, and, presumably, the introduction of artifacts by preparative techniques also is lessened.

B. METHODOLOGY

1. Specimens that are unfixed or lightly fixed with glutaraldehyde are infiltrated with glycerol and then mounted on metal blocks.
2. Next, the specimens are rapidly cooled in liquid nitrogen; the rapid freezing of the water-glycerol solution in the cells prevents ice crystal formation and reduces the destructive effects of freezing.
3. The frozen specimen then is placed in an apparatus that can fracture the tissue with a razor blade inside a vacuum of approximately 10^{-8} mm Hg. The blade strikes the specimen and causes a fracture to pass through it, much like breaking glass. The fracture plane passes between the inner and outer leaflets of the unit membranes of cells and along the cells' surfaces, often revealing internal details of membrane structure and allowing high-resolution views of the surfaces of cells.
4. After the specimen has been fractured, it is left for a variable amount of time in the vacuum to allow sublimation of some of the ice, revealing deeper surface details of the specimen.
5. Next, a layer of heavy metal is evaporated onto the exposed surface of the specimen from a point source filament at an angle to it. This shadows the surfaces and makes the details of relief more visible.
6. The specimen then is removed from the FFE apparatus and thawed, and its tissue is digested chemically, leaving behind a metal replica.

7. The metal replica of surface details then is examined with the TEM without other treatment.

VII. AUTORADIOGRAPHY

A. ISOTOPE INCORPORATION

1. Radioactive isotopes of cell components can be introduced into a cell and subsequently will be used in the metabolic events within the cell.
2. For example, to detect cells involved in DNA synthesis in a particular time window within an animal, a labeled precursor of DNA (e.g., tritiated thymidine) could be supplied in which tritium, the radioactive isotope of hydrogen, has been chemically substituted for normal hydrogen. The tritiated thymidine will be treated like normal hydrogenated thymidine by the cell's metabolic machinery, and the labeled precursor will become incorporated into the DNA of the cell.

B. SPECIMEN PREPARATION

1. The labeled macromolecule then can be fixed in place, the specimen sectioned, and the sections coated in the dark with a silver halide emulsion that is sensitive to the β particles released by the radioactive decay in the tritium atoms.
2. After an exposure period ranging from days to months, the exposed emulsions can be developed in much the same manner as any other photographic material, and metallic silver grains will be deposited in the emulsion layer directly above the site of the decaying isotope.
3. With this technique, cells involved in DNA synthesis can be localized in any part of the body.
4. It also is possible to follow the movement of labeled cells from one place to another in an organism by simply varying the amount of time that elapsed between the initial labeling period, called a "pulse," and the fixation needed to terminate the experiment.
5. During this "chase" period, labeled cells might migrate from the site of mitosis (e.g., in the intestinal crypts) to the eventual designation (e.g., on the tips of the intestinal villi).
6. Autoradiographs can be examined using electron microscopy, which provides much higher resolution than light microscopy, to determine the initial site of synthesis, pathways through the cellular organelles, and the eventual destination of cellular secretion products.

STUDY QUESTIONS

Directions: The question below contains five suggested answers. Choose the **one best** response to the question.

1. All of the following statements concerning the periodic acid-Schiff (PAS) reaction are true EXCEPT

(A) it requires oxidation of a tissue section before treatment with Schiff's reagent to be effective
(B) it gives a magenta stain to a wide variety of glycoconjugates
(C) it will stain the glycocalyx
(D) it will stain the basement membrane in many locations
(E) it can be used with α-amylase to demonstrate DNA

Directions: The group of questions below consists of lettered choices followed by several numbered items. For each numbered item select the **one** lettered choice with which it is **most** closely associated. Each lettered choice may be used once, more than once, or not at all.

Questions 2–6

For each of the following descriptions of parts of a light microscope, select the appropriate lettered structure shown in the adjacent photograph.

2. This structure is used to focus the image

3. This structure, which is involved in image formation, has a characteristic numerical aperture and gathers light rays

4. A specimen rests on this structure, which is used also to move the specimen about

5. This structure forms an image from diffracted rays in the back focal plane of the microscope

6. This structure supplies the light for illumination of a specimen

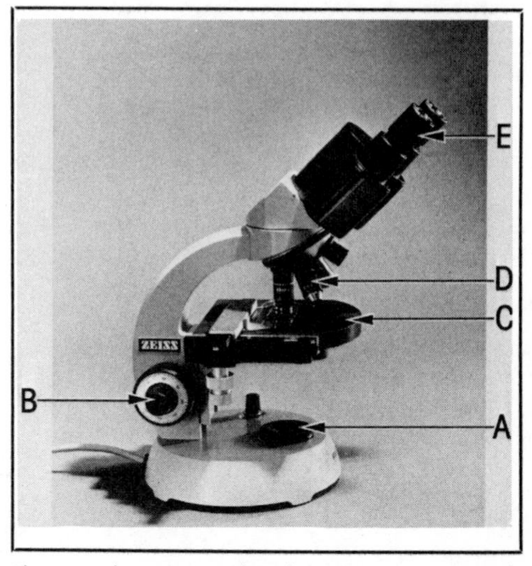

Photograph courtesy of Carl Zeiss, Inc., New York.

ANSWERS AND EXPLANATIONS

1. The answer is E. (*III E*) The periodic acid-Schiff (PAS) reaction is used to identify glycoconjugates by exposing a tissue section to oxidation followed by staining with Schiff's reagent. If the test is positive, it yields a magenta condensation product. PAS-positive substances include the glycocalyx and the basement membrane. The PAS reaction combined with α-amylase provides a specific histochemical test to demonstrate glycogen; the Feulgen reaction uses Schiff's reagent to demonstrate DNA.

2–6. The answers are: 2-B, 3-D, 4-C, 5-E, 6-A. (*IV A; IV C 1–2*) In a microscope, light coming from the light source (A) is gathered by a substage condenser. The light then passes through a specimen, which is placed on the stage (C), and the light rays are diffracted. Next, diffracted rays are collected by one of the objectives (D) and projected into the back focal plane of the microscope. The ocular (E) is located above the back focal plane. The complex diffraction pattern formed in the back focal plane is reformed into a visible magnified image by the ocular; the image is focused with (B).

2
Cell Biology

I. INTRODUCTION

A. RECENT ADVANCES. Intracellular components, called **organelles**, are understood now on a structure-function level that was only dimly appreciated before the modern era of cell biology.

B. TECHNIQUES OF STUDY

1. Electron microscopy
2. Cell fractionation by differential centrifugation
3. Biochemical analysis of cell fractions
4. Autoradiography
5. Enzyme histochemistry

C. PLASMA MEMBRANE

1. Each eukaryotic cell is surrounded by a semipermeable, lipoprotein bilayer called the plasma membrane (also called cell membrane or unit membrane).
2. This membrane represents the boundary between an individual cell and the outside world.
3. Organelles deep within the cell communicate with the plasma membrane either directly or functionally.

D. ENDOPLASMIC RETICULUM

1. The endoplasmic reticulum is a network of anastomosing sheets and tubes of membrane; its structure is similar to that of plasma membrane and other membranous organelles in the cell.
2. A rough endoplasmic reticulum, studded with ribosomes, is the principal site of synthesis for proteins that are exported from the cell.
3. A smooth endoplasmic reticulum, with no ribosomes attached, contains an array of enzymes involved in different functions, depending on physiologic state and cell type.

E. RIBOSOMES

1. Ribosomes exist freely within the cytoplasm, either alone or in small clusters called **polysomes**.
2. These free ribosomes are especially prominent in cells that manufacture protein destined for intracellular use rather than extracellular secretion.

F. GOLGI APPARATUS

1. The Golgi apparatus is another membranous organelle composed of many small vesicles and several flattened sacs of membrane.
2. This site of final processing of many glycosylated secretory or surface proteins is involved in the synthesis and turnover of the cell surface.

G. NUCLEAR ENVELOPE

1. The nuclear envelope is a double-membrane structure with nuclear pores.
2. It encloses the **nucleus** and so contains the genetic apparatus.
3. It is also the site of **cell division**.

H. MITOCHONDRIA

1. Mitochondria are bound by a double-membrane system.
2. They produce adenosine triphosphate (ATP) for the oxidative phosphorylation of reduced nucleotides produced by metabolism of glucose and other oxidizable substrates.

I. LYSOSOMES

1. Lysosomes are membrane-bound packets of hydrolytic enzymes.
2. They are used to degrade foreign ingested materials or to break down intracellular structures.

J. MICROTUBULES AND MICROFILAMENTS

1. Microtubules
2. Microfilaments
3. Intermediate filaments

K. CENTRIOLES

L. CYTOPLASMIC INCLUSIONS

1. Glycogen (a polymeric storage form of glucose)
2. Melanosomes (pigment granules)
3. Lipofuscin granules (poorly understood cell-degeneration products)

M. CELL SPECIALIZATION

II. PLASMA MEMBRANE

A. STRUCTURE

1. The plasma membrane looks relatively simple in the electron microscope, but recently it has been appreciated as extremely complex.
2. All cellular membranes are composed of lipids and proteins.
   a. The lipid components are predominately phospholipids such as phosphatidyl choline and cholesterol.
      (1) The phospholipids are **amphipathic**; that is, they are molecules with both **hydrophilic** and **hydrophobic** portions.
      (2) The phospholipids prevent free diffusion of ions and other small molecules.
      (3) The basic structure of a plasma membrane is a lipid bilayer.
         (a) The hydrophilic portions of the phospholipid are oriented toward the external environment and the internal cell sap.
         (b) The hydrophobic portions of the phospholipid bilayer face one another and interact strongly.
         (c) Cholesterol molecules are dispersed throughout the hydrophobic phospholipids.
   b. The proteins of plasma membranes also are amphipathic.
      (1) Their hydrophilic portions face the external and internal aqueous environments and are relatively rich in charged amino acids such as glutamic acid. The external hydrophilic portions of many plasma-membrane proteins also are glycosylated to a variable extent.
      (2) In contrast, the hydrophobic portions are embedded in the internal hydrophobic portions of the phospholipid bilayer and are relatively rich in nonpolar amino acids such as leucine.
   c. Recent evidence suggests that some proteins penetrate the entire thickness of the phospholipid bilayer while others penetrate only partly through it.
3. A fluid mosaic model of the membrane structure shows large protein molecules floating in a phospholipid lake, free to diffuse laterally in the plane of the membrane. This picture helps explain recent observations that various surface receptors diffuse in the membrane plane. Surface receptors also are actively moved about on the cell surface by the internal cytoskeletal elements.
4. Certain blood group substances actually are glycolipids with their lipid portion buried in the hydrophobic domain of the membrane and their carbohydrate portion projecting out into the aqueous pericellular environment.

B. FUNCTION

1. As the outer layer of the cell, the plasma membrane is a selective permeability barrier that excludes certain substances and transports others either into or out of the cell by energy-dependent processes.
2. It is a site of hormone receptors.
3. The membrane recognizes molecules in the environment of the cell such as antigens.
4. Specific cell-to-cell interactions occur at the plasma membrane.

C. **FREEZE-FRACTURE ETCHING (FFE)** technique has been instrumental in revealing new details of membrane structure (Fig. 2-1).

1. The fracture plane passes **between** the inner and outer leaflets of the plasma membrane and **through** the hydrophobic domain created by the phospholipids.
2. Studies, in which the external surface of a cell has been marked prior to freezing, reveal that the membrane face lying on the protoplasm (called the PF or **P-face**) is rich in intramembranous particles.
3. The membrane face lying next to the external world (called the EF or **E-face**) is sparsely dotted with intramembranous particles.
4. Intramembranous particles most likely represent proteins that pass more or less through the phospholipid bilayer.
5. They also may serve as anchorage sites for some components of the cytoplasmic filament networks.
6. The use of FFE terminology also has generated a name for the inner surface of the plasma membrane (PS or **P-surface**), which faces the protoplasm, as well as a name for the true external surface of the cell (ES or **E-surface**).
7. The structure of the plasma membrane as described above is also accurate, with minor exceptions, for the membranes of the endoplasmic reticulum, Golgi apparatus, nuclear envelope, mitochondria, and various membrane-delimited cytoplasmic bodies such as lysosomes.

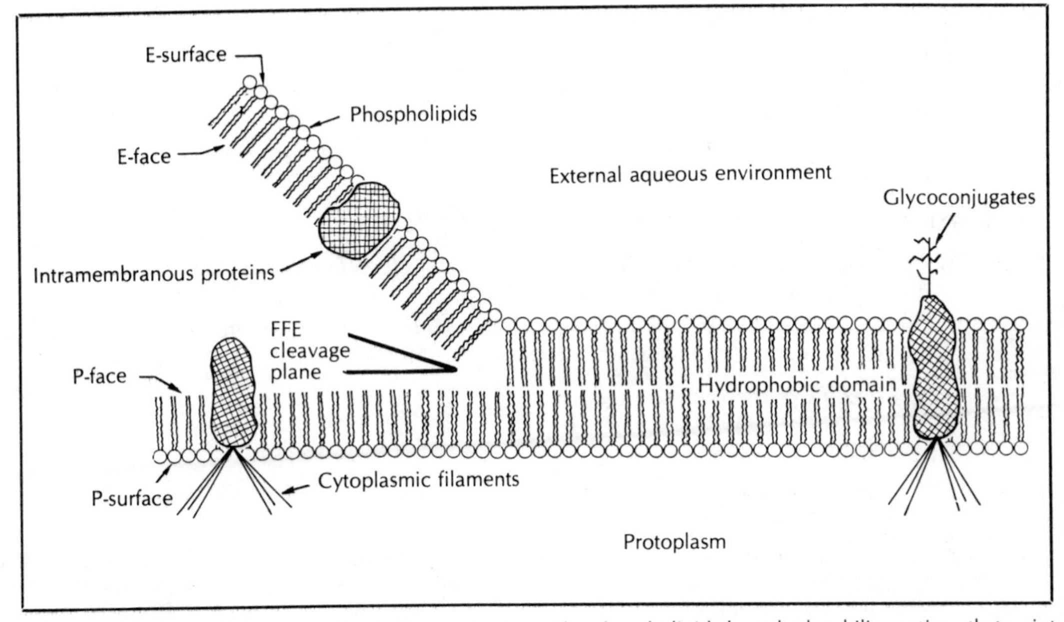

Figure 2-1. Diagram illustrating cell membrane structure. The phospholipids have hydrophilic portions that point toward the aqueous domains and hydrophobic portions that interact with one another. Integral membrane proteins have hydrophobic portions that interact with the hydrophobic portions of the phospholipids and hydrophilic portions, sometimes with glycoconjugate moieties added, that often project outward away from the cell structure. (Reprinted with permission from Johnson KE: *Histology: Microscopic Anatomy and Embryology*. New York, John Wiley, 1982 p 18.)

III. ENDOPLASMIC RETICULUM

A. STRUCTURE

1. The endoplasmic reticulum (**ER**) is a widely distributed system of tubes, sheets, and cavities which all are bounded by membranes.
2. This network within the cytoplasm may be studded with ribosomes (rough ER or **RER**) or may lack ribosomes (smooth ER or **SER**).
3. This complex organelle is highly developed in cells that are specialized for certain differentiated functions.
4. In pancreatic acinar cells, the RER and the nuclear envelope are in direct communication.

B. FUNCTION

1. In cells with a prominent RER, protein for secretion is actively synthesized. Pancreatic acinar cells, which manufacture many of the digestive enzymes for the small intestine, contain a prominent RER.
2. Similarly, hepatic parenchymal cells and goblet cells also contain a prominent RER.
3. In cells with a prominent SER, active steroid synthesis or drug detoxification occurs.
4. Enzymes for cholesterol metabolism and drug detoxification are located on the surface of SER membranes.
5. The RER often is continuous with and may be involved with the synthesis of other membrane systems within the cell, such as the SER and the Golgi apparatus.

IV. RIBOSOMES

A. STRUCTURE AND DISTRIBUTION

1. Individual ribosomes are not visible in the light microscope, but cells rich in ribosomes have basophilic cytoplasm due to the large amount of RNA in the ribosomes.
2. With the electron microscope, ribosomes commonly are found on the RER in cells that produce large amounts of export protein (e.g., pancreatic acinar cells or liver parenchymal cells); or they may be found in great numbers freely in the cytoplasm, especially in cells that produce intracellular proteins (e.g., erythroblasts).
3. A ribosome measures approximately 15×25 nm.
4. Each has one large and one small subunit.
   a. In membrane-associated ribosomes of the RER, the large subunit rests on the membrane.
   b. The small subunit sits on the large one.
5. There is a channel through the ribosomes, probably a hydrophilic core, through which passes the nascent chain of polypeptides.

B. FUNCTION

1. Proteins for export leave the cytoplasmic domain for the lumen of the RER by passing through ribosomes while they are being synthesized.
2. The ribosome is composed of ribosomal RNA and scores of ribosomal proteins which bind the ribosomes to membranes and are involved in the control of protein synthesis.

V. GOLGI APPARATUS

A. STRUCTURE

1. This organelle is actually a collection of membrane-bound, flattened sacs and vesicles of variable diameter.
2. Typically, three to five broad, flat lamellae are stacked together near the nucleus, like frisbees, with their concave surfaces pointing away from the nucleus and their convex surfaces pointing toward the nucleus.
3. The central portions of the lamellae are smooth and flat, while the peripheral portions of these structures may be dilated variably.

4. The proximal surface often is referred to as the **forming face** and the distal surface, the **maturation face**. The maturation face is composed of membranes that are somewhat thicker than those of the forming face.

5. Also, the maturation face often is closely associated with vesicles containing dense material that may be end products of Golgi processing.

B. FUNCTION

1. The Golgi apparatus currently is viewed as a dynamic organelle involved in **processing** and **modifying** secretion products initially produced in the ER.

2. The Golgi apparatus is especially prominent in cells producing large amounts of glycoprotein secretions.

3. It may be polarized and point in the direction of the release of secretion product. For example, goblet cells have a basal nucleus and an apical mass of mucous secretions.

4. The Golgi apparatus is involved in the addition of various carbohydrate moieties to the proteinaceous components of the mucus and typically lies in a polarized position between the basal nucleus and the apical mucous droplets.

5. The enzymes for addition of carbohydrate residues to proteins and lipids, known collectively as glycosyltransferases, recently have been shown as highly active in purified Golgi membranes.

6. Lipoproteins also are manufactured, in part, in the Golgi.

7. The lumen of the RER cisternae and the lumen of the Golgi cisternae are in functional communication.
   a. Proteins pass through the ribosomes and membranes of the ER and the lumen of the RER cisternae during their synthesis.
   b. They are transferred to the Golgi apparatus presumably by numerous rapid membrane-fusion events.

VI. COMPONENTS OF THE NUCLEAR APPARATUS

A. NUCLEUS

1. The nucleus of the adult human cell is a tangled mass of DNA threads in association with RNA and various nuclear proteins. The exact structure of this chemical complex is highly controversial.

2. Most adult human cells, with the exception of certain permanently proliferative cells in the skin, gastrointestinal tract, and reproductive system, have little or no mitotic capacity.

3. The great lengths of DNA structures, which may be many thousand times greater than the nuclear diameter, are packaged into multiple coils and supercoils of unknown structure. These coils are not evident in electron-microscopic sections of the nucleus.

4. DNA coiling is especially enigmatic in those cells in the body where proliferation is a normal part of the functioning of the organ (e.g., in the stratum germinativum of the skin). Here, cells undergo a cyclic coiling and uncoiling of DNA strands.

B. NUCLEOLUS

1. This RNA-rich structure, located in the nucleus, is prominent in cells involved in protein synthesis.

2. With the electron microscope, the nucleolus shows a rounded profile in cross section with a **fibrillar** central portion and a **granular** peripheral portion.
   a. The fibrillar portion represents the nuclear DNA sequences encoded for ribosomal RNA.
   b. The granular portion represents the granular intranuclear precursors of cytoplasmic ribosomes which are transported from the nucleus into the cytoplasm.

3. The mass transport of macromolecules between nucleus and cytoplasm suggests a rationale for the existence of nuclear pores through the double-membrane nuclear envelope.

C. NUCLEAR PORES

1. Nuclear pores have an octagonal shape and centrally located, electron-dense granules.

2. In cross section, nuclear pores are closed by a thin diaphragm of unknown significance.

D. NUCLEAR ENVELOPE

1. The nuclear envelope, enclosing the nucleus, is composed of an outer and an inner membrane, each similar in electron-microscopic appearance to the single membrane surrounding the entire cell.
2. The inner membrane is continuous with the outer membrane at nuclear pores, giving rise to an interrupted double membrane.
3. The outer membrane also is continuous in some locations with the ER and is thought to be in functional relationship to the ER.

VII. CELL CYCLE

1. Many differentiated adult cells have extreme variability in the length of their cell cycle.
2. For example, many of the neuronal elements of the central nervous system are fixed in an extremely long **G_1 phase** (Gap) following cell division. These extended G_1 phases commonly are called **G_0**.
3. The G_1 phase may be as brief as several hours in highly proliferative cells, or it may be much longer in cells that are less mitotically active.
4. After G_1 follows the **S phase**, during which the DNA content doubles from diploid to tetraploid, and the number of DNA strands doubles.
5. Following S there is a short period called **G_2**, so-called because it is the second gap in the cycle after the S phase. G_2 precedes the **M phase**, during which a cell undergoes DNA coiling in the formation of visible chromosomes and the complex events of mitosis occur.
6. During the relatively brief M phase, the nuclear envelope disintegrates, and chromosomes align on the metaphase plate and move to opposite poles of the mitotic spindle. The cytoplasm in the cell divides to form a pair of daughter cells. After M, a cell enters a new G_1 phase.

VIII. MITOCHONDRIA

A. STRUCTURE

1. Mitochondria are either ovoid or rod-shaped and are just large enough to be seen with the light microscope.
2. The electron microscope reveals their double-membrane structure. The outer membrane surrounds the entire organelle, compartmentalizing it from the rest of the cell.
3. The inner membrane is folded into a series of **cristae**, which protrude into an internal amorphous matrix. Usually these cristae interdigitate, and in some cells they may be tubular rather than lamellar.

B. FUNCTION

1. Mitochondria produce the energy-rich compound ATP from oxidizable substrates brought to the cells from the gastrointestinal tract via the general circulation and active transport.
2. Enzymes that convert compounds, like glucose, into energy-rich ATP bonds fall into two broad groups.
   a. Enzymes of the tricarboxylic acid (TCA) cycle reduce nucleotides produced from the oxidation of three carbon-breakdown products of glucose. There is substantial evidence that the TCA-cycle enzymes are located in the mitochondrial matrix surrounding the cristae.
   b. Enzymes generate ATP from adenosine diphosphate (ADP) by the oxidation phosphorylation process in the electron-transport chain. These enzymes are located in particulate substructures in the walls of the membranes in the cristae.

IX. LYSOSOMES

A. STRUCTURE AND DISTRIBUTION

1. These small (0.25–0.50 μm diameter), membrane-bound inclusions are prominent in macrophages—cells involved in the destruction of ingested foreign materials.
2. Lysosomes also are found in metabolically active cells whose high cellular activity causes substantial turnover of intracellular structures.

B. FUNCTION

1. Lysosomes contain nucleases, proteases, and glycosidases, all of which degrade macromolecules.

2. When a phagocyte engulfs a foreign object, it brings the foreign object into the cell in a membrane-bound structure called the **phagosome**.

3. When a lysosome meets a phagosome, their membranes fuse; the hydrolytic enzymes of the lysosome then are released and begin to attack and degrade the foreign object.

4. Lysosomes also may engulf and digest some of their own organelles, presumably after the organelles are "worn out" in the normal course of cell aging.

X. MICROTUBULES AND MICROFILAMENTS.

The complex architecture of cells is thought to be supported by an intracellular cytoskeleton composed of microtubules and microfilaments.

A. MICROTUBULES

1. **Structure.**
   a. Microtubules are present in most human cells and are thought to lend a degree of rigidity and structural integrity to cells.
   b. They are hollow, pipe-like structures with a diameter of approximately 24 nm and an indeterminate length. In certain highly asymmetric cells, individual microtubules may be many micrometers long.
   c. In cross section at high magnification, microtubules are composed of 13 **protofilaments** wrapped around one another in a loose spiral.
      (1) The protofilaments are composed of small globular subunits of a protein known as **tubulin**.
      (2) Tubulin subunits have the ability tb assemble into protofilaments and microtubules under appropriate conditions.

2. **Function.**
   a. Current belief is that microtubule assembly and disassembly are under precise physiologic control within the cell.
   b. A cell can assemble and disassemble a mitotic spindle, composed primarily of microtubules, at the appropriate time during the cell cycle.

B. MICROFILAMENTS

1. **Structure.**
   a. Microfilaments are about 5 nm in diameter and reveal no particular substructure.
   b. They are prominent in the cortex of many cells and accumulate in the cleavage furrow during cytokinesis.
   c. Microfilaments are believed to contain **actin**, a ubiquitous protein involved in cell motility.

2. **Function.**
   a. Recent studies show that these cortical microfilaments contain a myosin subfragment called **heavy meromyosin (HMM)**.
   b. HMM contains the actin-binding site of myosin and can be used as a histochemical reagent in cells made permeable to this macromolecule by glycerine treatment.

C. INTERMEDIATE FILAMENTS

1. Intermediate filaments are widely distributed throughout cells, have a diameter of about 10 nm and are not obviously associated with cell motility.

2. They contribute to the stiffness of certain cell types.
   a. For example, **tonofilaments**, which radiate from **desmosomes** found in intercellular junctions, are a kind of intermediate filament.
   b. Also, certain other highly rigid cell types are loaded with intermediate filaments.

XI. CENTRIOLES

A. STRUCTURE

1. Centrioles are cylindrical, with a diameter of 0.1–0.2 μm and a length of approximately 0.2–2.0 μm.

2. Most diploid cells contain a pair of centrioles, but several kinds of multinucleated cells contain many centrioles.

3. Centriole walls are composed primarily of structures closely related to microtubules.

4. Nine triplet structures radially arrange about the core of the centriole like the blades in a turbine.

5. The inner structure of a triplet resembles a microtubule. The outer two parts of a triplet look like parts of incomplete microtubules plastered against the adjacent member of the triplet.

B. FUNCTION

1. Centrioles are involved in self-duplication, in production of cilia and flagella, and in control of mitotic apparatus formation.

2. All these functions require rapid elaboration of large arrays of highly ordered microtubules or microtubule-containing structures.

3. Centrioles may serve as an initiation and orientation center for the formation of microtubules.

XII. CELL SPECIALIZATION BY ELABORATION OF ORGANELLES

A. GENERAL FEATURES

1. Many embryonic cells share a common fine structure.

2. Cell proliferation and protein synthesis are prominent during embryogenesis, with relatively little cell specialization and little cytodifferentiation.

a. Proliferative Embryonic Cell Characteristics.

(1) Prominent and enlarged nucleolus
(2) Many free ribosomes and polysomes in the cytoplasm
(3) Even distribution of mitochondria
(4) Enlarged Golgi apparatus in some cells

b. Striated Muscle Differentiation.

(1) Multinucleated sacs of contractile protein originate as myoblasts with a single large nucleus and a prominent nucleolus.
(2) The cytoplasm of these cells is particularly rich in polyribosomes but has few globular mitochondria and a sparse ER.
(3) Myoblasts fuse to form multinuclear myotubes at the same time that the cytoarchitecture of the myoblasts is changing greatly. As the cell becomes committed to a particular differentiation, specific proteins accumulate in the cell.
(4) The proteins of the contractile apparatus (actin, myosin, and other accessory proteins involved in the regulation of the muscle contraction-relaxation cycle) accumulate in bulk and alter the cell structure. The bulk of the cytoplasm becomes loaded with regular arrays of thick and thin filaments.
(5) Filaments of actin and myosin appear and assemble rapidly into sarcomeres. With the formation of definitive sarcomeres, nuclei become pushed to the cell periphery, and the ER becomes highly modified to form the **sarcoplasmic reticulum** (a membranous labyrinth for sequestration of calcium).
(6) Glycogen granules appear in great number, and mitochondria elongate and orient parallel to the long axis of the sarcomere.

B. CONSEQUENCES OF CYTODIFFERENTIATION

1. Cytodifferentiation involves the accumulation of large quantities of tissue-specific proteins.

2. It also involves a concomitant modification of the detailed cell structure.

STUDY QUESTIONS

Directions: Each question below contains five suggested answers. Choose the **one best** response to each question.

1. Which one of the following subcellular components has the largest single dimension?

(A) Microtubule
(B) Lysosome
(C) Nuclear pore
(D) Golgi apparatus
(E) Centriole

2. All of the following organelles incorporate a unit membrane EXCEPT

(A) ribosome
(B) rough endoplasmic reticulum
(C) lysosome
(D) nuclear envelope
(E) Golgi apparatus

Directions: Each question below contains four suggested answers of which **one or more** is correct. Choose the answer

A if **1, 2, and 3** are correct
B if **1 and 3** are correct
C if **2 and 4** are correct
D if **4** is correct
E if **1, 2, 3, and 4** are correct

3. Differentiated adult human cells incorporate which of the following structures?

(1) Plasma membranes
(2) Endoplasmic reticulum
(3) Golgi apparatus
(4) Sarcomeres

4. Organelles commonly found in ciliated epithelial cells include

(1) mitochondria
(2) smooth endoplasmic reticulum
(3) microtubules
(4) nucleolus

5. Prominent features in mucus-secreting goblet cells include which of the following?

(1) Intermediate filaments
(2) Golgi apparatus
(3) Centrioles
(4) Rough endoplasmic reticulum

ANSWERS AND EXPLANATIONS

1. The answer is A. (*X A 1*) All of these different organelles have several different dimensions (e.g., thickness and length). The microtubule has a small diameter relative to the other organelles, but it has a very great length.

2. The answer is A. (*III A; IV A; V A; VI D; IX A*) Most of the organelles in the body are composed, at least in part, of unit membranes. The ribosome, however, is not composed of unit membranes per se. Ribosomes can attach to the unit membranes of the rough endoplasmic reticulum.

3. The answer is A (1, 2, 3). (*XII A 5*) Most differentiated adult cells contain plasma membranes, an endoplasmic reticulum, and a Golgi apparatus. The qualifier ''most'' is important here, because these organelles are not universally present in adult cells. For example, the adult erythrocyte lacks an endoplasmic reticulum and Golgi apparatus. Sarcomeres are arrays of actin and myosin filaments and are peculiar to skeletal muscle and cardiac muscle tissue.

4. The answer is B (1, 3). (*VII A; X A*) Mitochondria and microtubules are common in ciliated epithelial cells. The mitochondria produce adenosine triphosphate (ATP) for ciliary beating. Cilia are composed of many microtubules surrounding a core.

5. The answer is C (2, 4). (*III B; V B*) The rough endoplasmic reticulum and Golgi apparatus are involved, respectively, in the synthesis and packaging of mucus. Therefore, these organelles are prominent in goblet cells. Intermediate filaments and centrioles are not particularly prominent in cells that secrete mucus.

3
Epithelium

I. INTRODUCTION. There are four basic tissue types found in the human body: epithelial, con-nective, muscular, and nervous. This chapter will discuss epithelial tissue; the other three tissu types will be discussed in later chapters.

A. TISSUE TYPES

1. **Epithelium** covers surfaces and cavities. It comprises many secretory portions of organs tha reasonably can be visualized as complex invaginations of the surface of the organism.
2. **Connective tissue** joins various epithelia. The cartilages, bones, joints, and even the bloo are specialized connective tissues.
3. **Muscular tissue** moves the skeleton, heart muscle, and walls of the cardiovascula gastrointestinal, and urogenital systems.
4. **Nervous tissue** is derived from embryonic epithelia and in some cases retains epitheli characteristics, although it is traditionally considered to be a fourth distinctive type of tissue

B. TISSUE ARRANGEMENT. Most major organs have mixtures of all four tissues.

1. The stomach, for example, is both lined and coated by epithelium. Connective tissu underlies these epithelia and joins them to the smooth muscle in the wall of the digestiv tract.
2. In addition, nerve cells coordinate the peristaltic contractions of the fibromuscular bag an coordinate the action of sphincters of the stomach.

II. FEATURES OF EPITHELIUM (Fig. 3-1)

A. LININGS

1. Virtually all of the free surfaces of the human body are lined by epithelium.
2. A cavity in the body not lined by epithelium is rare. Two examples are cavities in joints an the anterior surface of the iris, which is a naked connective tissue domain. Most other fre surfaces, however, are coated and protected by epithelium.
   - **a.** The skin and its derivatives (e.g., hair follicles, mammary glands, sweat glands, an sebaceous glands) all are covered by epithelium.
   - **b.** The entire gastrointestinal tract and all of its diverticula (e.g., the respiratory system, live pancreas, and gallbladder) are lined by epithelium.
   - **c.** The entire cardiovascular system and the body cavities derived from the extraembryon coelom, that is, the pericardial cavity, thoracic cavity, and peritoneal cavity, all are line by epithelium.
   - **d.** The entire urogenital system is lined by a layer of epithelial cells.

B. SHEETS

1. Epithelium is a group of cells with one or more layers tightly joined.
2. Firm adhesions bind these tissues into sheets.

C. POLARIZATION

1. Epithelium usually has an **apical surface**, which faces a free surface of the body or a lumen c an organ or gland, and a **basal surface**, which rests on an extracellular collagenous layer c fibrils and glycoproteins.
2. This collagenous layer, called the **basement membrane** (basal lamina), serves as a boundar between epithelium and underlying connective tissue.

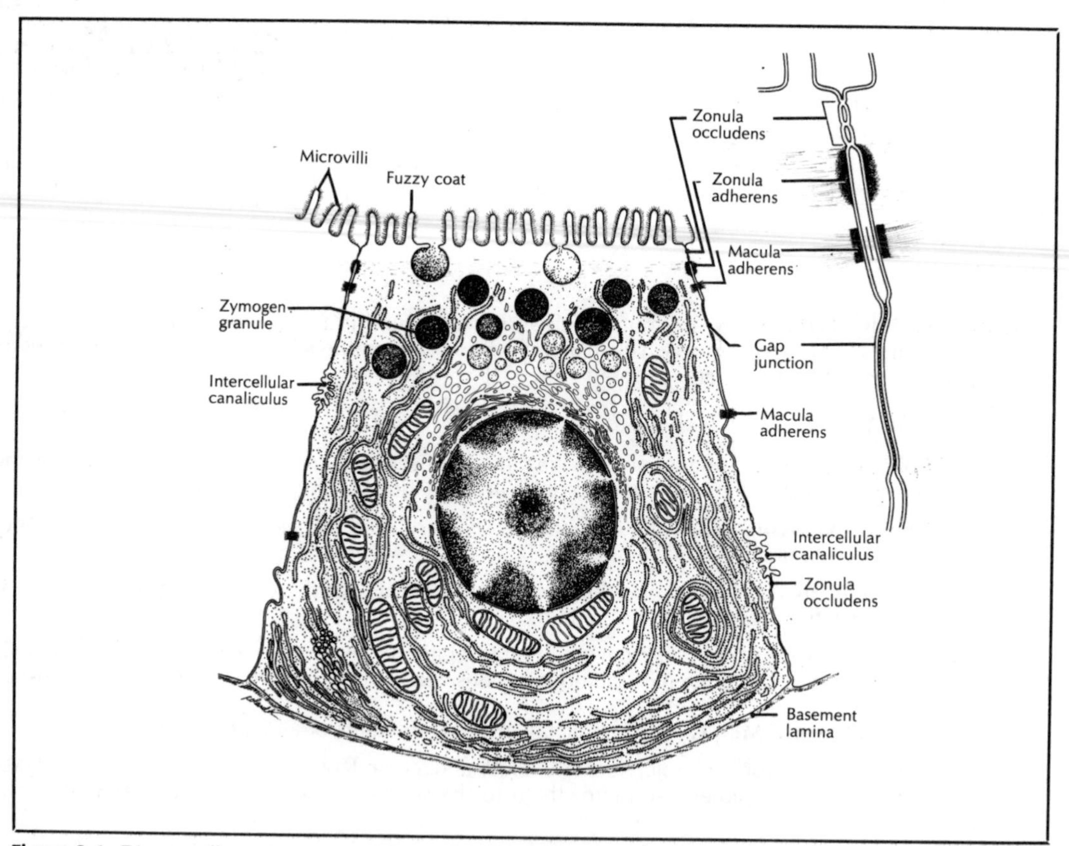

Figure 3-1. Diagram illustrating many of the cardinal features of epithelial organization. No real epithelial cells exhibit all of the features shown here, but most cells have many of these features. (Reprinted with permission from Greep R, Weiss L: *Histology*. New York, McGraw-Hill, 1977, p 129.)

D. AVASCULARITY. Blood vessels do not penetrate the basement membrane. Even in such tissues as the stria vascularis in the inner ear—where blood vessels penetrate deeply into the epithelial lining of the endolymphatic cavity to produce endolymph—blood vessels are surrounded by a basement membrane and, therefore, are technically outside the epithelium.

E. BLOOD SUPPLY

1. In most organs, the connective tissue below and surrounding the epithelium is filled with blood vessels and lymphatics.
2. Epithelium is nourished by **diffusion**.

F. APICAL SPECIALIZATIONS

1. In many cases the apical surface of epithelium is highly specialized.
2. Apical surfaces may be ciliated or may contain surface projections of variable length and frequency.

G. JUNCTIONAL COMPLEX

1. The apical junctions between cells in many epithelia are equipped with a microscopic ceiling and adhesive structure known as the **junctional complex**.
2. The junctional complex isolates the sensitive internal milieu of the organism from the noxious, toxic, and infectious outside world. It also joins the cells tightly.

III. TYPES OF EPITHELIA

A. GENERAL CLASSIFICATION. There are three broad classes of epithelia.

1. **Simple** epithelium has one cell layer, with all cells resting on the basal lamina and reaching the apical surface.
2. **Stratified** epithelium has more than one cell layer, and all cells do not rest on the basal lamina or reach the apical surface.
3. **Pseudostratified** epithelium has all cells resting on the basal lamina but not all cells reaching the apical surface. This epithelium **appears** to be stratified because nuclei lie at different levels; it is not, however, because all cells rest on the basal lamina.

B. SPECIFIC CLASSIFICATION. Table 3-1 lists types of epithelia in the body and gives an example of where each can be found.

Table 3-1. Types of Epithelia in the Human Body

Type	Site
Simple squamous	Endothelium of blood vessels
Simple cuboidal	Renal proximal convoluted tubule
Simple columnar	Small intestine
Pseudostratified columnar with cilia	Trachea
Pseudostratified columnar with stereocilia	Epididymis
Stratified squamous	Vagina
Stratified squamous, keratinized	Epidermis
Stratified cuboidal	Sweat glands
Stratified columnar	Prostatic urethra
Transitional	Urinary bladder

Note.—Reprinted with permission from Johnson KE: *Histology: Microscopic Anatomy and Embryology*. New York, John Wiley, 1982, p 31.

1. **Squamous** epithelium is **flat**.
2. **Cuboidal** epithelium is approximately equal in height and width.
3. **Columnar** epithelium is taller than it is wide.
4. In a **stratified** epithelium, the apical layer of cells is used to classify the epithelium.
   a. Epithelium of the skin, called **epidermis**, is stratified (i.e., clearly many layers), squamous (i.e., outer layer very flattened), and keratinized (i.e., no nuclei visible in the outer layers of the cell).
   b. The epithelium lining the trachea is pseudostratified, ciliated, and columnar.
   c. The epithelium lining the vagina is stratified squamous but is not keratinized.
   d. The epithelium lining the proximal convoluted tubules of the kidney commonly is described as simple cuboidal. An example of cuboidal epithelium from the kidney is shown in Figure 3-2. Some epithelia in the body are highly specialized and defy simple classification or description.
      (1) Each **chorionic villus** in the **placenta** is covered by a **syncytial epithelium (syncytiotrophoblast)** which rests on a simple cuboidal epithelium **(cytotrophoblast)** which in turn rests on a basement membrane.
      (2) **Seminiferous epithelium** in the male is stratified, highly proliferative, and after puberty sheds live haploid gametes from its apical surface.
      (3) Many sensory epithelia are complex miniature energy transducers and probably are more complicated than the chips in a hand-held calculator.

IV. EPITHELIAL APICES

A. FUNCTIONAL FEATURES

1. Epithelial apices are not adhesive.
   a. When hands are clapped they do not adhere, even if they are clapped very hard or left together for many days.
   b. Similarly, the apices of the mesothelial lining in the thoracic cavity, the pericardial cavity, and peritoneal cavities do not adhere. Because of mesothelium, organs are free to glide past one another without impediment or pain. If mesothelium is destroyed by infection or during surgery, organs will stick together and painful **adhesions** will result.
   c. The lumen of even the smallest tubule in the body does not collapse on itself and adhere because of this property of the epithelial apex.
   d. In many cases, the apex is coated with a thick, glycoconjugate-rich layer external to the outer leaflet of the plasma membrane. Called a **glycocalyx**, this layer probably is an adden-

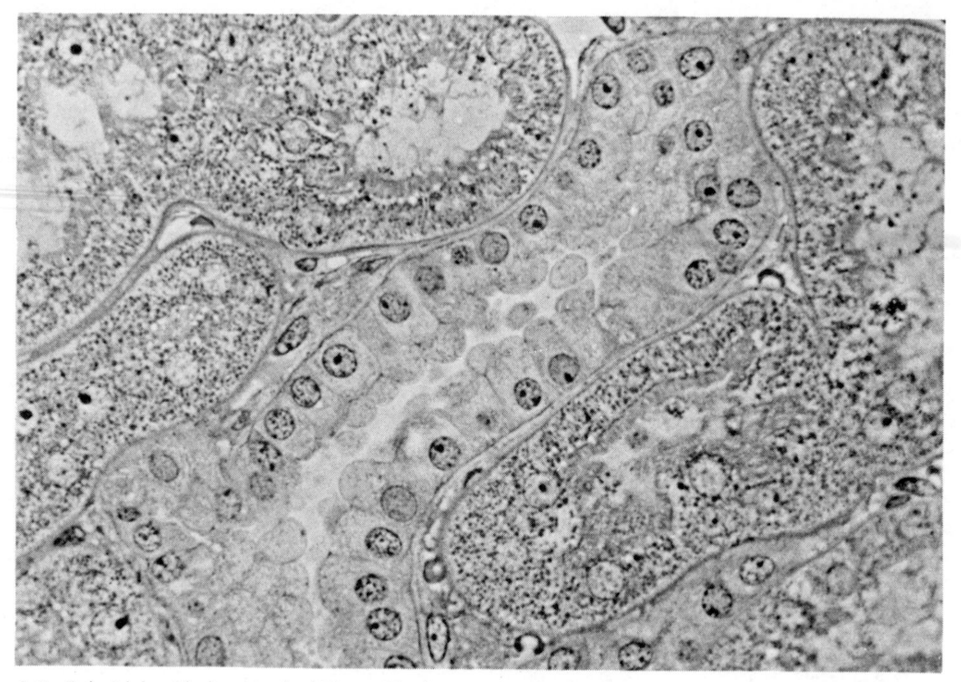

Figure 3-2. Cuboidal epithelium in the kidney. The large strip of cuboidal epithelium in the center of the picture is flanked on either side by coils of the proximal convoluted tubules.

dum to an integral membrane protein and may be related to the nonadhesive character of the apical surface.

2. Epithelial apices are specialized.
   a. The apex may contain scattered **microvilli** (e.g., in mesothelium) or a brush border of microvilli to increase absorptive area. Brush borders are found in the kidney and the gastrointestinal tract.
   b. The apex may have a forest of **stereocilia** which are elongated, nonmotile microvilli. Such apices are found in the epididymis.
   c. The apex may have **cilia**, as do those found in the respiratory tract.

B. FINE STRUCTURAL FEATURES

1. **Flagella** are long modified cilia which occur as one flagellum per cell. (The tail of a mature spermatozoon is a single flagellum).
2. **Cilia** are shorter and much more numerous per cell than flagella.
   a. They are 5–10 μm long, protrude from the surface of the cell, and characteristically are covered by a plasma membrane.
   b. In cross section, each circular cilium measures 0.2 μm in diameter and reveals nine pairs of microtubules radially arranged about a central pair of microtubules. The outer pairs actually are composed of one complete microtubule with one partial microtubule pressed against it.
      (1) The incomplete microtubule has a pair of arms which are composed of a 500,000 molecular weight protein known as **dynein**.
      (2) The microtubules are composed of two kinds of **tubulin subunits** called α- and β-tubulin.
   c. The mechanism of ciliary beating remains a topic of intense research, but many similarities between the movement of cilia and the movement of muscle cells are noted.
3. **Polarized Ultrastructure.** Often an epithelial cell has a polarized fine structure in addition to dramatic differences between its apex and base.
   a. A ciliated cell usually has many mitochondria in its apex, just below the cilia.
   b. A **goblet cell** secreting **mucus** into the lumen of the trachea has a basal nucleus, which is capped by an active Golgi apparatus on its apical side, and large numbers of apical mucous droplets.

V. BASEMENT MEMBRANE

A. CHARACTERISTICS AND DISTRIBUTION

1. In small capillaries the basement membrane is thin or fenestrated, whereas in other areas it is quite thick.
2. In the trachea, the basement membrane serves as a part of the defense against bacterial invasion.
3. In the skin, the epidermal basement membrane is strong and serves as an attachment site for the constantly stressed epidermis.
4. Lymphatic capillaries have no basement membrane.

B. SUBSTRUCTURE

1. The basement membrane has an amorphous **basement lamina** immediately adjacent to the base of the epithelium.
2. Below the basement lamina is a fibrous **reticular lamina**.

C. FINE STRUCTURE

1. **Collagen.** Basement membrane collagen is highly glycosylated **type IV collagen**. It is rich in the amino acid **hydroxylysine**, and it is composed of three $\alpha 1$ (IV) subunits. This type of collagen is highly cross-linked and is rich in carbohydrate units.
2. Other glycoproteins and proteoglycans also are present in many basement membranes. (These biochemical characteristics have been established for some of the thicker and more unusual basement membranes and may not apply to all that are found in humans.)
3. Overall, however, the detailed fine structure of the basement membrane varies considerably with its location and function.
   a. The basement membrane of a capillary is very thin.
   b. The basement membrane of the double epithelium in Bowman's capsule is quite thick but lacks a reticular lamina.
   c. In the epidermis, the basal cells in the stratum germinativum are anchored firmly to the thick basement membrane by rivet-like intracellular specializations known as **hemidesmosomes**.
      (1) These structures are similar to the desmosomes found in many junctional complexes.
      (2) Desmosomes and hemidesmosomes are considered strong sites of cell adhesion either to another cell or to an extracellular substrate, such as basement membrane collagen.

VI. JUNCTIONAL COMPLEX

A. FUNCTION. As mentioned previously, epithelium serves as a boundary between functional compartments in the body. The lumen of the small intestine, for example, contains a mixture of digestive enzymes which could virtually dissolve the wall of the gastrointestinal tract. This does not happen, however, because the noxious contents of the lumen are isolated from the sensitive gut wall by an apical epithelial specialization called a **junctional complex**.

B. STRUCTURE. There are four components of junctional complexes. Their structural characteristics vary and so will be discussed separately.

1. **Zonula Occludens.** Most adherent cells are held together by simple appositions where there is a gap of 10–15 nm between the outer leaflets of the plasma membrane. In the small intestine, however, there is no gap between cells where they join at the bases of the organ's numerous apical microvilli. Instead, the outer leaflets of the membrane appear to fuse for short distances into an occluding junction called the **zonula occludens** or tight junction.
   a. This structure extends completely around the apex of the columnar epithelial cells in a belt.
      (1) Electron-dense tracers placed in the intestinal lumen, for example, do not pass between cells.
      (2) Electron-dense tracers placed in the bloodstream will percolate between cells up to the zonula occludens but will not pass between cells into the intestinal lumen.
   b. With freeze-fracture etch, the zonula occludens in some epithelial tissues is seen as an anastomosing network of ridges (i.e., points of membrane fusion) which represent multiple barriers to molecular movement from the lumen to the lateral extracellular compartment.
2. **Zonula Adherens.** Just below the zonula occludens is a divergence of the plasma membrane

with a clear separation of 10–15 nm. This structure is called the **zonula adherens** or intermediate junction.

a. It consists of a simple membrane apposition with variable amounts of electron-dense material in the intervening gap.

b. In addition, 10-nm filaments radiate away from the zonula adherens into the cytoplasmic matrix of apposed cells.

c. This structure is thought to represent an adhesive junction.

3. **Macula Adherens.** Below the zonula adherens is a third structure, the **macula adherens** or desmosome. Each apposed cell contributes half of one desmosome. When an epithelial cell abuts on a basement membrane, sometimes it forms a **hemidesmosome**.

a. At the macula adherens, the plasma membranes diverge slightly and are 25–30 nm apart.

b. An intermediate dense line runs immediately between the cells. On the inner aspect of each apposed plasma membrane there is punctate electron-dense material.

c. Long bundles of 10-nm filaments called **tonofilaments** radiate away from the plaques of electron-dense material.

d. As its name implies, the macula adherens is considered a structure that holds cells together.

4. **Gap Junction.** Another structure, the **gap junction** or nexus, also can be found in many epithelia.

a. The gap junction is a specialized region where the outer leaflets come as close as 2 nm from each other, but a small gap remains.

b. This junction appears to be composed of a hexagonal array of barrel-shaped structures, with six subunits, arranged around an electron-lucid central core.

(1) This core is thought to represent an aqueous channel between closely apposed cells.

(2) There is clear evidence that ions and other small molecules can pass freely between closely apposed epithelial cells, presumably via these aqueous channels in the gap junction.

5. Not all features of the junctional complex are equally prominent in different epithelia. For example, the luminal-junctional specializations at the bile canaliculi of liver parenchymal cells have well-formed zonulae occludentes and prominent gap junctions but relatively few maculae adherentes.

VII. PRIMARY GERM LAYERS AND EPITHELIUM

A. **Three primary germ layers** are established during gastrulation.

1. **Ectoderm** gives rise to the skin and its associated glands, the lining of oral and anal canals, the central nervous system, and various neural crest derivatives.

2. **Mesoderm** gives rise to most of the connective tissues including cartilage and bone—although some cartilage and bone in the head is derived from neural crest and therefore is of ectodermal origin—muscle, the cardiovascular system, blood, the lining of the body cavities, and much of the urogenital system.

3. **Endoderm** gives rise to the epithelium of the gastrointestinal tract and its diverticula such as the respiratory system, liver and gallbladder, and pancreas. Parts of the urinary system also are endodermal derivatives.

B. **Epithelium** is derived from all three primary germ layers. For example, epithelium of the skin is derived from ectoderm, epithelium lining the body cavities (i.e., mesothelium) and blood vessels (i.e., endothelium) is derived from mesoderm, and epithelium lining the gut tube is derived from endoderm.

VIII. NEURAL CREST

A. NEURAL TUBE FORMATION

1. The primitive nervous system in the embryo is established as a tubular structure, the neural tube, by curling and fusion of the flat **neural plate**.

2. During this neurulation process, cells on the edge of the folding plate emigrate from the epithelium just prior to fusion.

3. These **neural crest** cells migrate ventrally and become dispersed widely throughout the embryo.

B. **NEURAL CREST DERIVATIVES** include

1. **Melanocytes**
2. The **adrenal medulla** and **Schwann cells**
3. **Ganglia** of the autonomic nervous system, dorsal root ganglia, and part of the myenteric plexus
4. Some **cartilage** and **bone** as well as other connective tissue elements of the head and odontoblasts

STUDY QUESTIONS

Directions: Each question below contains four suggested answers of which **one or more** is correct. Choose the answer

- **A** if **1, 2, and 3** are correct
- **B** if **1 and 3** are correct
- **C** if **2 and 4** are correct
- **D** if **4** is correct
- **E** if **1, 2, 3, and 4** are correct

1. Where is simple columnar epithelium found?

(1) On a basal lamina
(2) In the epididymis
(3) In the small intestine
(4) In the prostatic urethra

2. True statements concerning stratified epithelium include which of the following?

(1) Some cells rest on the basement membrane
(2) The cells rarely are joined by desmosomes
(3) The outer cells can be keratinized
(4) It is found in the trachea and duodenum

Directions: The group of questions below consists of lettered choices followed by several numbered items. For each numbered item select the **one** lettered choice with which it is **most** closely associated. Each lettered choice may be used once, more than once, or not at all.

Questions 3–7

Match the following.

(A) Zonula occludens
(B) Zonula adherens
(C) Both
(D) Neither

3. Present in cuboidal epithelium
4. Present in pseudostratified epithelium
5. Present in the basal lamina
6. Present in desmosomes
7. Forms an anastomosing network of ridges in freeze-fracture etch

ANSWERS AND EXPLANATIONS

1. The answer is B (1, 3). (*III A, B*) Simple columnar epithelium is found in areas of active absorption, such as the small intestine. Simple columnar epithelium does not have stereocilia. These structures are found in pseudostratified epithelium (e.g., in the epididymis). The cells of all simple epithelia rest on a basal lamina (also known as a basement membrane).

2. The answer is B (1, 3). (*III B 4*) In all epithelia at least some cells rest on a basement membrane. In many stratified epithelia, cells are joined by desmosomes (e.g., in the epidermis). In some instances stratified epithelia do have keratinized cells (e.g., in the skin). Stratified epithelium is not found in the trachea where there is pseudostratified epithelium, and it is not found in the duodenum where there is simple columnar epithelium.

3–7. The answers are: 3-C, 4-C, 5-D, 6-D, 7-A. (*VI B*) Both the zonula occludens and the zonula adherens are parts of the junctional complex and commonly are found in cuboidal and pseudostratified epithelia. The zonula occludens forms an anastomosing network of ridges in freeze-fracture etch. The basal lamina underlies epithelia but is not part of the junctional complex. The macula adherens is another name for a desmosome.

4
Connective Tissue

I. INTRODUCTION

A. COMPOSITION OF CONNECTIVE TISSUE

1. **Cellular components** include
   a. Fibroblasts, osteoblasts, and chondroblasts
   b. Osteocytes and chondrocytes
   c. Reticular cells and mast cells
   d. Formed elements of the blood

2. **Fibrous components** include
   a. Collagenous fibers
   b. Elastic fibers
   c. Reticular fibers

3. **Amorphous Ground Substance.** This third component of connective tissue is a mixture of proteoglycans and glycoproteins.
   a. **Proteoglycans** are macromolecules composed predominantly of carbohydrates. Proteoglycans also consist of polymers of sugars, amino sugars, and uronic acids (known as glycosaminoglycans) all attached to a central core of protein.
   b. **Glycoproteins** are macromolecules composed predominantly of protein with highly variable carbohydrate chains attached covalently to the protein backbone.
   c. **Minerals**. In many cases there also is variable mineralization of amorphous ground substance.

B. GENERAL CONNECTIVE TISSUE includes

1. **Loose** (areolar) connective tissue

2. **Dense** connective tissue, which may be regular or irregular

C. SPECIFIC CONNECTIVE TISSUE includes

1. Adipose tissue

2. Reticular tissue

3. Cartilage

4. Bone

II. CONNECTIVE TISSUE TYPES

A. CLASSIFICATION

1. Histologists have developed a classification system to help describe some of the varieties of connective tissues (CT).

2. This classification system involves three variables of CT.
   a. The number of cells per unit volume
   b. The ratio of cells to fibers
   c. The orientation of fibers

B. DESCRIPTION

1. **Loose Connective Tissue.**
   a. This CT type is found in the subcutaneous fascia, the mesenteries, and the **lamina propria**—the CT domain underlying moist epithelia in the gastrointestinal tract and elsewhere.

b. Loose CT is relatively rich in cells and is not very fibrous. Cells of loose CT include
   (1) Fibroblasts
   (2) Macrophages
   (3) Mast cells
   (4) All formed elements of peripheral blood except erythrocytes and platelets

c. The collagen and elastic fibers of loose CT are arranged randomly, are scattered throughout, and are relatively sparse.

2. Dense **regular** CT has fibers that either run in register for great lengths or are arranged in overlapping layers. This CT type is found in
   a. Ligaments
   b. Tendons
   c. The cornea

3. Dense **irregular** CT has a large number of fibers oriented in many different directions. This CT type is found in
   a. Dermis
   b. Capsules surrounding various organs
   c. Periosteum and perichondrium

III. FIBROBLASTS

A. STRUCTURE

1. This stellate or elongated fusiform cell has a deceptively simple appearance.
2. It has an elliptic nucleus with several prominent nucleoli.
3. The cytoplasm is rich in rough endoplasmic reticulum, and there is a prominent Golgi apparatus.
4. Mitochondria are elongated and prominent, and lysosomes are seen occasionally.
5. The fine structure of the fibroblast is typical for a cell involved in the secretion of extracellular proteins.

B. FUNCTION

1. Fibroblasts synthesize and secrete proteins such as collagen, elastin, and various proteoglycans, the latter in large quantities.
2. Fibroblasts are highly differentiated cells which retain the capacity to become proliferative during wound repair.
3. When removed from an organism, fibroblasts are highly proliferative and highly motile, although they probably are much less active in situ.

C. VARIATION.
Fibroblasts are found in all types of CT, but their detailed morphology may vary.

1. In loose areolar CT, fibroblasts are relatively stellate.
2. In dense regular CT, fibroblasts are elongated and considerably compressed between collagen fibers.

IV. MACROPHAGES

A. STRUCTURE

1. When macrophages are inactive, it is difficult to distinguish them with the light microscope.
2. Active (i.e., phagocytic or moving) macrophages, however, characteristically show numerous protrusions and folds projecting from the cell surface. This morphology results from the active movement of macrophages.
3. The nucleus of alveolar macrophages usually is irregular and lobulated. The cytoplasm may be rich in ingested foreign material as well as in lysosomes and phagosomes.

B. FUNCTION

1. Macrophages are important components of the immune system and so are abundant in the
   a. Lymph nodes
   b. Spleen
   c. Bone marrow

2. Alveolar macrophages have several functions.
   a. They phagocytose and destroy inspired debris.
   b. They move about the surface of the alveoli and engulf foreign material.
   c. Often, they are marked by particulate debris and are commonly called "dust cells" in the lung.

C. DISTRIBUTION

1. These phagocytic CT cells are part of the **mononuclear phagocyte system** (MPS). Monocyte precursors in bone marrow and monocytes in bone marrow and peripheral blood all are precursors of macrophages and can be converted rapidly into macrophages when needed.

2. In the adult human there is a complex macrophage system located in the
   a. CT, where macrophages are known as **histiocytes**
   b. Lungs
   c. Pleural and peritoneal cavities
   d. Sinusoids in the liver, where macrophages are known as **Kupffer cells**

V. MAST CELLS

A. STRUCTURE. These large CT cells contain numerous granules in the cytoplasm which stain metachromatically with certain basic dyes (i.e., toluidine blue stains them a metachromatic purple color rather than an orthochromatic blue).

1. Individual mast cells contain hundreds of these membrane-delimited organelles.

2. Their metachromasia is a result of the presence of large amounts of a sulfated proteoglycan called **heparin**, which is an anticoagulant.

B. FUNCTION. Mast cells are involved in inflammatory reactions and immediate hypersensitivity allergic reactions.

1. Mast cells avidly bind a portion of the immunoglobulin E (Ig E) molecules that are released into the serum when a person is exposed to antigens such as ragweed pollen.

2. When a person is reexposed to the antigens, the IgE bound to the surface of the mast cells facilitates release of several substances, including
   a. Histamine, which promotes capillary leakage and edema
   b. Slow-reacting substance of anaphylaxis (SRS-A), which promotes contraction of smooth muscle and leakage from blood vessels
   c. Eosinophil chemotactic factor (ECF)
   d. Heparin

C. DISTRIBUTION

1. Mast cells are relatively common in the **lamina propria** (i.e., the loose CT underlying the luminal epithelium) of the respiratory and digestive systems.

2. Mast cells occur in peripheral blood as basophils.

VI. COLLAGEN FIBERS

A. STRUCTURE. With the electron microscope, collagen fibers show an alternating dark and light banding pattern with an overall periodicity of 67 nm. Collagen fibers are composed of numerous smaller fibers, which in turn are made up of subunits called tropocollagen.

1. **Tropocollagen.**
   a. These subunits assemble themselves into stable fibers under appropriate conditions. Intermolecular cross-links also are created between the subunits to increase their tensile strength.
   b. Tropocollagen is composed of three polypeptide chains which coil around one another to produce a single long spiral.
   c. There are at least four different molecular types of tropocollagen. They are labeled I–IV to denote differences in amino acid sequences. Each type has slight but definite differences in its structure, and all vary greatly in their distribution in the body.
      (1) **Type I** collagen is composed of two $\alpha 1$ (I) chains and one $\alpha 2$ chain. The $\alpha 1$ and $\alpha 2$ chains react differently in certain chromatographic conditions. Type I collagen is found predominantly in bone, tendons, ligaments, the dermis, and dentin of teeth.
      (2) **Type II** collagen is composed of three $\alpha 1$ (II) chains and is found in hyaline cartilage.

(3) **Type III** collagen is composed of three $\alpha 1$ (III) chains and is associated with smooth muscle cells in the cardiovascular system, gastrointestinal tract, and uterus.
(4) **Type IV** collagen is composed of three $\alpha 1$ (IV) chains and is found in basement membranes.

d. The tropocollagen molecule is very slender, measuring 300 nm in length and 1.5 nm in width.
(1) In the collagen fiber, tropocollagen molecules are arranged end-to-end in long rows.
(2) Between rows they are aligned with a quarter stagger, resulting in an overall 67-nm periodicity.

2. **Alpha Chains.**
a. Alpha chains are approximately equal in molecular weight (95,000) and are coiled in left-handed helices. Together, however, they constitute a right-handed helix in the tropocollagen molecule.
b. Alpha chains have a peculiar amino acid composition in that every third residue is glycine and many of the remaining residues are proline and hydroxyproline. Hydroxylysine also is present and serves as an attachment site for carbohydrate residues.
c. The amino and carboxyl termini each contain a brief section of nonhelical peptides called **telopeptides**, which contain residues involved in covalent cross-linking between individual tropocollagen molecules.

B. **SYNTHESIS** (Figure 4-1). The complex process of collagen synthesis is conducted not only by mesodermally derived fibroblasts but by osteoblasts, chondroblasts, smooth muscle cells, and epithelial cells as well.

1. Collagen alpha chains are assembled from appropriate aminoacylated transfer RNAs on messenger RNAs which are bound to the rough endoplasmic reticulum.

2. After translation, the nascent chain is hydroxylated at prolyl and lysyl residues by prolyl and lysyl hydroxylases and is glycosylated at selected hydroxylysine residues by glucosyl and galactosyl transferases.

3. Once synthesis is complete, pro alpha chains—which contain extra polypeptides at both the carboxy and amino termini in addition to the telopeptides (registration peptides)—assemble into procollagen. Specific peptidases then cleave the extra peptides, leaving a fully assembled tropocollagen molecule.

4. Individual tropocollagen molecules become extensively cross-linked by covalent bonds **after** they self-assemble into fibers. Initially these fibers are held together by weaker, noncovalent interactions.

5. In CT, fibroblasts secrete extracellular **lysyl oxidases** which cause oxidative deamination of lysyl and hydroxylysyl residues in the telopeptides to create aldehyde groups. These aldehydes then condense with terminal amino groups in lysyl and hydroxylysyl residues in adjacent molecules, forming a stable covalent bond.

6. Different collagen types have different degrees of cross-linking. As a result, their tensile strengths, susceptibilities to proteolytic degradation, and thermal stabilities vary.

C. **DISTRIBUTION.** Collagen is the most widely distributed and abundant protein in the human body. It is a major constituent of the CT proper as well as cartilage and bone.

VII. ELASTIC FIBERS

A. **STRUCTURE.** Elastic fibers have an amorphous component, which contains the protein **elastin**, and an 11-nm fibrillar component.

1. Like collagen, elastin is rich in glycine and proline; unlike collagen, elastin has an abundance of hydrophobic amino acids.

2. Also, elastin differs from collagen in that it uniquely contains the amino acids **desmosine** and **isodesmosine**.

B. **SYNTHESIS.** The synthetic mechanisms of elastin are similar to those of collagen.

1. Proelastin molecules cross-link to one another when lysyl oxidases create aldehyde residues on three different proelastin chains.

2. These three chains combine with a fourth lysyl residue, forming a heterocyclic carbon and nitrogen ring of desmosine and thereby joining the four proelastin molecules into a large macromolecule. In this manner, huge arrays of elastic macromolecules are created.

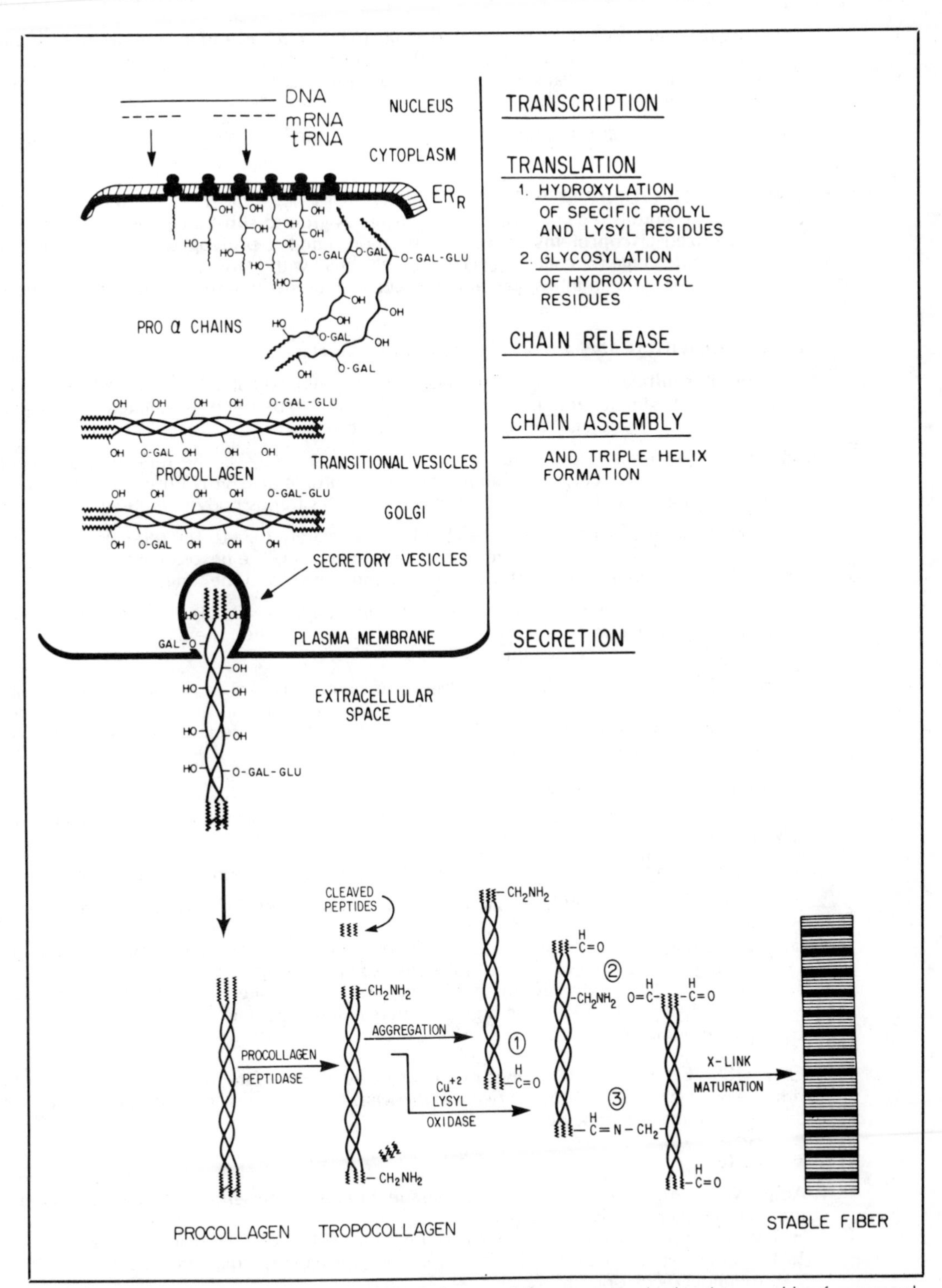

Figure 4-1. Diagram illustrating the intracellular and extracellular events involved in the assembly of mature collagenous fibers. These events occur not only in fibroblasts but in chondroblasts and osteoblasts as well. (Reprinted with permission from Weiss L, Greep R: *Histology*. New York, McGraw-Hill, 1977, p 164–165.)

C. DISTRIBUTION

1. Because these fibers have the remarkable ability to be stretched and then return to their initial length without permanent deformation, they are common in areas subjected to constant stretching or periodic expansion and relaxation, such as the cardiovascular and respiratory systems. The skin also has an abundance of elastic fibers.

2. When studying the distribution of elastic fibers with the light microscope, special stains are needed to reveal them.

VIII. AMORPHOUS GROUND SUBSTANCE.

All CT types have variable amounts of chemical substances with no fibrous organization. The **amorphous ground substance** contains a variety of poorly characterized **glycoproteins** as well as the more familiar **proteoglycans**. The latter have polysaccharide side chains called glycosaminoglycans with large numbers of repeating disaccharides—usually an amino sugar and uronic acid or neutral sugar—attached to protein backbones.

A. GLYCOSAMINOGLYCANS are divided into five major classes.

1. **Chondroitin sulfate** has a molecular weight of approximately 30,000 and a repeating unit of *N*-acetylgalactosamine and D-glucuronic acid. The hexosamine is sulfated. Chondroitin sulfate is found in cartilage, bone, skin, and the cornea.

2. **Hyaluronic acid** has a molecular weight of at least 1,000,000. The repeating unit in this macromolecule is *N*-acetylglucosamine and D-glucuronic acid. It is unsulfated. Hyaluronic acid is found in cartilage, the umbilical cord, and the vitreous body of the eye.

3. **Dermatan sulfate** has a molecular weight of approximately 30,000. The repeating unit of dermatan sulfate is *N*-acetylgalactosamine and L-iduronic acid. The hexosamine is sulfated. Dermatan sulfate is found in skin, blood vessels, heart valves, and the lungs.

4. **Keratan sulfate** has a molecular weight of approximately 10,000. The repeating unit is *N*-acetylglucosamine and galactose. The hexosamine is sulfated. Keratan sulfate is found in the cornea, cartilage, and nucleus pulposus.

5. **Heparan Sulfate and Heparin.** Each of these has a molecular weight of approximately 15,000 and a repeating unit of *N*-acetylglucosamine and D-glucuronic acid. The hexosamine is sulfated—more so in heparin than in heparan sulfate. Both are found in the aorta, the liver, the lungs, and mast cell granules.

B. THE PROTEIN COMPONENTS of proteoglycans are not so well-known.

The proteoglycans of cartilage are thought to be coupled together in a large macromolecular complex called a **proteoglycan aggregate**.

1. This aggregate is thought to be like a bottle brush with a backbone of hyaluronic acid.

2. To this backbone, linker proteins attach radially arrayed glycosaminoglycans, which resemble bristles on the brush.
   - **a.** The glycosaminoglycans give a substantial net negative charge to the complex. This allows for strong interactions with other molecules, such as collagen, in the extracellular matrix.
   - **b.** In addition, glycosaminoglycans bind large amounts of water, allow for free diffusion of small molecules through their highly hydrated domains, and contribute to the gel-like structure of many CT types.

IX. ADIPOSE TISSUE.

There are two types of adipose tissue: **white fat**, which contains **unilocular** (i.e., with one large fat vacuole) **adipocytes**, and **brown fat**, which contains **multilocular** (i.e., with many small fat vacuoles) **adipocytes**.

A. STRUCTURE

1. **Adipocytes** comprise the bulk of adipose tissue, but they represent only about 20 percent of the total number of cells in the tissue.
   - **a.** The adipocyte of white fat is a large cell with a single, centrally placed fat vacuole.
   - **b.** The adipocyte of brown fat is a large cell with many small fat vacuoles scattered throughout its cytoplasm.
   - **c.** The cytoplasm of either adipocyte type is pushed peripherally but otherwise contains the usual array of organelles.
   - **d.** The nucleus is flattened considerably.
   - **e.** The fat vacuole is not bounded by a phospholipoprotein bilayer but instead by a regularly spaced array of 9-nm diameter filaments of unknown length. The fat vacuole itself is amor-

phous and is composed primarily of triglycerides with traces of cholesterol, cholesterol esters, monoglycerides, and phospholipids.

2. In addition to adipocytes, adipose tissue contains scattered collagen fibers, a rich blood supply, fibroblasts, leukocytes, and macrophages.

B. FUNCTION

1. **General.**
   a. Food storage, when a human consumes excess food
   b. Protective padding
   c. Thermoregulation, particularly in the infant

2. **Specific.**
   a. The adipocyte contains enzymes that catalyze fatty acid synthesis from glucose during times of nutritional excess. In the mitochondria and endoplasmic reticulum the fatty acids are synthesized into triglycerides, and these are transported into the central fat vacuole either directly or via liposomes.
   b. During fasting, the large fat vacuole either decreases in size, breaks into several smaller vacuoles, or disappears altogether. In addition, there appears to be a proliferation of the smooth endoplasmic reticulum and a tremendous increase in cell surface area. These changes presumably are related to the adipocyte's attempts to mobilize and export stored triglycerides.
   c. Adipocytes without their central fat vacuoles are quite similar to fibroblasts, but recent evidence suggests that they may be separate cell types.
      (1) When human adipocytes are placed in tissue culture in appropriate media, they lose their lipid vacuoles and assume a fibroblastic morphology. When fed media supplemented with serum taken from obese patients, these cells rapidly assume the characteristic adipocyte morphology.
      (2) In contrast, human fibroblastic cells in media supplemented with serum from nonobese patients do not assume an adipocyte-like morphology; instead, they remain fibroblastic.

C. LOSS AND GAIN OF ADIPOSE TISSUE

1. When a person loses a large amount of body weight, there is a decrease in the volume of fat and contained adipocytes as the lipid vacuoles disappear. The total number of adipocytes, however, changes very little.

2. Fat accumulates in the body in two ways: by an increase in the number of adipocytes, called **hyperplasia**, or by an increase in the size of adipocytes, called **hypertrophy**.
   a. In humans, there are two periods of hyperplastic and hypertrophic fat growth. The first occurs soon after birth, and the second is prepubertal.
   b. Children with an early onset of obesity may have many more adipocytes than an adult. Many adults gain weight in later life not by an increase in the number of fat cells but by an increase in the size of fat cells. Whatever the case, however, it appears that once fat cells have formed, they are fixed in number; only their size is variable.
   c. Only when caloric output exceeds caloric intake in an adult does the amount of fat decrease.

D. DISTRIBUTION

1. White adipose tissue is common in the entire subcutaneous compartment and is particularly prevalent over the abdomen and around the hips and buttocks.

2. There is a sexual dimorphism in fat deposition. Approximately 22 percent of the body weight of a normal female is adipose tissue, while a normal male has approximately 15 percent of his body weight as fat.

3. Fat serves a very important function as anatomic packing material. The kidneys and eyes are protected by substantial fat padding. The mesenteries and the omentum also are infiltrated with adipose tissue to a variable degree, depending on the nutritional status of the individual.

X. RETICULAR CONNECTIVE TISSUE

A. STRUCTURE

1. Reticular CT is composed of a lacework reticulum of collagen fibers. These collagen fibers are associated with a rich coat of glycoproteins, but this association is poorly understood. These **reticular fibers** have the ability to reduce silver salts and to cause the deposition of silver

metal, thus staining the fibers black. Because of this property, these fibers often are described as **argyrophilic fibers**.

2. In addition to reticular fibers, reticular CT contains cells that in many cases look similar to fibroblasts. These reticular cells are phagocytic in some instances, and they may have a filtering or surveillance function in certain organs.

B. DISTRIBUTION. Reticular CT is prevalent around the small vascular channels of the liver, spleen, lymph nodes, and bone marrow.

XI. EMBRYOLOGY OF CONNECTIVE TISSUE

A. DERIVATION

1. The bulk of the body's CT, including cartilage and bone, is derived largely from mesoderm.
2. Some of the CT in the head and neck, however, may be derived from the neural crest. There is little direct evidence of this in humans, but evidence from animal studies is compelling.

B. FORMATION

1. Once the primary germ layers are established, a series of **somites** form as paired blocks lateral to the neural tube.
2. Somites further subdivide into several structures, two of which are
   a. A **sclerotome**, which contributes to the vertebrae
   b. A **dermomyotome**, which later forms dermal CT and muscles
3. Many of the spaces between the major structures in the early embryo are packed with a space-filling embryonic CT known as **mesenchyme**. An example of mesenchyme is shown in Figure 4-2.
   a. Blood vessels and the formed elements of the blood arise directly from mesenchyme.
   b. Mesenchymal cells condense and form bones directly (e.g., in the vault of the skull) or differentiate into cartilage, which ossifies later to become bones (e.g., of the axial and appendicular skeletons).
   c. Mesenchymal cells invade the shafts of developing bones and later differentiate into hematopoietic cells.
   d. In addition, mesenchyme can differentiate into several types of fibroblasts, adipocytes, and reticular cells.
   e. All of these differentiation steps involve an elaboration of certain organelles and the production of cell-specific extracellular matrix proteins peculiar to each cell type.

XII. EPITHELIAL-MESENCHYMAL INTERACTIONS

A. THE DEVELOPMENT OF COMPLEX ORGANS within the body requires two processes.

1. **Cytodifferentiation**, which is acquisition of novel cytologic and biochemical functions
2. **Morphogenesis**, which is the rearrangement of cell groups into new configurations

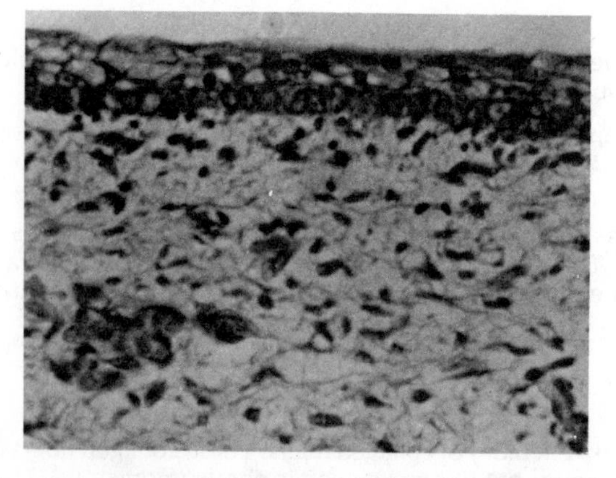

Figure 4-2. A section through human embryonic skin. The epidermis consists of three cell layers, which run across the top of the photograph. Beneath the epidermis, a large field of loosely arranged cells and fibers make up mesenchyme. (Reprinted with permission from Johnson KE: *Histology: Microscopic Anatomy and Embryology.* New York, John Wiley, 1982, p. 258.)

B. RECIPROCAL INDUCTIVE INTERACTIONS

1. It would be useless to develop alveoli in the lung or the acini in salivary glands without airways or ducts to connect the parenchyma to the surface of the body. In many instances, there is a reciprocal interaction that occurs between the developing epithelium of an organ and the mesenchymally derived connective tissue surrounding that epithelium.

2. It is possible to separate the epithelial and mesenchymal components of many organs and then recombine and culture them in artificial media.
   a. When pulmonary epithelium is cultured alone it will not mature in vitro, but when cultured in combination with mesenchyme it will show extensive development.
   b. Pulmonary epithelium in combination with pulmonary mesenchyme will show considerable development of a dichotomously branching pattern that is characteristic of the lung.
   c. When the same epithelium is combined with salivary mesenchyme, it will develop budding that is characteristic of the salivary gland rather than the lung.

3. The distribution of collagen fibers, basal laminae, and other components of the extracellular matrix has a clear bearing on the course of organ morphogenesis. The distribution of these extracellular matrix components probably is under complex control of both epithelium and mesenchyme.

STUDY QUESTIONS

Directions: Each question below contains five suggested answers. Choose the **one best** response to each question.

1. Dense regular connective tissue contains all of the following components EXCEPT

(A) a relatively large number of extracellular fibers
(B) a relatively small number of cells
(C) many collagen fibers in a regular array
(D) some fibroblasts
(E) no amorphous ground substance

2. A defect in collagen cross-linking would result in which of the following symptoms?

(A) Fever
(B) Edema
(C) Poor wound healing
(D) High blood pressure
(E) Blindness

Directions: The group of questions below consists of lettered choices followed by several numbered items. For each numbered item select the **one** lettered choice with which it is **most** closely associated. Each lettered choice may be used once, more than once, or not at all.

Questions 3–7

For each of the following connective tissue locations, select the tissue type most likely to be found there.

(A) Loose areolar tissue
(B) Adipose tissue
(C) Dense irregular connective tissue
(D) Dense regular connective tissue
(E) Reticular connective tissue

3. Dermis
4. Orbital space of the skull
5. Lymphatic system
6. Tendons
7. Lamina propria

ANSWERS AND EXPLANATIONS

1. The answer is E. (*I A 1–3*) Almost all connective tissue types have some sort of amorphous ground substance. Dense regular connective tissue has many extracellular fibrils, is relatively hypocellular, has regular arrays of collagen fibers, and has fibroblasts.

2. The answer is C. (*VI B 4*) During wound healing, fibroblasts secrete new collagen. Wound healing will be poor if new collagen is poorly cross-linked.

3–7. The answers are: 3-C, 4-B, 5-E, 6-D, 7-A. (*II A, B*) The dermis is a dense irregular connective tissue. Adipose tissue in the orbit cushions the eyeball. Reticular connective tissue provides part of the structural framework for a lymph node. Tendons have great tensile strength due to their large numbers of parallel connective tissue fibers. The lamina propria is a loose areolar connective tissue.

5
Cartilage and Bone

I. INTRODUCTION

A. FUNCTION. Cartilage and bone are among the major components of the human body, and they serve mainly to support it.

1. Cartilage provides flexible support.
2. Bone is the body's chief support element.
   - **a.** It provides rigid structures for muscle attachment and constitutes a system of levers that turn muscle contraction into a series of useful body movements.
   - **b.** Bone protects vital organs in the skull and thoracic cavity.
   - **c.** Bone largely stores calcium, which can be mobilized when necessary, and it provides a compartment for hematopoiesis.
3. The histologic organization of cartilage and bone provides a unique combination of rigid, yet plastic, living tissue.

B. COMPONENTS of cartilage and bone include

1. **Cells** such as chondroblasts or osteoblasts, chondrocytes or osteocytes, osteoclasts, hematopoietic elements, and—in bone only—blood vessels
2. **Fibers** such as collagen fibers
3. **Amorphous ground substance**, which is rich in proteoglycans and glycoproteins and, in bone, contains a mineralized matrix

II. CARTILAGE

A. STRUCTURE

1. Components of cartilage include
   - **a.** Cells (e.g., chondrocytes)
   - **b.** Fibers (e.g., collagen)
   - **c.** Amorphous ground substance (e.g., chondroitin sulfate and hyaluronate)
2. Cartilage is dominated by the acellular elements, and it is devoid of blood vessels and nerves.

B. GROWTH

1. Cartilage can grow rapidly and still maintain a degree of support. These properties make it an excellent "skeletal" tissue for the fetus.
2. Most bones in the adult first existed as cartilaginous models during fetal life.

C. DISTRIBUTION. Cartilage is found on the articular surfaces of long bones and in the trachea, bronchi, nose, ears, larynx, and intervertebral discs.

D. VARIETIES

1. Hyaline cartilage
2. Elastic cartilage
3. Fibrocartilage

III. HYALINE CARTILAGE

A. DISTRIBUTION

1. This bluish, opalescent tissue is found in adults on the ventral ends of the ribs.
2. Hyaline cartilage also is found in the respiratory tract—in the larynx, trachea, and bronchi—and on the articular surfaces of bones.

B. FORMATION

1. Cartilage forms in the embryo by the condensation of undifferentiated mesenchymal cells into **centers of chondrification**.
2. These cells begin to secrete collagen and amorphous ground substance and soon isolate themselves into separate compartments called **lacunae**.
3. Cells with the potential to secrete the extracellular matrix of cartilage are called **chondroblasts**. (Collagen and amorphous ground substance together comprise the extracellular matrix.)
4. Once chondroblasts have walled themselves off with matrix they are called **chondrocytes**.

C. GROWTH

1. Growth resulting from the division of cells in the middle of cartilage is known as **interstitial growth**.
2. Growth by proliferation and differentiation of new chondroblasts at the periphery of the growing piece of tissue is known as **appositional growth**.

D. FURTHER DEVELOPMENT

1. The mesenchyme surrounding a growing piece of cartilage differentiates into a dense irregular connective tissue layer called the **perichondrium**.
2. The innermost layer of the perichondrium, lying adjacent to the cartilage itself, contains a layer of chondroblasts, which are able to form more cartilage.

E. STRUCTURE

1. **Chondrocytes** (Fig. 5-1).
   a. With the electron microscope, the chondroblasts and chondrocytes of growing cartilage reveal a prominent nucleolus and a basophilic cytoplasm with abundant rough endoplasmic reticulum.
   b. Chondrocytes also have a prominent Golgi apparatus.
2. **Amorphous Ground Substance.**
   a. The amorphous ground substance of cartilage contains glycoproteins and proteoglycans.
   b. One of the proteoglycans predominates in the matrix of cartilage. It is a copolymer of protein with three glycosaminoglycans—**chondroitin sulfate**, hyaluronate, and keratan sulfate.
3. **Fibrous Components.**
   a. As much as 40 percent of the dry weight of cartilage is composed of **collagen** fibrils that are 10–20 nm in diameter.
   b. The **avascular** cartilage is nourished by diffusion through the aqueous compartment of the extracellular matrix of cartilage.

IV. ELASTIC CARTILAGE AND FIBROCARTILAGE

A. ELASTIC CARTILAGE

1. Without special stains, elastic cartilage looks much like hyaline cartilage. Stains for the elastic protein **elastin**, however, show an anastomosing network of elastic fibers.
2. Elastic cartilage is found in the external ear—in the walls of the external auditory meatus and auditory tubes—the epiglottis, and parts of the larynx.

B. FIBROCARTILAGE

1. When compared to hyaline cartilage and elastic cartilage, fibrocartilage has a relatively abundant extracellular matrix and relatively few cells.

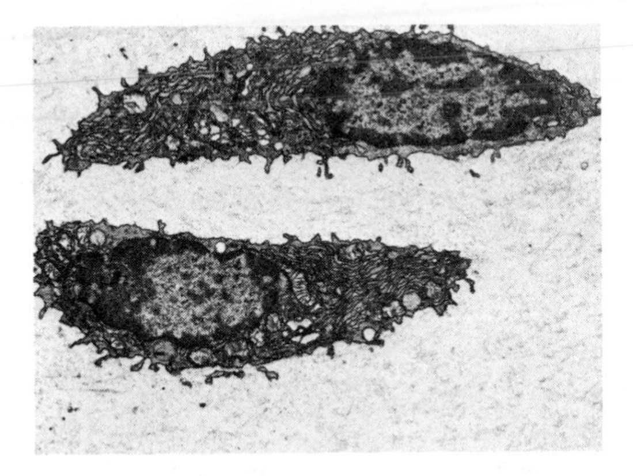

Figure 5-1. Two chondrocytes from the hyaline cartilage between two bones in a rat's foot. Each chondrocyte is a secretory cell that has been surrounded by the collagen fibers and amorphous ground substance that it secretes. (Courtesy of Dr. Daniel P. DeSimone, George Washington University School of Medicine.)

2. Fibrocartilage occurs in the intervertebral discs, symphysis pubis, and in certain articular cartilage and tendons. Often it is found in zones located between an inserting tendon and a large piece of cartilage that is on the end of a bone.

V. MACROSCOPIC STRUCTURE OF BONE

A. **REGIONS.** Bone may contain as many as three different regions. A typical long bone, for example, has all three.

1. The tubular middle portion, known as the **diaphysis**, is made of dense, **compact bone.**
2. The **metaphysis** is a region where compact bone forms a shell around a mass of **spongy** or **cancellous bone**.
3. A long bone typically is capped by an **epiphysis**. Between the epiphysis and the metaphysis is the **epiphyseal plate**, where a bone grows in length. Long bones, such as the humerus, have epiphyseal plates at both ends.

B. ARTICULAR SURFACES

1. On the articular surfaces of a long bone, a thin layer of hyaline cartilage covers the top of the compact bone.
2. Normally, the surface of this articular cartilage is extremely slippery.

C. PERIOSTEUM AND ENDOSTEUM

1. Except on the articular surfaces, bone is covered with a dense irregular tissue layer called the **periosteum**. The cells of the inner layer of the periosteum, called **osteoblasts**, secrete bone constituents, giving the periosteum the ability to form new bone.
2. Both the marrow cavity and the surfaces on spicules of spongy bone are lined with a thin layer of cells called the **endosteum**. Like the periosteum, the endosteum has osteogenic potential.
3. The periosteum and endosteum are responsible for growth in the diameter of a bone.

D. **VARIATIONS.** Not all bones have two epiphyseal plates. For example, most tarsal bones have a single ossification center. Others have complex ossification centers that reflect their complicated final gross morphology.

VI. MICROSCOPIC STRUCTURE OF BONE

A. GENERAL APPEARANCE

1. With the light microscope, a section of bone reveals a large amount of mineralized extracellular matrix arranged in **lamellae** (plates).
2. Within the lamellae exist small **lacunae** and an anastomosing network of miniature **canaliculi**. The lacunae are occupied by living **osteocytes**, and delicate cell processes from these osteocytes fill the canaliculi.
3. The mineralized matrix prevents free diffusion so that the osteocytes are connected to the blood vessels in the bone via the canaliculi.

B. LAMELLAR PATTERNS. Bone lamellae in compact bone are arranged in three common patterns.

1. Along the periosteal and endosteal surfaces of a long bone there are **circumferential lamellae**.
2. Throughout the mass of compact bone, lamellae are arranged in 4–20 concentric layers around a vascular space. These cylindrical units, called **haversian systems (osteons)**, run parallel to the long axis of the bone.
3. **Interstitial systems** are irregular arrangements of lamellae. These often are roughly triangular or quandrangular.

C. VASCULAR CHANNELS. There are two types of these in compact bone.

1. **Haversian canals** always are surrounded by concentrically arranged lamellae, where the long axis of the cylinder of lamellae is parallel to the long axis of the blood vessel. Even in cases where haversian canals branch into Y-shaped haversian systems, lamellae maintain their concentric arrangement.
2. **Volkmann's canals** occur where blood vessels from the periosteum penetrate compact bone, crossing haversian systems in their descent into it. Therefore, Volkmann's canals have their long axes perpendicular or nearly perpendicular to the lamellae of haversian systems.

D. CANCELLOUS BONE is composed of small numbers of lamellae which are not closely arranged into formal haversian systems. Instead, lamellae lie close to blood vessels and the endosteum, and they receive their nutrition by direct diffusion.

E. THE PERIOSTEUM varies in microscopic structure during bone development.

1. While a bone grows in diameter, the inner layer of the periosteum is a cuboidal layer of cells loosely arranged in a sheet. These **osteoblasts** are active in bone formation; they secrete uncalcified fibers and amorphous matrix, which subsequently calcify to form compact bone.
2. Once a bone reaches its full diameter, osteoblasts become quiescent and are indistinguishable from the other densely packed fibroblasts that comprise the periosteum. (When needed to repair a fracture, however, quiescent osteoblasts express their osteogenic potential and resume active bone secretion.)
3. Dense bundles of collagen fibers become incarcerated in the bony matrix of a growing bone. These **Sharpey's fibers** anchor the periosteum firmly to the underlying bone.

F. THE ENDOSTEUM is similar to the layer of osteoblasts but is somewhat thinner. Like osteoblasts, the cells of the endosteum can secrete bone.

VII. CYTOLOGIC STRUCTURE OF BONE

A. CELL TYPES IN GROWING BONE

1. Osteoprogenitor Cells.

a. These are relatively undifferentiated cells that commonly undergo mitosis.
b. Each has a pale-staining oval nucleus and an acidophilic or faintly basophilic cytoplasm.
c. Osteoprogenitor cells are found near all free surfaces of bone including the periosteum and endosteum. Also, they line haversian canals and are found on trabeculae of degenerating cartilage at the epiphyseal plate of a growing bone.

2. Osteoblasts.

a. These cells actively secrete bone matrix. Typical of cells that make collagen for export, osteoblasts have a large amount of rough endoplasmic reticulum and, therefore, basophilic cytoplasm.
b. The nucleus of the osteoblast is prominent and contains a large basophilic nucleolus.
c. The Golgi apparatus also is well-developed. Its functions include packaging and glycosylating collagen for export as well as producing glycosaminoglycans for the matrix.

3. Osteocytes.

a. These cells are somewhat similar to osteoblasts. Because the osteocyte is less active in matrix secretion, however, its rough endoplasmic reticulum and Golgi apparatus are less prominent than those of the osteoblast.
b. Although less active, osteocytes are not inert. They respond rapidly to parathyroid hormone during regulation of serum calcium concentration, and they can secrete matrix and cause deposition of new inorganic matrix.

c. Osteocytes occupy lacunae in the solid matrix, and they are attached to one another by slender cellular processes within the canaliculi.
   (1) The processes of osteocytes are joined by **gap junctions** that allow exchange of small molecules.
   (2) The processes may be involved in rapid hormone conduction from the haversian canals to the osteocytes in lamellae that are far removed from the blood supply.

4. **Osteoclasts.**
   a. These are large multinuclear cells formed by the fusion of many mononuclear cells.
   b. The cytoplasm is acidophilic in routine hematoxylin and eosin preparations and so stains intensely with eosin.
   c. There are numerous mitochondria and many prominent Golgi apparatus in the cytoplasm.
   d. These cells are affected by parathyroid hormone and calcitonin and are involved in the regulation of serum calcium concentration.
   e. Osteoclasts secrete acid hydrolases and certain ions that cause the breakdown of bone. The osteoclast forms a conspicuous ruffled membrane that is applied closely to bone fragments; this tremendous elaboration of osteoclast cell surface probably is involved in bone resorption. As osteoclasts erode bone lamellae, they create small pockets in the bone called Howship's lacunae.

B. INTERCONVERSION OF CELL TYPES

1. In the embryo and possibly in the adult, **osteoprogenitor cells** exist on or near free bone surfaces. These cells can cytodifferentiate into **osteoblasts**.

2. Osteoblasts also can differentiate. As they secrete the extracellular matrix of bone, they can become **osteocytes**.

3. Either there is a reserve population of osteoprogenitor cells in the adult or dedifferentiation takes place, because new osteoblasts and osteocytes appear and take part in the formation of new bone in a healing fracture site.

VIII. CHEMICAL COMPOSITION OF BONE MATRIX

A. INORGANIC SALTS

1. The extracellular matrix of bone is dominated by inorganic salts. The bone mineral is composed mainly of a complex calcium phosphate which is similar to hydroxyapatite $[Ca_{10}(PO_4)_6(OH)_2]$. Other components of the mineral are calcium carbonate, citrate, fluoride, magnesium, and sodium.

2. If the organic matrix is removed from a bone by ashing it, the remaining mineralized bone is extremely brittle. If the mineral component of a bone is removed by prolonged exposure to chelating agents, the bone becomes rubbery.

3. It is thought that both components work synergistically to give bone its extraordinary tensile strength and flexibility.

B. GLYCOCONJUGATES

1. Indirect histochemical evidence suggests that bone matrix contains some poorly characterized glycoproteins.

2. Bone also contains such glycosaminoglycans as
   a. Keratan sulfate
   b. Chondroitin sulfate
   c. Hyaluronic acid

C. COLLAGEN.
Bone matrix contains a large amount of type I collagen in the form of cross-banded fibers. The periodicity of collagen fibers is 67 nm, and the diameter of the fibers is 50–70 nm.

IX. FORMATION OF BONE

A. GENERAL DESCRIPTION

1. The chemical constituents of bone are secreted by osteoblasts. During development of the embryo, bone forms in either of two ways.
   a. Directly in undifferentiated mesenchymal fields **(intramembranous ossification)**
   b. After the destruction of a preexisting cartilaginous model **(endochondral ossification)**

2. Actually, bone forms in only one way; that is, osteoblasts secrete bone organic matrix which later is mineralized.

3. Endochondral ossification is complicated by the need to degrade a cartilage model of the bone before the formation of the bone itself.

B. INTRAMEMBRANOUS OSSIFICATION

1. The flat bones of the skull—the frontal, parietal, occipital, and temporal bones, parts of the occipital and sphenoid bones, and parts of the mandibula—develop by intramembranous ossification and are commonly called **membrane bones**. In this type of bone formation, condensations of mesenchymal cells occur near blood vessels.

2. For unknown reasons, osteoprogenitor cells differentiate into osteoblasts, and these cells begin to secrete **osteoid**. Almost immediately after secretion, the osteoid becomes mineralized. These small spicules of bone begin to grow around many blood vessels.

3. As they continue to grow, these bone fragments coalesce to form large spongy networks of trabeculae. Within the growing bone fragments are many lamellae of bone and entrapped osteocytes with canaliculi.

4. The trabeculae continue to thicken by addition of many new layers of lamellae.
   a. Where compact bone forms, trabeculae become arranged into haversian systems.
   b. Where spongy bone persists, trabeculae cease growth and the vascular tissue differentiates into hematopoietic tissue.

5. The osteoprogenitor cells on the outer surface of the growing bone continue to differentiate into osteoblasts which, in turn, secrete more osteoid.

6. When the bone reaches its final size, these osteoblasts revert to quiescent osteoprogenitor cells.

C. ENDOCHONDRAL OSSIFICATION

1. Long bones in the extremities and bones in the pelvis, vertebral column, and base of the skull are formed by endochondral ossification. Cartilaginous models of these bones appear in the embryo. For this reason, these bones sometimes are called **cartilage bones**. (Recall that a piece of cartilage has a dense irregular connective tissue capsule called the perichondrium.)

2. Cartilage can grow by **interstitial growth** or by **appositional growth**. Bone, being rigid, cannot grow interstitially but can grow appositionally.
   a. In growing cartilage, the inner layer of the perichondrium has chondroblasts. These cells secrete cartilage matrix and entrap themselves in their own secretion product, thereby becoming chondrocytes.
   b. In the center of a cartilage model of a long bone, such as the tibia, the cartilage hypertrophies and degenerates, creating a cavity.
   c. Simultaneously, in the diaphyseal region of the cartilage model, the perichondrium differentiates into a layer of osteoprogenitor cells. These, in turn, differentiate into osteoblasts which secrete a collar of osteoid around the cartilage model. This collar of osteoid is calcified immediately, forming a **bony collar**.
   d. Soon after the bony collar is formed, a **vascular bud** burrows through it. The cavity formed by hypertrophy and degeneration of the cartilage is invaded rapidly by growing blood vessels, hematopoietic stem cells, and osteoprogenitor cells. Blood vessels proliferate toward both ends of the bone and stimulate further hypertrophy and degeneration of cartilage.
   e. The bony collar increases in thickness by appositional growth. The cartilage at the ends of the bone continues to proliferate, hypertrophy, degenerate, and calcify. As this cartilage transforms into bony spicules, spongy bone forms at the ends of the bone.
   f. The epiphyses are invaded secondarily by new vascular buds, and the process repeats—cartilage degenerates, ossifies, and then forms more spongy bone.

3. Throughout life, cartilage persists on the articular surfaces of the bone. In children with growing bones, cartilage persists at the epiphyseal plate.

4. During childhood and puberty, this cycle of proliferation, hypertrophy, degeneration, and calcification of cartilage continues at the epiphyseal plate and is followed by more spongy or cancellous bone formation. Once full adult stature is attained, the epiphyseal plate stops growing and becomes completely ossified, and the bone stops growing in length.

5. Achondroplastic dwarfs have abnormalities in endochondral bone formation.

X. HISTOPHYSIOLOGY OF BONE

A. CALCIUM STORAGE

1. In addition to its obvious mechanical and protective functions, bone serves as a ready store of calcium and phosphate.

2. Among other things, calcium is essential for cell adhesion, membrane permeability, muscle contraction, and blood clotting.

3. There is a rapid exchange of calcium between blood and bones. It has been estimated that, in every minute during life, one out of every four calcium atoms in blood exchanges with a calcium atom in bone.

B. ENDOCRINE REGULATION OF SERUM CALCIUM

1. **Parathormone (PTH).**
   a. PTH acutely stimulates osteocytic osteolysis and chronically stimulates osteoclastic osteolysis, providing calcium for the blood by way of solubilization of calcium phosphates in bone.
   b. PTH is produced by the parathyroid gland.
   c. PTH and calcitonin have opposite and antagonistic effects on bone resorption and deposition.

2. **Calcitonin (Thyrocalcitonin).**
   a. Calcitonin depresses bone resorption and is antagonistic to the effects of PTH.
   b. Calcitonin is produced by C cells in the thyroid gland.

C. REGULATION OF MATURATION AND GROWTH

1. **Gonadal steroids** seem to regulate the rate of maturation and the form of the skeleton. Many sexual dimorphisms of the skeleton are controlled by gonadal steroids (e.g., the male pelvis and female pelvis are dramatically different).

2. The anterior pituitary gland secretes **growth hormone**, which has a direct effect on the epiphyseal plate. Hypophysectomy results in cessation of growth at the epiphyseal plate and can be counteracted by the administration of growth hormone.

XI. JOINTS

A. SYNARTHROSES are joints with little mobility.

1. Where a bone joins directly to another bone (e.g., in the skull), the joint is called a **synostosis**.

2. Where bones are joined by cartilage, as in the symphysis pubis, the joint is called a **synchondrosis**.

3. In the vertebral column are examples of bones that are joined by ligaments. These joints are called **syndesmoses**.

B. DIARTHROSES are joints with great mobility.

1. These joints have a fluid-filled cavity enclosed by a fibrous connective tissue capsule.

2. The joint cavity is not lined by an epithelium but does have a peculiar layer of cells known as **synovial cells**.

3. The capsule may be thickened into a **synovial membrane** with a rich vascular supply.

4. These blood vessels are thought to leak synovial lubricating fluid into the joint cavity.

STUDY QUESTIONS

Directions: Each question below contains five suggested answers. Choose the **one best** response to each question.

1. The characteristic that best describes osteocyte processes is that they

(A) secrete elastin
(B) occupy canaliculi and are joined by gap junctions
(C) contain a Golgi apparatus
(D) contain many tonofilaments
(E) disappear when osteocytes divide

2. The chondroblast is best described as

(A) a mesenchymal derivative that secretes a basement membrane
(B) an endodermal derivative that secretes proteoglycan
(C) a mesodermal derivative that secretes collagen and proteoglycan
(D) a mesothelial derivative that secretes collagen and glycoprotein
(E) a neural crest derivative that secretes reticular fibers

Directions: The question below contains four suggested answers of which **one or more** is correct. Choose the answer

A if **1, 2, and 3** are correct
B if **1 and 3** are correct
C if **2 and 4** are correct
D if **4** is correct
E if **1, 2, 3, and 4** are correct

3. Bones that would be of normal size in an adult male achondroplastic dwarf include the

(1) tibia
(2) frontal bone
(3) metatarsal bones
(4) mandibula

Directions: The group of questions below consists of lettered choices followed by several numbered items. For each numbered item select the **one** lettered choice with which it is **most** closely associated. Each lettered choice may be used once, more than once, or not at all.

Questions 4–8

For each description of regions in a developing bone, choose the appropriate lettered area shown in the adjacent micrograph.

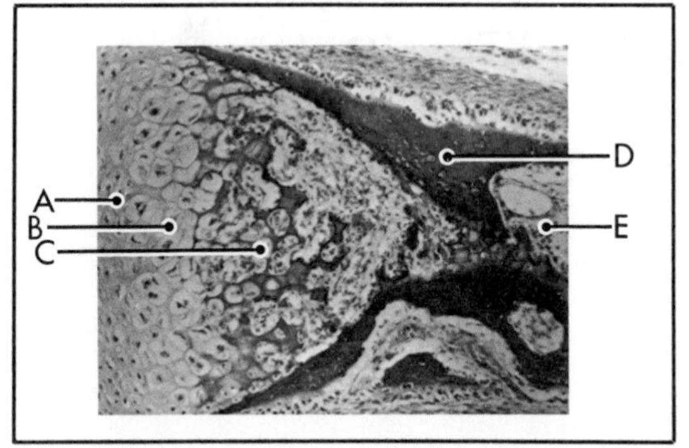

Reprinted with permission from Johnson KE: *Histology: Microscopic Anatomy and Embryology*. New York, John Wiley, 1982, p 68.

4. Part of the bony collar, which later will become the metaphysis of bone

5. Zone of cartilage calcification

6. Zone of chondrocyte multiplication

7. Marrow cavity, which is a compartment for hematopoiesis

8. Zone of chondrocyte hypertrophy

ANSWERS AND EXPLANATIONS

1. The answer is B. (*VII A 3*) Osteocytes do not secrete elastin. Osteocytes contain a Golgi apparatus, but it is located in the cell body, not in slender processes. Tonofilaments are present where cells are joined by desmosomes not gap junctions. Osteocytes do not divide.

2. The answer is C. (*III B*) Chondroblasts usually are derived from mesoderm, although some chondroblasts in the head region appear to originate from the neural crest. Chondroblasts do not secrete reticular fibers and they do not secrete a basement membrane.

3. The answer is C (2, 4). (*IX C 12*) Achondroplastic dwarfs have abnormalities in the development of epiphyseal plates during endochondral bone formation. Therefore, the tibia and metatarsal bones would be abnormally shortened. The frontal bone and mandibula, both of which are formed by intramembranous processes, would be relatively normal in morphology.

4–8. The answers are: 4-D, 5-C, 6-A, 7-E, 8-B. (*IX C*) This is an example of endochondral bone formation in its early stages. Chondrocytes are proliferating (A) and undergoing hypertrophy (B) and calcification (C). A bony collar (D) surrounds a marrow cavity (E).

6
Muscular Tissue

I. INTRODUCTION. The capacity for autonomous locomotion is distributed widely among living organisms. Even bacteria often are equipped with flagella to propel them toward food sources or away from noxious stimuli. The ability to move also is distributed widely throughout the human body's tissues and cells. Polymorphonuclear leukocytes and macrophages move about actively in the body and continue to translocate in tissue culture. Great tissue masses in the body, that is, muscular tissue, also can be moved by the activity of cells specialized for locomotion.

A. ACTIN AND MYOSIN. All mobile cells within the body contain substantial amounts of the proteins, actin and myosin.

1. The microfilaments observed frequently in the cortex of different cell types are made of **actin**. Actin constricts the cleavage furrow during mitosis and moves microvilli up and down slowly.
2. **Myosin** also is distributed widely and is not restricted to muscle tissue.
3. In cells that have become highly specialized to move rapidly, repeatedly, or forcefully, the relative concentrations of actin and myosin are high at the expense of other proteins.

B. TYPES OF MUSCULAR TISSUE

1. **Smooth muscle** cells are distributed widely throughout the cardiovascular, gastrointestinal, urogenital, and respiratory systems. The movement of smooth muscle cells is controlled largely by the autonomic nervous system, and this movement is **involuntary**.
2. **Skeletal muscle** is striated because of the organization of the contractile proteins within the individual cells.
   a. Individual skeletal muscle fibers (cells) are **syncytial**; that is, many nuclei are contained within the plasma membrane of a single cell.
   b. Skeletal muscles move the bones relative to one another. The **voluntary** contractions and relaxations of skeletal muscles allow the body to move about and engage in physical activities.
   c. Most but not all skeletal muscles have at least partial voluntary control.
3. **Cardiac muscle** also is striated because the contractile proteins are arranged in paracrystalline arrays similar to those found in skeletal muscle.
   a. Individual cardiac muscle fibers are uninuclear.
   b. Cardiac muscle is restricted to the heart and those portions of the aorta and venae cavae directly adjacent to the heart.

II. ACTIN AND MYOSIN

A. ACTIN

1. This 42,000 molecular weight protein is distributed widely in nature and is prominent in muscular tissue.
2. Actin exists as a globular molecule called **G-actin**, which can polymerize into a filamentous structure called **F-actin**. F-actin is composed of two beaded chains of G-actin subunits which are wound around one another in a long, gentle spiral.
3. Thin filaments in striated muscular tissue are composed of F-actin chains associated with the regulatory proteins, **troponin** and **tropomyosin**.

B. MYOSIN

1. A myosin molecule has a molecular weight of 470,000 and can be divided into subfragments of **light meromyosin** and **heavy meromyosin** (LMM and HMM, respectively).

2. Thick filaments in muscle fibers are thought to be groups of myosin molecules, where strong sideways interactions between the relatively straight LMM chains cause the formation of rod-like multimolecular aggregates.

III. SMOOTH MUSCLE

A. MICROSCOPIC ANATOMY

1. Typical smooth muscle cells stain deeply and often are eosinophilic due to their high protein concentration.
2. These elongated, tapering cells are 10–20 μm in diameter and range in length from 20 μm in blood vessels to 1 mm and more in the uterine wall during pregnancy.
3. With the light microscope, smooth muscle masses show numerous cells, with one centrally placed nucleus per cell.
4. In sections cut perpendicular to the long axis of smooth muscle fibers, the nucleus is not always visible in the cytoplasm of each cell. This results from the fusiform cells being packed together so that the fat portions of one cell mesh with the thin portions of another.
5. With special stains, such as the periodic acid-Schiff reaction, a basement membrane can be seen surrounding individual smooth muscle cells.
6. With the electron microscope, a smooth muscle cell shows a substantial collection of mitochondria and modest amounts of Golgi apparatus in the perinuclear region.
7. The peripheral portions of these cells are dominated by filamentous structures.
   a. In thin sections cut perpendicular to the long axis of smooth muscle cells, numerous thin, actin-containing filaments and a few thick, myosin-containing filaments are visible.
   b. Smooth muscle cells are not striated, indicating that the actin-containing and myosin-containing filaments are arrayed less formally than are those of striated skeletal and cardiac muscle cells.

B. FUNCTION

1. Masses of smooth muscle fibers are essential for controlling the size and motility of the lumina in the cardiovascular, gastrointestinal, urogenital, and respiratory systems.
2. Smooth muscle cells in the walls of muscular arteries partially control the distribution of blood to different parts of the body.
   a. Ingestion of food triggers rhythmic peristalsis of the smooth muscle layers in the gut tube. This movement aids in the transport and digestion of food.
   b. In the reproductive system, smooth muscle cells propel gametes toward each other for fertilization.

IV. SKELETAL MUSCLE

A. MICROSCOPIC ANATOMY (Figs. 6-1 and 6-2)

1. With the light microscope, muscles (e.g., the biceps) show clear striations in sections cut parallel to the long axis of individual muscle cells.
2. The prefix "sarco-" commonly is applied to components of skeletal and cardiac muscle cells. The plasma membrane of muscle cells is called the **sarcolemma**, and the endoplasmic reticulum is called the **sarcoplasmic reticulum**.
3. Skeletal muscle cells are formed by the fusion of many uninucleated myoblasts into a single multinucleated, syncytial skeletal muscle cell.
4. The nucleus of a skeletal muscle cell is displaced to the cell periphery by masses of regularly arranged actin and myosin filament bundles called **myofibrils**.
5. Mitochondria and glycogen granules also are arranged peripherally to the central myofibrils.
6. Histologically, a gross muscle such as the biceps is surrounded by a connective tissue capsule called the **epimysium**.
7. A gross muscle is subdivided further into a number of **fascicles**, each surrounded by a connective tissue sheath known as the **perimysium**.
8. Individual fascicles contain many multinucleated muscle cells, each surrounded and bound to its neighbors by a delicate connective tissue sheath known as the **endomysium.** All of these investments are composed of fibroblasts and collagen fibers.

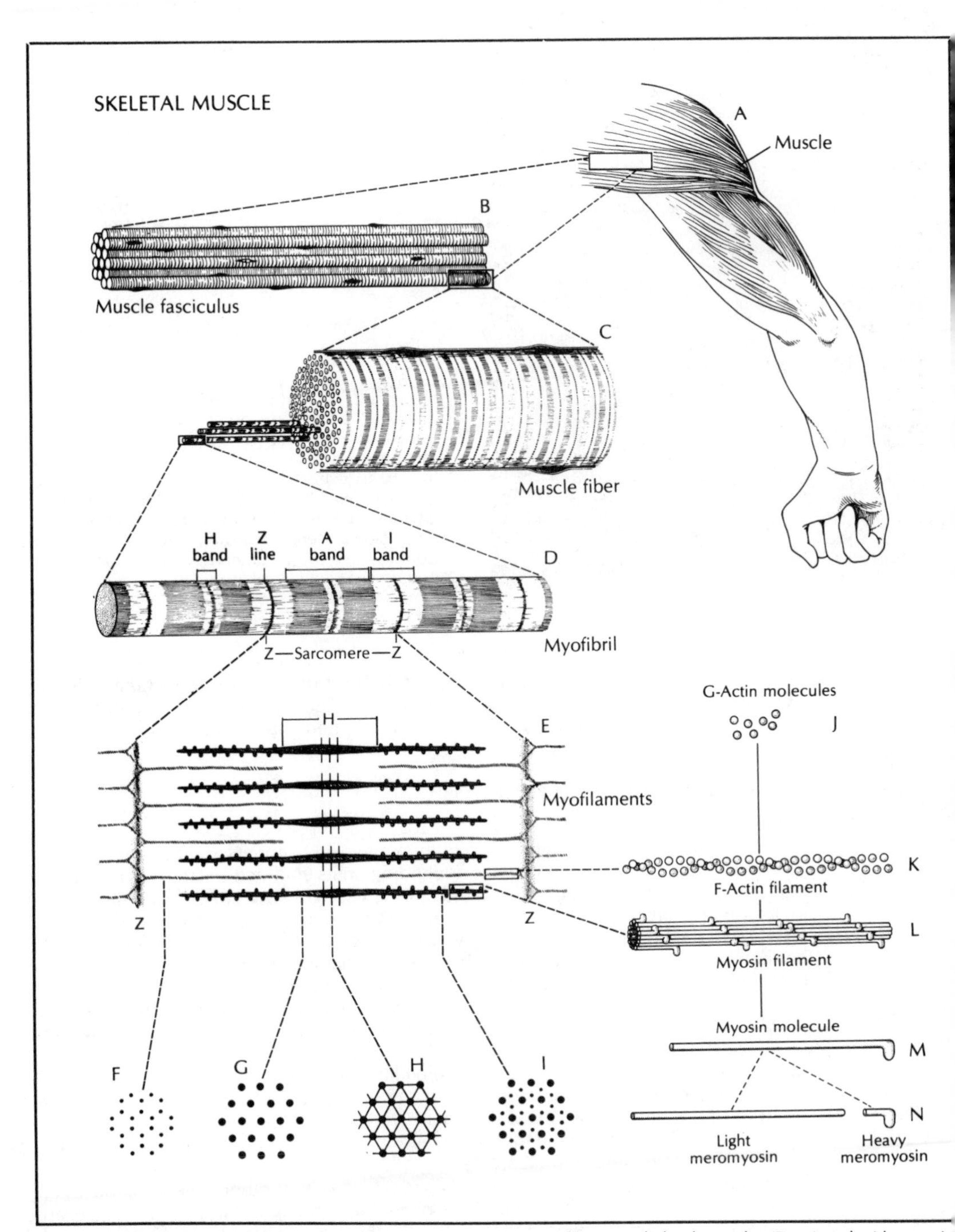

Figure 6-1. Diagram illustrating the basic microscopic anatomy of human skeletal muscle. (Reprinted with permission from Bloom W, Fawcett D: *A Textbook of Histology.* Philadelphia, WB Saunders, 1975, p 306.)

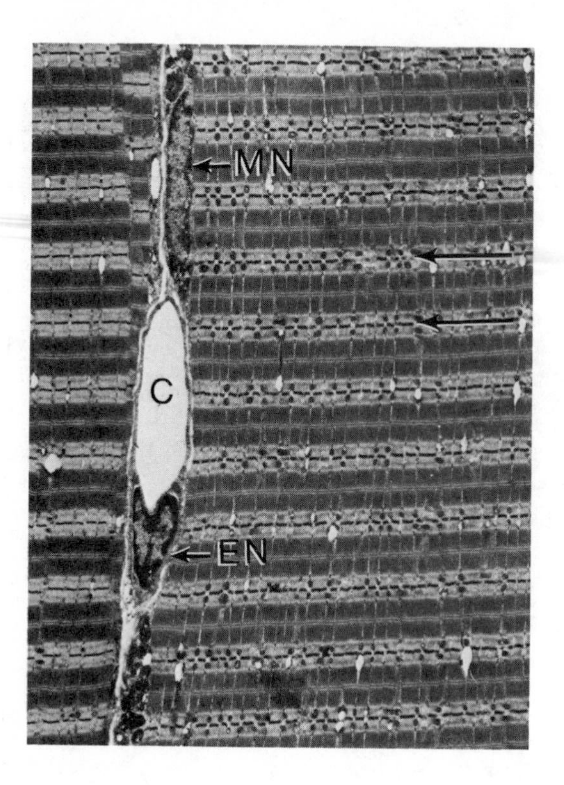

Figure 6-2. Low-power electron micrograph of skeletal muscle. A white muscle fiber is on the left, and a red muscle fiber is on the right, with a capillary (C) between them. Notice that there are many more mitochondria at the I-band level in the red fiber (*at arrows*). The nuclei of a muscle fiber (*MN*) and of a capillary endothelial cell (*EN*) also are indicated. (Prepared by Dr. Carlton Sexton, Department of Radiology, Yale University. Courtesy of Dr. Michael K. Reedy, Department of Anatomy, Duke University.)

9. If skeletal muscle tissue is homogenized, thereby destroying the sarcolemma surrounding muscle cells, large numbers of striated myofibrils will be released. These myofibrils are composed of numerous **sarcomeres**, which are the fundamental structural subunit of all striated muscles.

B. SARCOMERE STRUCTURE AND FUNCTION

1. Structure of Sarcomeres.

a. Each sarcomere essentially is an array of thick, myosin filaments and thin, actin filaments.

b. The total length of a sarcomere varies depending on its contractile status. As individual sarcomeres change length, the entire gross muscle changes length. For example, shortening in sarcomeres produces a shortening of myofibrils, muscle fibers, and gross muscles.

c. Banding Pattern (Fig. 6-3). Individual sarcomeres are divided into definite regions or **bands**, the largest of which are the A band and the I band. These two bands alternate with each other in a repeating pattern along the length of a myofibril.

(1) The **A band** is a region of electron density; it represents the entire length of the thick filaments in a sarcomere. The **M band**, at the center of each sarcomere, represents the thickest point of the thick filaments.

(2) The **I band** is a region exclusively of thin filaments. These thin filaments insert at **Z lines** and project toward the center of each sarcomere. (An individual sarcomere extends from one Z line to the next Z line inclusively.)

(3) The **H band**, located in the middle of the A band, represents a region where there is no overlap of thick and thin filaments.

(4) When a muscle contracts, filaments slide past one another. The H band becomes shorter and I band shortens, but the A band length remains constant. This is because in the A band, everywhere but in the H band, there is an overlap between thick and thin filaments.

2. Regulation of Sarcomere Contraction.

a. The dominant proteins of the sarcomere are actin and myosin. In the presence of adenosine triphosphate (ATP) and at high calcium concentrations, actin and myosin form a labile **actomyosin complex**. Portions of myosin molecules are present in the cross-bridges that span the gap between thick and thin filaments.

b. Sarcomeres also contain regulatory proteins, but in lesser amounts.

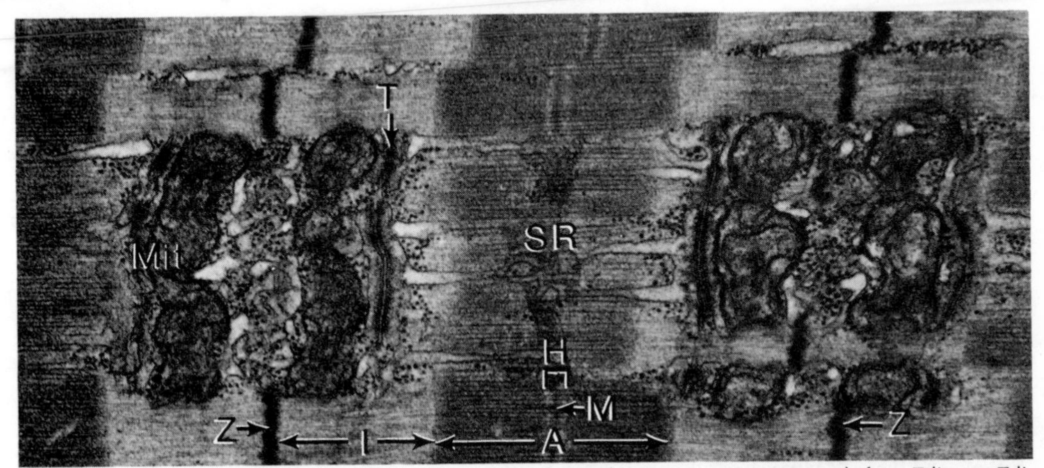

Figure 6-3. High-resolution electron micrograph of skeletal muscle. One sarcomere extends from Z line to Z line and is centered about the narrow M band. This sarcomere is contracted. The H band is narrow due to extensive interdigitation between thin filaments of actin (in the I band) and thick filaments of myosin (in the A band). Mitochondria (*Mit*) lie over the I band in close association with the T tubules (*T*) and sarcoplasmic reticulum (*SR*). (Courtesy of Dr. Helen A. Padykula, Department of Anatomy, University of Massachusetts. Reprinted, in a slightly different version, with permission from Weiss L, Greep R: *Histology*. New York, McGraw-Hill, 1977.)

(1) **Troponin** is a globular regulatory protein that is distributed along the thin filaments at regular intervals.
(2) **Tropomyosin**, another regulatory protein, lies in the groove along the helical thin filaments.
(3) At low calcium concentrations, troponin and tropomyosin together prevent interaction between actin and myosin.

3. Sliding Filaments and Swinging Cross-Bridges.

a. In the sarcomere, the thick and thin filaments are intimately interdigitated.
b. If a cut were made through the A band, perpendicular to the long axis of the sarcomere, a hexagonal array of thin filaments would be seen around each thick filament. In the I band, where there are no thick filaments, only thin filaments would be seen.
c. During a contraction and relaxation cycle, calcium concentration around the myofibrils increases suddenly. This causes a conformational change in the troponin molecule so that the cross-bridge (S-1) binding site of actin becomes exposed and an ATP-actomyosin complex forms.
d. Another conformational change then occurs. The S-1 fragment, still in association with the actin-containing thin filament, swings like an oar in an oarlock and causes the thin filament to slide relative to the thick filament. When this occurs at millions of cross-bridges, the entire sarcomere is shortened suddenly. (The A band remains the same length, but the H band and the I band shorten.)
e. The calcium concentration falls rapidly following the hydrolysis of ATP and the swinging of the cross-bridges. This drop in calcium severs the association between the actin and myosin, and so the contraction stops.
f. ATP is hydrolyzed to adenosine diphosphate (ADP), which subsequently is phosphorylated again to form ATP. Muscle contraction involves substantial ATP hydrolysis. This provides an explanation for the plethora of mitochondria in skeletal and cardiac muscle.

4. The Triad and Contraction.

a. Muscle cells have a highly modified endoplasmic reticulum, called the **sarcoplasmic reticulum**, which is an integral part of the mechanism for regulating calcium concentration around the myofibrils. The sarcoplasmic reticulum forms an anastomosing network of interconnecting cisternae (i.e., flattened, membrane-delimited sacs) which communicate directly with dilated **terminal cisternae**.
b. **T tubules** invaginate from the sarcolemma, near the junction between the A band and the I band. In many locations, T tubules penetrate deeply into the mass of myofibrils, so that each sarcomere is served by two T tubules.
c. The T tubules form intimate contacts with the dilated terminal cisternae of the sarcoplasmic reticulum. One T tubule and two cisternae make up a **triad**. The remainder of the I

bands are covered by collections of mitochondria, which run somewhat perpendicular to the long axis of the myofibrils.

d. The sarcoplasmic reticulum is rich in **calsequestrin**, a protein that binds calcium in the sarcoplasmic reticulum.

(1) In a resting muscle, the calcium concentration in the sarcoplasmic reticulum is quite high. When a nerve impulse to initiate contraction reaches the neuromuscular junction, it causes a depolarization of the sarcolemma.

(2) A traveling wave of depolarization passes down the sarcolemma and the T tubule, similar to the way that an action potential is propagated along a nerve fiber. The depolarization of the sarcolemma also causes a depolarization in the sarcoplasmic reticulum.

(3) Suddenly, the membranes of the terminal cisternae become permeable to calcium ions. These ions rush into the region surrounding the myofibrils and intiate the formation of ATP-actomyosin complex.

(4) After contraction, the calsequestrin in the sarcoplasmic reticulum rapidly decreases the calcium ion concentration around the myofibrils by sequestering calcium ions within the sarcoplasmic reticulum. This completes the contraction cycle.

V. CARDIAC MUSCLE

A. MICROSCOPIC ANATOMY

1. Individual cardiac muscle cells have a single, centrally placed nucleus rather than many nuclei, as are found in each skeletal muscle cell.

2. Like skeletal muscle, cardiac muscle has striations arranged in thin and thick filaments.

3. Cardiac muscle cells may have branches.

4. With the electron microscope, cardiac muscle shows T tubules that are larger than those seen in skeletal muscle. The sarcoplasmic reticulum, however, is not as well-developed as that of skeletal muscle.

5. There are many more mitochondria present in cardiac muscle, and they run parallel to the long axis of the myofibrils in the I bands rather than perpendicular to it, as is found in skeletal muscle.

6. Cells are joined tightly at complex junctions. These junctions are visible with the light microscope as **intercalated discs** (Fig. 6-4).

a. With the electron microscope, intercalated discs are complex interdigitations between adjacent cells.

b. On the inner aspect of apposed cells, numerous filamentous masses serve as anchoring sites for thin filaments of the sarcomere nearest to the end of the cell. These are thought to be similar to the macula adherens in the junctional complex but are more extensive and so are called **fasciae adherentes**.

c. The gap between cells is very small, both at the transverse portion of the intercalated disc and longitudinally between adjacent cells. Recently, it has been shown that extensive **gap junctions (nexus)** exist in these locations, so that cells are in direct ionic (i.e., electrical) communication.

B. DISTRIBUTION

1. Cardiac muscle is restricted largely to the myocardium.

2. It occurs also in proximal portions of the aorta and the venae cavae.

VI. NEUROMUSCULAR JUNCTION

A. STRUCTURE

1. Muscle cells have scattered surface invaginations called **primary synaptic clefts**. The axon for a motor nerve terminates in this cleft.

2. There are many deep folds in the sarcolemma of the primary synaptic cleft. These folds are called either secondary synaptic clefts or **junctional folds**.

3. The contours of the membranes of the nerve and muscle cells follow one another, but the nerve and muscle cells always are separated by a synaptic cleft containing the basement membrane surrounding the entire muscle fiber.

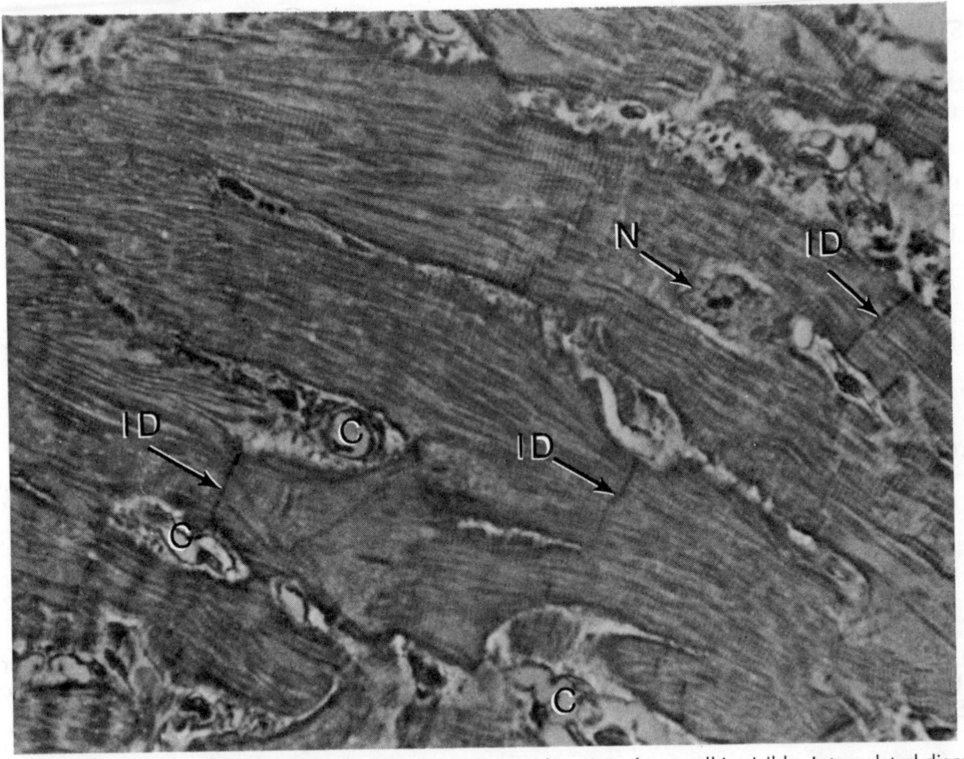

Figure 6-4. Light micrograph of cardiac muscle tissue. The nucleus (*N*) of one cell is visible. Intercalated discs (*ID*) are intercellular junctions between individual muscle cells. Cardiac muscle tissue has many capillaries (*C*). (Reprinted with permission from Johnson KE: *Histology: Microscopic Anatomy and Embryology*. New York, John Wiley, 1982, p 82.)

4. The muscle cell cytoplasm in the region of the neuromuscular junction has an abundance of mitochondria, rough endoplasmic reticulum, and free ribosomes, perhaps for the synthesis of acetylcholine receptors in the membrane of the muscle cell.

5. The axon innervating a neuromuscular junction lacks myelin. Instead, its associated Schwann cell forms a protective cap over the junction.

B. FUNCTION. Skeletal muscle fibers interact with motor nerves at the neuromuscular junction.

1. A wave of depolarization passes down the nerve fiber to the neuromuscular junction.

2. The nerve fiber contains **synaptic vesicles** filled with **acetylcholine**. When released into the synaptic cleft, this neurotransmitter causes a depolarization of the sarcolemma which, in turn, is transmitted to the T tubule.

3. The action potential passes down the T tubule to the terminal cisternae of the sarcoplasmic reticulum, causing a sudden increase in calcium ion permeability and thus triggering a contraction.

STUDY QUESTIONS

Directions: The question below contains five suggested answers. Choose the **one best** response to the question.

1. In a muscle fiber that is stretched beyond the point where it can generate any tension upon stimulation, the

(A) thin filaments overlap the thick filaments so much that the H band is narrow
(B) thin filaments do not overlap the thick filaments, so the H band is broad
(C) thick filaments overlap the thin filaments so little that the H band is narrow
(D) thick and thin filaments overlap extensively, causing the I band to shorten
(E) thick and thin filaments overlap hardly at all, resulting in a small M band

Directions: The groups of questions below consist of lettered choices followed by several numbered items. For each numbered item select the **one** lettered choice with which it is **most** closely associated. Each lettered choice may be used once, more than once, or not at all.

Questions 2–6

Match the following.

(A) Troponin
(B) Tropomyosin
(C) Both
(D) Neither

2. Present in thick filaments
3. Present in thin filaments
4. Present in the I band
5. Present in the H band
6. Present in the A band

Questions 7–9

Match the following.

(A) Thick filaments
(B) Thin filaments
(C) Both
(D) Neither

7. Are absent in the A band
8. Insert into the Z line
9. Are attached permanently to cross-bridges

ANSWERS AND EXPLANATIONS

1. The answer is B. (*IV B*) As a sarcomere is stretched, Z lines move further apart, and eventually the thin filaments no longer interdigitate with thick filaments. When this happens, tension can no longer be generated.

2–6. The answers are: 2-D, 3-C, 4-C, 5-D, 6-C. (*II A 3*) Troponin and tropomyosin are regulatory proteins that are closely associated with the thin, actin-containing filaments. They are absent from thick filaments. The I and A bands contain thin filaments and therefore contain both proteins. No thin filaments are present in the H band, and therefore neither protein is present here.

7–9. The answers are: 7-D, 8-B, 9-A. (*IV B*) Thick and thin filaments overlap in the A band. Thin filaments insert into Z lines. Cross-bridges are attached permanently to myosin-containing thick filaments; they attach only transiently to thin filaments during contraction.

7
Nervous Tissue

I. INTRODUCTION

A. DIVISIONS

1. The nervous system incorporates the central nervous system (CNS), which includes the brain and spinal cord, and the peripheral nervous system (PNS), which is the collection of nerves and ganglia scattered throughout the body.
2. These two systems are interconnected both structurally and functionally.

B. COMPONENTS

1. The **neuron** is the basic cellular element of the nervous system, and its structure varies widely throughout the nervous system.
   a. In the spinal cord, for example, motor neurons have their cell bodies in the anterior (ventral) horn and their axonal terminations on muscle fibers that are as much as a meter away. These motor neurons conduct motor impulses from the spinal cord to the skeletal muscles.
   b. Other neurons in the CNS may have an extremely complex dendritic arborization or may be small interconnecting cells.
2. **Glial Cells.** The nervous system has an assortment of glial cells, which usually have a subsidiary role unrelated to communication. Astrocytes and Schwann cells are examples of glial cells.
3. The **brain** is the central organ for the coordination and regulation of the human body. It is the seat of thought, emotion, and creativity. It controls speech, movement, behavior, and a broad range of intellectual and emotional functions.
   a. The nervous activity of the brain is transmitted to the body by way of the spinal cord and PNS.
   b. Sensory input concerning the state of the body and the effects of various stimuli are relayed back to the brain via the PNS and the spinal cord.
4. The **autonomic nervous system** (ANS) regulates many automatic nervous functions that control blood flow under different physiologic states. The ANS also coordinates digestion and other visceral muscular activity and controls certain involuntary aspects of sexual activity and waste elimination.

C. NEURON STRUCTURE

1. Neurons have a **cell body**, usually one **axon**, and a collection of **dendrites** that vary in number and morphology. Both the axon and the dendrites are long cytoplasmic processes which radiate from the cell body.
2. The cytoplasm within the cell body is called the **perikaryon**.
   a. The nucleus and associated organelles, most notably the rough endoplasmic reticulum (RER), usually are prominent in the perikaryon.
   b. Proteins are synthesized actively in the perikaryon and then are transported radially, especially to the axon. This transport is related to a bulk cytoplasmic transport.
   c. With the electron microscope the perikaryon reveals crowded clusters of RER and unbound ribosomes. With the light microscope these clusters are seen as **Nissl bodies**, which stain intensely with various basic dyes and are the reason for the extremely basophilic cytoplasm.
   d. Numerous mitochondria are interspersed between clusters of RER and ribosomes. Also in the perikaryon are abundant cytoplasmic microtubules and microfilaments and a prominent Golgi apparatus.

D. MYELINATION. Myelin is an insulating material that surrounds axons in the nervous system.

1. Myelin in the CNS is produced by oligodendroglial cells.
2. Myelin in the PNS is produced by Schwann cells.

E. DEVELOPMENT

1. The nervous system develops from the **neural plate**, an ectodermal thickening on the dorsal surface of the embryo. This plate rapidly folds into a closed tubular structure.
2. Just before the neural folds fuse to form the **neural tube, neural crest** cells leave the apices of the folds. These cells then disperse widely throughout the body and give rise to a diverse collection of cells such as pigment cells and those found in some skull bones and the adrenal medulla.
3. The neural tube develops cranially to give rise to the brain and caudally to give rise to the spinal cord.
4. Much of the ANS is derived from the neural crest.

II. NEURONAL VARIATION IN THE CNS

A. GENERAL CONSIDERATIONS

1. In the brain and spinal cord are found **gray matter** and **white matter**. The gray matter is relatively rich in the cell bodies of neurons; the white matter is relatively rich in myelin.
2. Throughout the CNS, cell bodies of neuronal elements cluster to form macroscopic **nuclei**, and large bundles of myelinated nerve fibers run together in gross **tracts**.
   a. In the spinal cord the gray matter is centrally placed and divided into an anterior horn, an intermediate horn, and a posterior horn. These horns are associated with motor neurons, autonomic neurons, and sensory neurons, respectively. In the spinal cord the gray matter is surrounded by white matter, which is composed of nerve axons that in many cases contain myelin.
   b. In the cerebral cortex and cerebellum there is an addition of a superficial layer of cortical gray matter.

B. HISTOLOGIC CONSIDERATIONS. Morphologic variations of neurons involve such factors as the length of the axon, number of dendrites, shape of the cell body, and size of the entire cell.

1. Neurons vary tremendously in shape. Also, they can vary in diameter from somewhat less than that of an erythrocyte to more than 100 μm.
2. Neurons are classified as multipolar, bipolar, or unipolar.
   a. Most neurons in the CNS are multipolar; except in sensory epithelia, bipolar neurons are rather uncommon.
   b. In embryos, however, bipolar neurons are somewhat common. Many of these neurons later undergo a process where the axon and dendrite become closely associated. This results in cells that appear unipolar but in fact are modified bipolar neurons called **pseudounipolar neurons**. These cells often are encountered in ganglia associated with satellite cells.
3. **Neuropil.** Each human has more than ten billion neurons and probably five times as many glial cells. Within the various functional nuclei in the brain are found large collections of nerve cell bodies and associated axons, dendrites, and glial cells. Each of these complex meshworks of processes is called a **neuropil**.
   a. The regional variations throughout the CNS partially are a result of variations in both the types of neurons present in the region and the organization of the communicating processes in the surrounding neuropil.
   b. It has been estimated that nearly half of the neuronal cytoplasm in the CNS is in the neuropil; the remainder is in the nerve cell bodies.

C. NEURON TYPES

1. **Pyramidal Cells** (Fig. 7-1).
   a. Each of these cells in the cerebral cortex has a roughly pyramidal cell body. A slender axon projects from the base of this body.
   b. Pyramidal cells also have extensive dendrites, each dendrite having many long branches and multiple regions where hundreds of synaptic terminals from other cortical neurons make contact.

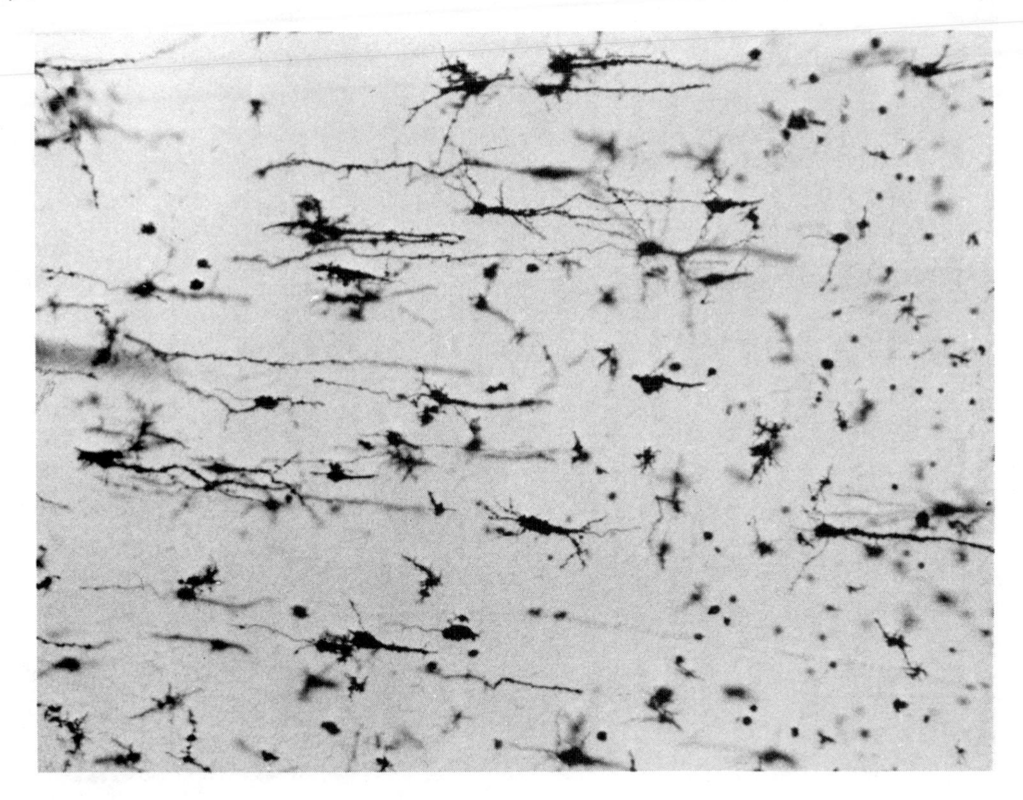

Figure 7-1. Pyramidal neurons are common in this light micrograph of a Golgi preparation of the cerebral cortex of a monkey. (Courtesy of Dr. Ernest N. Albert, Department of Anatomy, George Washington University.)

2. **Purkinje Cells** (Fig. 7-2).
   a. In the cerebellar cortex, at the boundary between the molecular and the granular layers, is found the Purkinje cell.
   b. This cell has a rounded cell body with a slender axon projecting through the granular layer (which is part of the cortical gray matter) into the white matter.
   c. It has three to five main dendrites which immediately break into a lavish dendritic arborization in one plane like a fan.
   d. Some estimate that there are several hundred thousand synaptic terminal swellings called **boutons** on the dendritic arborization of each Purkinje cell. Considering that there also are hundreds of thousands of Purkinje cells, the total number of synapses on **all** Purkinje cells is astronomical.

3. **Golgi type I neurons** have long axons that begin in the CNS and terminate at a distance in the CNS or peripherally. Motor neurons of the anterior horn and cerebral pyramidal cells both are Golgi type I neurons.

4. **Golgi type II neurons** have short axons that begin and end in restricted regions within the CNS. The bipolar cells of the retina and olfactory epithelium both are Golgi type II neurons.

D. **MOTOR NEURONS** form the basic structural unit of nervous tissue in the anterior horn of the spinal cord. The following are characteristic features of motor neurons.

1. **Structural Features.**
   a. In addition to a single axon, motor neurons have several dendrites as processes from the cell body. These dendrites communicate with other neurons by way of synapses.
   b. With the electron microscope, dendritic processes show reduced amounts of endoplasmic reticulum; Nissl bodies and Golgi apparatus, however, are absent. Instead, these processes are loaded with microtubules—which are 17 nm in diameter and called **neurotubules**—and microfilaments—which are 10 nm in diameter and called **neurofilaments**. Also scattered within these processes are small mitochondria.
   c. Especially in a large motor neuron, the nucleus is large, euchromatic, and has a prominent nucleolus.

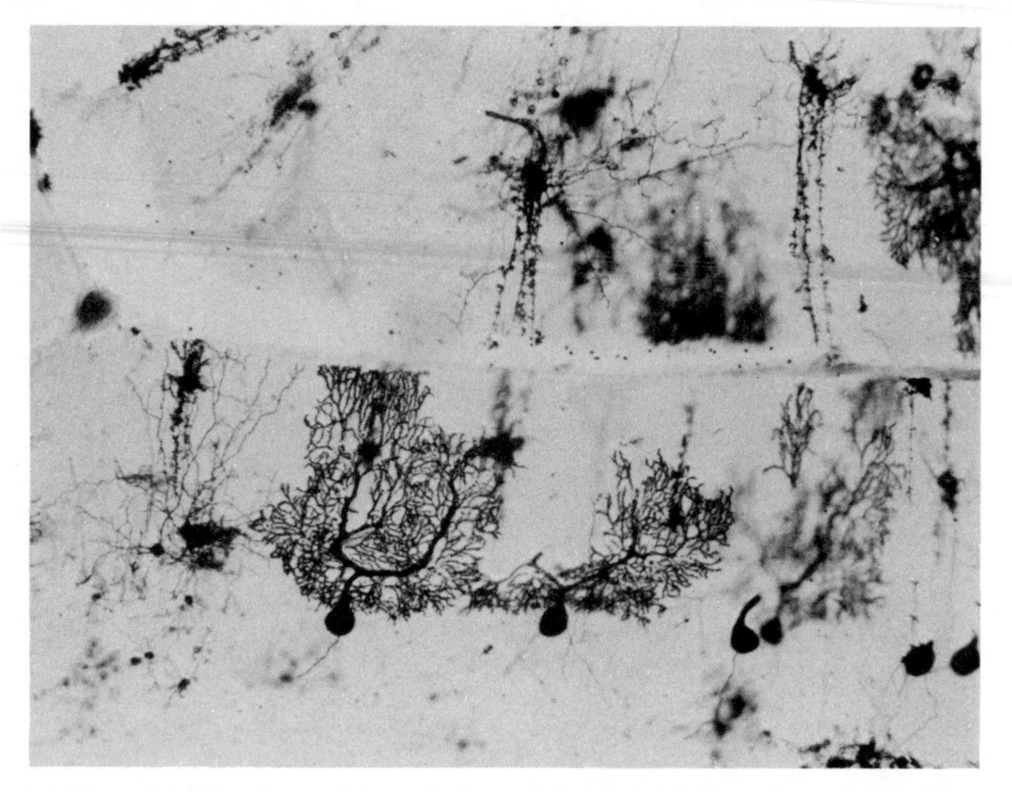

Figure 7-2. Several Purkinje cells are evident in the bottom half of this light micrograph of a Golgi preparation of the cerebellum of a monkey. The Purkinje cell has a rounded cell body, a complex dendrite arborization, and a slender axon that projects from the cell body. (Courtesy of Dr. Ernest N. Albert, Department of Anatomy, George Washington University.)

2. **Functional Features.**
   a. Like all neurons, motor neurons are active secretory cells. They secrete large numbers of synaptic vesicles (i.e., membrane-delimited packets of the neurotransmitter **acetylcholine**) at the motor plate.
   b. The neurotubules and neurofilaments are considered participants in both the rapid anterograde (i.e., toward the neuromuscular junction) and the slower retrograde (i.e., toward the body) transport of cytoplasm that occurs in the motor neuron.
   c. Motor neurons contact muscle fiber bundles that they innervate at the neuromuscular junctions. Here, the Schwann cells do not myelinate the axon, but they remain in close contact with it.

III. NEUROGLIA, or simply **glia**, is a term used to describe the collection of non-neuronal cells in the nervous system. Glia vary considerably in structure and function.

A. CNS GLIA

1. **Ependymal cells** are remnants of the highly proliferative elements of the neural tube. They line the brain ventricles and the lumen of the spinal cord. In addition, they form a layer of cells over the choroid plexus.

2. **Astrocytes** are relatively large cells with many processes. Many of these processes are closely associated with blood vessels, suggesting that astrocytes may be involved in controlling passage of substances from the blood into the brain parenchyma. There are two varieties of astrocytes in humans.
   a. Fibrous astrocytes are most abundant in white matter and have relatively few cytoplasmic filaments.
   b. Protoplasmic astrocytes are most abundant in gray matter and have numerous cytoplasmic filaments.

3. **Oligodendroglia** are smaller and have relatively fewer processes than astrocytes. Oligodendroglia function in myelination in the CNS.

4. **Microglia** are mesodermally derived phagocytic cells that are part of the mononuclear phagocyte system. These are the smallest CNS glia.

B. PNS GLIA

1. **Schwann cells** are neural crest derivatives that are responsible for myelination in the PNS.

2. Sometimes, **satellite cells** also are included in the list of glia. Satellite cells are associated with neurons in peripheral ganglia, where they form a kind of cellular capsule around the neuronal cell bodies.

IV. PERIPHERAL NERVES.

Large nerves encountered in the gross anatomy laboratory are collections of many axons of motor and sensory nerves running to and from a particular part of the body into the spinal cord or brain. These axons, with their myelinating Schwann cells, are called **nerve fibers**.

A. MICROSCOPIC ANATOMY

1. In a peripheral nerve, such as the ulnar nerve, each nerve fiber is surrounded by a delicate connective tissue (CT) sheath called the **endoneurium**. These CT sheaths are not nervous tissue. (Only neurons and glia are properly considered nervous tissue.) The endoneurium surrounds the basement membrane of Schwann cells with a delicate sheath of collagen fibers and flattened fibroblastic cells.
2. Bundles of axons with Schwann cells and endoneurial fibroblasts are gathered into collections known as **fascicles**. These bundles are encapsulated by a CT sheath of collagen fibers and fibroblasts known as the **perineurium**.

3. In a peripheral nerve, many fascicles then are gathered and surrounded by a CT capsule of collagen fibers, adipocytes, and blood vessels. This CT capsule is called the **epineurium**. When passing from the CNS into progressively smaller branches of nerves, these capsules are reduced in thickness as the overall diameter of the entire nerve trunk decreases.

B. FREE NERVE ENDINGS

1. In tendons, fine branches of nerve fibers mix with collagenous fibers and probably are responsible for sensations of pain.

2. Free nerve endings occur also in the skin, in the mucosa of the respiratory system, and in the cornea, and they are responsible for the high degree of tactile sensitivity found in these areas.

C. SYNAPSE MORPHOLOGY

1. Synapses are sites of anatomic and functional interaction between individual neurons.

2. There is no cytoplasmic continuity between individual neurons at synapses. Instead, neurons are segregated by the **neurolemma** that surrounds each neuronal element or process.

3. At synapses, axonal processes typically show terminal swellings known as **boutons**, and the neurofilaments that are abundant in the rest of the axon terminate abruptly.

4. The boutons are loaded with mitochondria and numerous small **synaptic vesicles** that are 40–65 nm in diameter. The presynaptic and postsynaptic membranes are closely apposed here but remain separated by a definite 20-nm **synaptic cleft**.

5. The synaptic vesicles contain **neurotransmitters** such as **acetylcholine**. When an action potential reaches the presynaptic membrane, it stimulates the release of these vesicles; the neurotransmitter substance, in turn, initiates a new wave of depolarization at the postsynaptic membrane of the next neuron in the chain. In this manner, nerve impulses are passed from one nerve cell to the next, even though these individual cells are not in direct anatomic communication.

6. **Norepinephrine**, **dopamine**, and **serotonin** also serve as neurotransmitters.

V. MENINGES.

The brain and spinal cord are surrounded by meninges derived from mesoderm and from the neural crest. Although meninges are not true nervous tissue, it is appropriate to discuss them here.

A. LAYERS

1. **Dura Mater.**
   a. The outermost layer is called the dura mater. It is a tough, thick CT capsule surrounding the entire CNS.

b. In the spinal canal the dura is separated from the periosteum of the vertebrae; within the skull the dura is loosely associated with the skull bones and their periosteum.
c. Blood vessels and, in places, large venous sinuses are present in the dura.
d. The inner aspect of the dura is lined by a mesothelium, and it faces onto a thin serous cavity called the **subdural space**.

2. Arachnoid.
a. The arachnoid is an avascular intermediate layer with a thin capsule of CT cells and fibers as well as a series of anastomosing CT trabeculae that project away from the arachnoid toward the surface of the brain.
b. The large area between the arachnoid and the inner **pia mater** is called the **subarachnoid space**.

3. Pia Mater.
a. The main blood supply of the CNS is carried in the pia mater.
b. The arachnoid and pia mater are so intimately linked that many histologists refer to them together as the **pia-arachnoid layer**.
c. The subarachnoid space is quite wide along the length of the spinal cord and over most of the brain.
(1) In several locations over the brain, the subarachnoid space is greatly enlarged to form **cisternae**.
(2) The largest of these, the **cisterna magna**, lies just posterior to the cerebellum and has three holes through which the fourth ventricle of the brain communicates with the subarachnoid space.

B. CEREBROSPINAL FLUID AND CHOROID PLEXUS

1. The fluid in the subarachnoid space, the ventricles of the brain, and the central canal of the spinal cord is called the **cerebrospinal fluid** (CSF).

2. The walls of the brain near the cisterna magna are extremely reduced in thickness. They are composed solely of a highly convoluted layer of ependymal cells coating a complex anastomosing network of capillaries from the pia-arachnoid.

3. These specialized pial capillaries, called the **choroid plexus**, are fenestrated and modified to produce CSF, a blood filtrate important for the metabolism and protection of the brain parenchyma.

4. CSF is produced largely at the choroid plexus and drains from the CNS back into the systemic circulation at the **arachnoid villi**. From these projections of the subarachnoid space, CSF drains into the enlarged subdural spaces that run along the superior midline of the brain.

VI. DEVELOPMENT OF THE NERVOUS SYSTEM

A. EARLY DEVELOPMENT

1. Nervous tissue proper is derived wholly from the ectoderm of the embryo.

2. During the third week gestation, after the establishment of the primary germ layers, the dorsal ectoderm of the embryo thickens considerably to form a **neural plate**.

3. At the end of the third week, this neural plate begins to fold upon itself—first in the future cervical region and later craniad and caudad to that point—to form a **neural groove** and a **neural tube**.

4. As the edges of the neural groove come together but just before these folds meet and fuse, **neural crest cells** migrate from the neuroepithelium.

5. After the neural folds fuse to form a tube, the surface ectoderm of the embryo grows over the entire neural tube. Thus, the entire nervous system comes to lie beneath the surface ectoderm.

B. NEUROEPITHELIUM

1. The neural plate and neural tube form into a pseudostratified epithelium called the **neuroepithelium**.

2. Recent autoradiographic studies have shown that the height of cells in this epithelium and their relationship to the basement membrane depend on their cell-cycle kinetics.
a. Tall, thin columnar cells engage in DNA synthesis and become progressively shorter as they prepare for mitosis. The nuclei migrate toward the lumen of the neural tube prior to cell division; following cell division, the daughter cells again become elongated, and their

nuclei migrate away from the lumen toward the external limiting membrane (i.e., basement membrane).

b. Different cells in different parts of the cell cycle have their nuclei at different levels above the basement membrane.

3. After division, some neuroepithelial cells differentiate into **neuroblasts** and **glioblasts**. The former differentiate into a wide variety of neurons, and the latter differentiate into glial cells.

4. As neuroblasts in the spinal cord and brain grow axons that project outward from the neuroepithelial layer and as these axons become progressively myelinated, the neuroepithelium that lines the lumen of the CNS persists and becomes surrounded by an increasing thickness of neuroblast derivatives.

C. MANTLE AND MARGINAL LAYERS

1. In the spinal cord the cell bodies of developing neurons accumulate in a cellular **mantle layer**—the precursor of gray matter; the mantle layer is surrounded by a peripheral **marginal layer**—the precursor of white matter.

2. Later, the mantle is subdivided by the **sulcus limitans** into a posterior (dorsal) **alar plate** and an anterior (ventral) **basal plate**. These represent, respectively, the future posterior (sensory) and anterior (motor) horns of the spinal cord.

D. BRAIN DEVELOPMENT

1. **General Description.**
   a. The **prosencephalon** or forebrain subdivides into the
      (1) Telencephalon, which forms cerebral hemispheres
      (2) Diencephalon, which forms the optic cup, infundibulum, thalamus, and hypothalamus
   b. The **mesencephalon** or midbrain develops only slightly but forms brain structures that receive sensory input from the head and that control movement of certain structures (e.g., the eyes through the oculomotor muscles).
   c. The **rhombencephalon** or hindbrain subdivides into the
      (1) Metencephalon, which forms the cerebellum and pons
      (2) Myelencephalon, which forms the medulla oblongata

2. **Detailed Description.** Many of the developmental events that lead to the formation of the spinal cord occur also in the formation of the brain. For example, in the hind-brain, a sulcus limitans forms as a dividing line between a posterior alar plate area and an anterior basal plate area.
   a. **Myelencephalon.**
      (1) The myelencephalon develops much like the spinal cord, except that the walls of the neural tube move apart dorsally, similar to the unfolding of a flower. Consequently, the roof plate becomes extremely thin.
      (2) In the myelencephalon, the basal plate develops several different nuclei. These become the motor innervation for voluntary muscles in the head and neck and for involuntary muscles in the viscera.
      (3) The alar plate also develops several nuclei which connect with sensory innervation for the special senses in the head and for the interoceptive sense in the viscera.
      (4) The thinned roof of the myelencephalon becomes richly vascularized in several areas, giving rise to the choroid plexus.
      (5) At four months gestation, two lateral **foramina of Luschka** and a single medial **foramen of Magendie** develop as perforations in this thin layer. These allow free communication between the brain ventricles and the subarachnoid space for the circulation of CSF.
   b. **Metencephalon.**
      (1) Through extensive proliferation, the primitive cerebellar neuroepithelium forms a mantle layer of neuroblasts and a marginal layer of numerous myelinated neuronal processes.
      (2) Some neuroblasts that form in the mantle layer have a continued capacity for proliferation. These migrate through the marginal layer and form an **external granular layer** of cells.
         (a) Cells in this layer retain their proliferative ability for a long time, unlike other neuronal elements that form from the neuroepithelium.
         (b) After their final division, these cells differentiate into nonproliferative neuroblasts.
      (3) While cells of the external granular layer are proliferating, Purkinje cells differentiate and produce dendritic arborizations.
      (4) Simultaneously, other cells in the external granular layer are differentiating into a variety of Golgi type II neurons, which migrate away from the free surface of the cerebellum.
      (5) The end result of this migration and differentiation is a striking stratification of the

cerebellum. The layers include: an **outer molecular layer** composed of Purkinje dendrites and some type II neurons derived from the external granular layer; a layer of **Purkinje cell bodies**; an **internal granular layer** of type II neurons that have migrated in from the external granular layer; and finally an inner mass of white matter, composed of many myelinated axons. Nearest to the lumen of the cerebellum are found other collections of nerve cell bodies in another layer of gray matter.

E. MYELINATION

1. Many neurons in the PNS have axons that are myelinated by a series of glial elements known as **Schwann cells**.
   a. Myelin is formed by multiple layers of Schwann cell membrane being wrapped around the axon during development of the nervous system. This "wrapping" process is caused either by the rotation of the entire Schwann cell or by the rotation of only the cell's **mesaxon**.
   b. As the mesaxon grows and becomes wrapped around the axon, it fuses with itself, leaving layer upon layer of membrane fusion sites.
   c. The thick coat of fused membranes provides an insulation around the axon, so that ions cannot flow out of the axon into the extracellular space around the axon.
   d. A long axon from a motor neuron, for example, is covered by thousands of Schwann cells. Where one myelinating Schwann cell ends and another begins, there is a **node of Ranvier** (Fig. 7-3). At this site, the axon is exposed directly to the fluid and ions surrounding it.

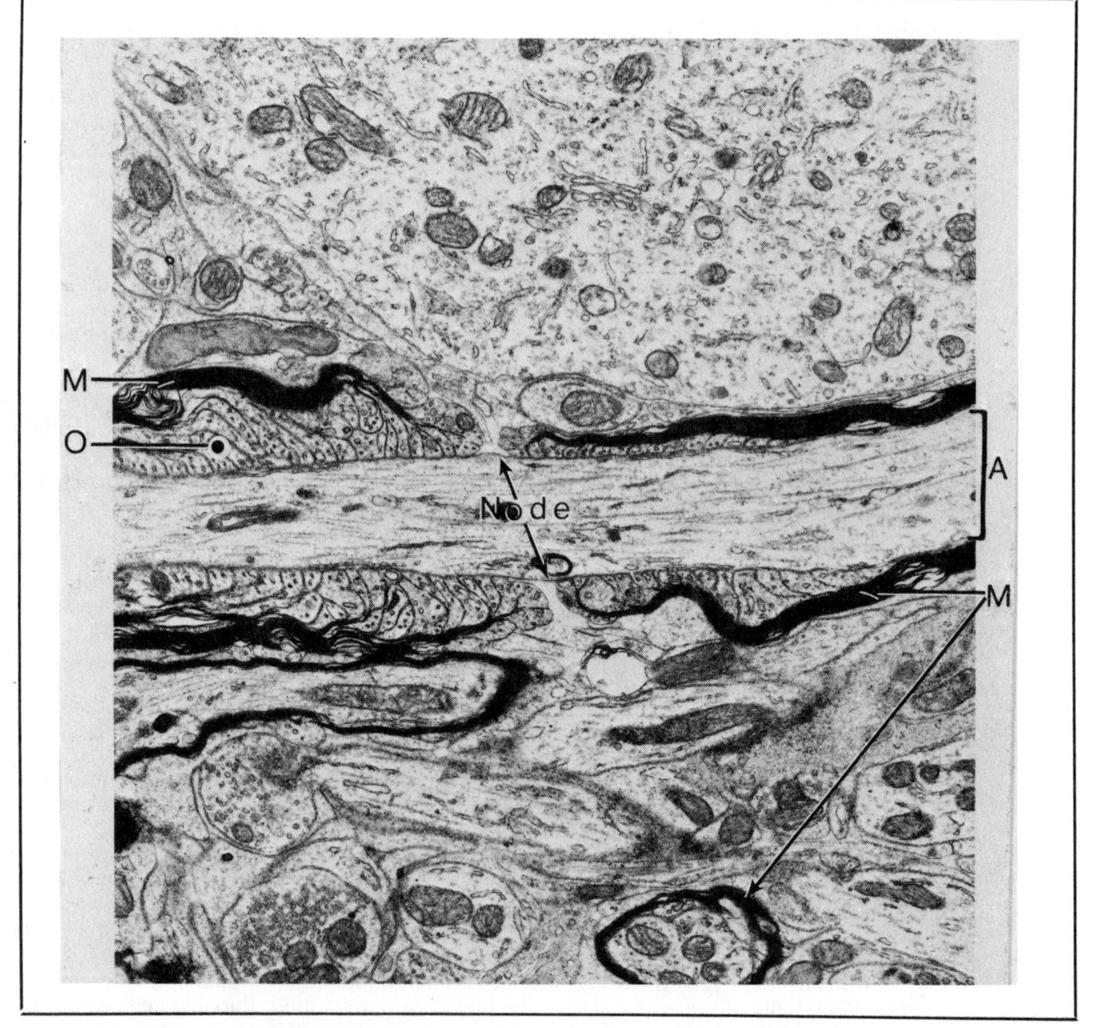

Figure 7-3. Electron micrograph of a node of Ranvier surrounding a neuronal axon (*A*) in the medulla oblongata of a rat. The myelin sheaths (*M*) surround axons and are produced by oligodendroglia cells (*O*). (Courtesy of Dr. Jeffrey M. Rosenstein, Department of Anatomy, George Washington University.)

e. As a result of these periodic breaks in insulation, nerve impulses, which actually are ionic fluxes, pass down the axon via saltatory conduction (i.e., skipping from node to node). Thus, nerve conduction velocity is much greater in myelinated nerve fibers than in unmyelinated fibers.
f. Generally, the thicker the myelin sheath, the greater the internodal distance and conduction velocity. Conversely, when the myelin sheath is thin, the internodal distance and the conduction velocity are small.

2. Even so-called unmyelinated nerve fibers in the PNS are associated with a Schwann cell. In these cases, however, several axons from many different unmyelinated fibers are found embedded within the cytoplasm but surrounded by the membrane of the Schwann cell.
3. **Schmidt-Lantermann clefts** are conical defects in the myelin sheath. These move up and down the Schwann cell, bringing Schwann cell cytoplasm and its nutrients to the living membranes of the myelin sheath.
4. Oligodendroglial cells are responsible for myelination in the CNS.

STUDY QUESTIONS

Directions: The question below contains four suggested answers of which **one or more** is correct. Choose the answer

A if **1, 2, and 3** are correct
B if **1 and 3** are correct
C if **2 and 4** are correct
D if **4** is correct
E if **1, 2, 3, and 4** are correct

1. Astrocytes are nervous tissue components that

(1) are glioblast derivatives
(2) are larger than microglia
(3) have processes closely related to blood vessels
(4) outnumber pyramidal neurons in the cerebrum

Directions: The group of questions below consists of lettered choices followed by several numbered items. For each numbered item select the **one** lettered choice with which it is **most** closely associated. Each lettered choice may be used once, more than once, or not at all.

Questions 2–6

For each description of a myelinated nerve fiber, choose the appropriate lettered structure shown in the micrograph below.

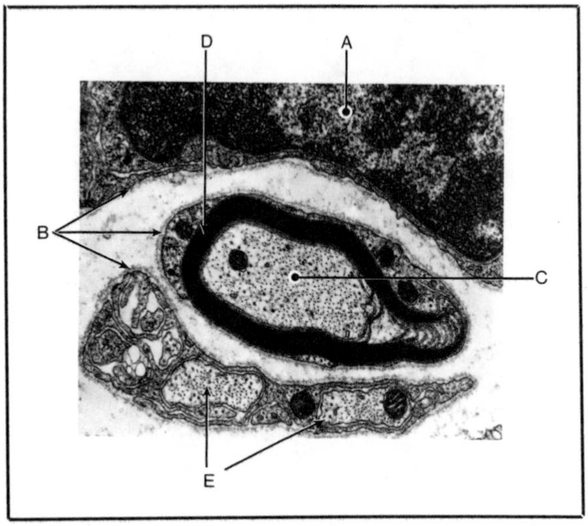

Courtesy of Dr. Jeffrey M. Rosenstein, Department of Anatomy, George Washington University.

2. Axon of myelinated nerve
3. Nucleus of Schwann cell
4. Axon of unmyelinated nerve
5. Basement membrane of Schwann cell
6. Multiple layers of Schwann cell membrane

ANSWERS AND EXPLANATIONS

1. The answer is E (all). (*II A 2*) Astrocytes are glial cells that are derived from glioblasts. They are larger than microglia and have processes close to blood vessels. There are many more glial cells than neuronal cells in the nervous system.

2–6. The answers are: 2-C, 3-A, 4-E, 5-B, 6-D. (*II E*) The myelin sheath (D) surrounds an axon (C). Myelin is produced by Schwann cells (A) in the peripheral nervous system. The Schwann cell has a basement membrane (B) around it.

8
Peripheral Blood

I. INTRODUCTION. Blood is a highly modified type of connective tissue (CT). It is mesodermally derived, is ubiquitous in the body, and is composed of cells, fibers, and an extracellular amorphous ground substance.

A. COMPONENTS

1. The **cells** in blood include
   a. Erythrocytes
   b. Leukocytes
   c. Platelets
2. The **fibers** in blood are potential fibers of **fibrinogen**, which is converted to fibrin fibers generated during **clotting.** Blood clots when platelets interact with and respond to **collagen fibers** that they encounter when they leave the circulatory system in response to a wound. Clotted blood is a solid CT which rapidly stops blood flow from a wound and provides some cells for wound healing.
3. The **fluid** and **proteins** in blood plasma represent the extracellular amorphous ground substance.

B. FUNCTIONS

1. Blood carries **oxygen** and **nutrients** to the cells of the body and carries **waste** materials away from cells to the kidneys and lungs.
2. Blood carries many cellular elements of the **immune system**.
3. Blood is involved in the **homeostasis** of the human body.

II. ERYTHROCYTES

A. STRUCTURE

1. An **erythrocyte** or red blood cell (RBC) is a biconcave disc that is approximately 7–8 μm in diameter and 2 μm thick at the edge.
2. There are approximately 5×10^6 RBCs/mm^3 of blood.
3. Each RBC is bound by a typical plasma membrane but otherwise lacks organelles.
4. About 33 percent of RBC mass is devoted to **hemoglobin**, a protein containing **heme**. Heme is an iron porphyrin with a high capacity to **bind oxygen**.
5. An individual RBC has a large surface area of approximately 130 μm^2. The total surface area of all RBCs in a normal adult's 6 liters of blood, however, is about 2000 times greater than the surface area of the body. This enormous surface area partly explains the large gas transport and binding capacity of blood.
6. RBCs represent about 45 percent of the total volume of blood. When a sample of blood is taken from a normal person and then is centrifuged briefly to sediment the RBCs, the **hematocrit** is approximately 45 (i.e., 45 percent of the blood's total volume is occupied by packed cells). Hematocrit can vary greatly depending on a person's state of health.

B. FUNCTION

1. RBCs function in oxygen and carbon dioxide transport.
2. They do most of their work within capillaries, the smallest branches of the cardiovascular system. In some instances, the diameter of the capillaries is considerably less than the

diameter of the RBCs that must pass through these capillaries (e.g., in the capillary bed in the lungs). Consequently, the RBCs are deformed when squeezed through these small channels.

3. This stress on RBCs in the microvasculature probably contributes to the short life of the cells. After a life span of approximately 120 days, RBCs are removed from circulation and destroyed. RBC destruction occurs in the spleen, liver, and bone marrow. The exact mechanism controlling the removal of RBCs from the circulation is still being studied.

III. LEUKOCYTES. In addition to the RBCs, peripheral blood contains a collection of leukocytes.

A. CLASSIFICATION. There are two types of leukocytes.

1. **Granulocytes** include
   - **a.** Neutrophils
   - **b.** Eosinophils
   - **c.** Basophils

2. **Agranulocytes** include
   - **a.** Lymphocytes
   - **b.** Monocytes

B. GRANULOCYTES. Blood films usually are stained with a mixture of stains such as Wright's stain, Giemsa stain, or May-Grünwald stain. These all are mixtures of charged and uncharged dyes which stain the granules of the different granulocytes so that they are distinct.

1. **Neutrophils.** The most common leukocyte in normal human peripheral blood is the neutrophil or polymorphonuclear leukocyte (PMN).
   a. Structure.
   - **(1)** These cells are 12–15 μm in diameter. They are the most numerous leukocytes, representing 40–60 percent of all leukocytes. Approximately 4500 neutrophils occur in every mm^3 of peripheral blood.
   - **(2)** Small, 0.2 μm-diameter granules comprise 80 percent of neutrophilic granules; large, 0.4 μm-diameter azurophilic granules comprise the remaining 20 percent.
   - **(3)** Neutrophils have from three to five nuclear lobes.

   b. Function.
   - **(1)** Neutrophils are essential for life. They are the major cells responsible for the **phagocytosis** and later destruction of bacteria.
     - **(a)** In neutrophils, azurophilic granule membranes fuse with the membranes of **phagosomes** to produce a special kind of **phagolysosome**.
     - **(b)** The azurophilic granules are modified lysosomes that contain the usual array of hydrolytic enzymes for the destruction of ingested, invading bacteria.
     - **(c)** Azurophilic granules also contain the enzyme **myeloperoxidase** which catalyzes the formation of bactericidal molecular oxygen from hydrogen peroxide.
     - **(d)** Neutrophils are chemotactically attracted to small peptides and other substances released by the bacteria.
   - **(2)** Sepsis or serious infectious disease often will result in a rapid decrease followed by an increase in the number of neutrophils in a differential count of a blood film. (A blood film with three neutrophils is shown in Figure 8-1.)

2. **Eosinophils** are less common than neutrophils; approximately 200 eosinophils occur in every mm^3 of peripheral blood, representing 5 percent of the leukocytes.
   a. Structure.
   - **(1)** These cells are about the same diameter as neutrophils but otherwise are quite distinct from them.
   - **(2)** The nucleus has two or three lobes, and there is a striking array of large (0.6–1.0 μm-diameter) red or orange **eosinophilic granules**.

   b. Function.
   - **(1)** Eosinophils are present in large numbers in the lungs of patients with bronchial asthma. Here they release specific granules resulting in the formation of numerous **Charcot-Leyden crystals**.
   - **(2)** More importantly, eosinophils are prominently involved in **parasitic infections.** For example, in **schistosomiasis**, an infection by parasitic blood flukes, specific antibodies are produced. These specific antibodies, in turn, potentiate the antiparasitic action of eosinophils.

3. **Basophils** are the rarest leukocyte; approximately five basophils occur in every mm^3 of peripheral blood, representing less than 1 percent of the leukocytes.

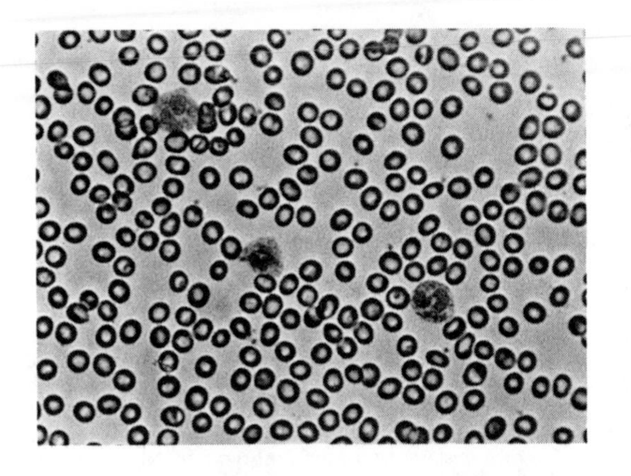

Figure 8-1. Human blood film with many normal erythrocytes and three normal neutrophils.

a. **Structure.**
   (1) Basophils have approximately the same diameter as neutrophils and eosinophils. Also, each basophil contains a nucleus with two or three lobes.
   (2) A mass of small, 0.5 μm-diameter granules often obscures the nucleus. These basophilic granules stain red-purple (i.e., metachromatically).
   (3) With the electron microscope, basophil granules are membrane bound; they appear to have crystalline regions as well, suggesting that they are modified lysosomes.

b. **Function.**
   (1) Basophil granules contain: **histamine**, a potent vasodilator; **heparin**, a glycosaminoglycan with anticoagulant activity; and **slow-reacting substance** (SRS), a slow-acting vasodilator.
   (2) Basophils and **mast cells** are similar both structurally and functionally.
      (a) When certain antigens enter the body, they stimulate the formation of a class of antibodies known as immunoglobulin E (IgE).
      (b) These IgEs bind to the surface of basophils and mast cells, but there is no immediate effect.
      (c) A basophil and mast cell response can result later, however, when the same antigen is present again. The response either is restricted (e.g., in bronchial asthma), or it is more severe and systemic (e.g., in anaphylactic shock from a bee sting).

C. AGRANULOCYTES

1. **Lymphocytes** are the most abundant of the agranulocytes; approximately 2500 occur in every mm^3 of peripheral blood, representing 20–40 percent of the leukocytes.
   a. **Structure.**
      (1) Lymphocytes have a round, densely stained nucleus which occupies most of the volume of the cell. There is only a thin shell of cytoplasm around the nucleus.
      (2) Lymphocytes have a highly variable diameter ranging from 5–8 μm to 15 μm. As the diameter of lymphocytes increases, the relative amount of cytoplasm increases greatly while the nuclear diameter increases only slightly.
      (3) The chromatin of lymphocytes is unevenly stained, exhibiting frequent, densely stained blocks of heterochromatin.
      (4) There are no specific granules in lymphocytes.
      (5) A blood film with two small lymphocytes is shown in Figure 8-2.
   b. **Function.**
      (1) Lymphocytes are key cells in the immune system. They differentiate into plasma cells to secrete antibodies or cells to ward off viral infections, yeast infections, and malignant cells (see Chapter 10 for a more detailed discussion).
      (2) Lymphocytes are not restricted to the peripheral blood but are widely distributed in the CT domains of the body as well as in the lymph nodes, spleen, tonsils, and bone marrow.

2. **Monocytes** occur at a frequency of approximately $300/mm^3$ of peripheral blood, representing about 7 percent of the leukocytes.
   a. **Structure.**
      (1) Monocytes are the largest leukocytes, ranging from 12–18 μm in diameter.
      (2) The cytoplasm is agranular and the nucleus is rounded, with a variable degree of indentation on one side.
      (3) The chromatin of monocytes, unlike that of lymphocytes, forms a somewhat uniformly

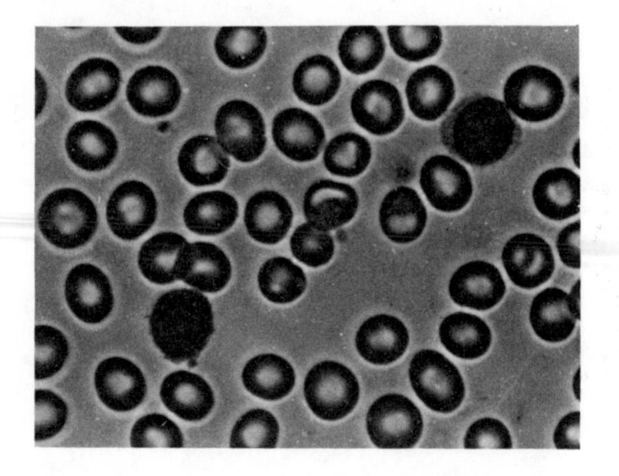

Figure 8-2. Human blood film with many normal erythrocytes and two small lymphocytes. The cell in the lower left has a very thin shell of cytoplasm around the centrally placed nucleus; the cell in the upper right has somewhat more cytoplasm associated with the nucleus. (Reprinted with permission from Johnson KE: *Histology: Microscopic Anatomy and Embryology*. New York, John Wiley, 1982, p 111.)

stained, delicate network.

(4) With the electron microscope, a monocyte reveals a large number of lysosomes and a prominent Golgi apparatus.

b. **Function.**

(1) Monocytes are direct precursors of **macrophages**.

(2) When monocytes leave the closed circulation, they can enter the lungs in the form of alveolar macrophages and remove inspired debris; or they can enter the CT where they also have a phagocytic function and are sometimes called **histiocytes**.

IV. PLATELETS

A. STRUCTURE

1. These are 2–4 μm in diameter and occur in tremendous numbers; approximately 200,000–400,000 platelets occur in every mm^3 of blood.

2. In blood films, platelets have a purple central **granulomere** and a poorly stained peripheral **hyalomere**, and the platelets may be clustered together either in small groups or in large masses.

3. With the electron microscope, a platelet reveals a plasma membrane and some intracellular organelles such as lysosomes and a peripheral band of microtubules, which maintain the platelet's shape of a biconvex lens. Platelets lack a nucleus, have few mitochondria, and have little endoplasmic reticulum.

4. Platelets are fragments of large, multinucleated bone marrow cells called **megakaryocytes**.

5. Platelets contain membrane-bound dense granules which probably contain **serotonin** and other vasomotor substances.

B. FUNCTION.

The primary function of platelets is to initiate **clot formation** to stop bleeding.

1. Platelets have **surface enzymes** that recognize collagen in the extracellular matrix. This recognition triggers a chain of events resulting in clot formation.

2. Platelets aggregate and set up the cascade of enzymatic reactions that convert the potential fibers of blood, called **fibrinogen**, to the actual fibers in a clot, called **fibrin**.

STUDY QUESTIONS

Directions: Each question below contains five suggested answers. Choose the **one best** response to each question.

1. Eosinophils are best described as being

(A) less numerous in peripheral blood than basophils
(B) more numerous in peripheral blood than neutrophils
(C) more numerous in peripheral blood than lymphocytes
(D) more numerous in patients with schistosomiasis
(E) incapable of phagocytosis

2. Platelets are best described as being

(A) inhibited by collagen from aggregating
(B) fragments of megakaryocytes
(C) seen in the peripheral blood with a small nucleus
(D) without a plasma membrane
(E) more abundant than erythrocytes in peripheral blood

Directions: The group of questions below consists of lettered choices followed by several numbered items. For each numbered item select the **one** lettered choice with which it is **most** closely associated. Each lettered choice may be used once, more than once, or not at all.

Questions 3–7

For each description of a peripheral blood film, choose the appropriate lettered structure shown in the adjacent micrograph.

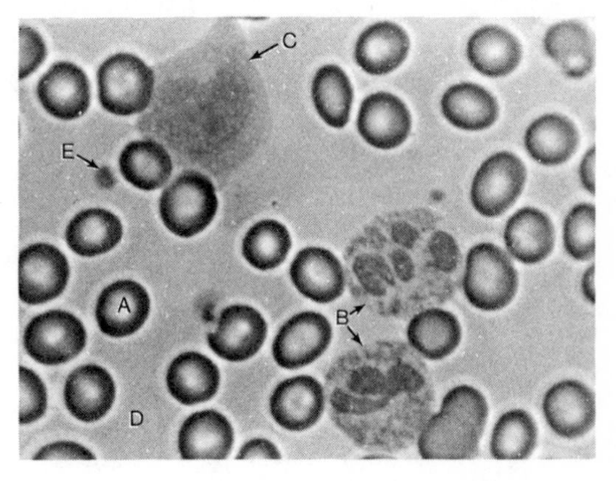

3. Contains hemoglobin, is produced in bone marrow, and is destroyed in the spleen

4. Bacterial phagocyte that contains azurophilic granules

5. Derived from bone marrow precursor and can become a macrophage

6. Granulocyte that can become a phagocyte in the gastrointestinal lamina propria

7. Agranulocyte that can become a phagocyte on the alveolar epithelium

ANSWERS AND EXPLANATIONS

1. The answer is D. (*III B 2*) Eosinophils are more numerous than basophils, less numerous than neutrophils, less numerous than lymphocytes, capable of phagocytosis, and elevated in schistosomiasis and other parasitic infections. They contain characteristic large red granules in their cytoplasm.

2. The answer is B. (*IV A, B*) Platelets are megakaryocyte fragments that are stimulated to aggregate when they are exposed to collagen. They have no nucleus but are surrounded by a plasma membrane. They are not as frequent in blood as erythrocytes.

3–7. The answer are: 3-A, 4-B, 5-C, 6-B, 7-C. (*II; III B, C*) The structure lettered (A) is a red blood cell. The structure lettered (B) is a neutrophil or polymorphonuclear leukocyte. Neutrophils can become phagocytes in many locations including the lamina propria of the gastrointestinal tract. (The lamina propria is a loose irregular connective tissue domain beneath the mucosal epithelium of many visceral organs.) The structure lettered (C) is a monocyte; this cell also can become an alveolar macrophage.

9
Bone Marrow and Hematopoiesis

I. INTRODUCTION. In the central portions of many bones there is a cavity filled with **bone marrow**. This highly specialized reticular connective tissue has both **hematopoietic** and **osteogenic** potential. During the fetal period, the bone marrow in most growing bones, along with the periosteum, contributes to the growth of the bones. This osteogenesis primarily is the work of endosteal bone marrow **osteoblasts**. Other bone marrow stem cells differentiate into **erythrocytes**, **leukocytes**, and **platelets**. In adults, most marrow has undergone a fatty regression, and the modest hematopoietic needs of a healthy adult are met by small amounts of active bone marrow.

II. BONE MARROW

A. TYPES

1. **Red marrow** is characterized by active hematopoiesis.
2. **Yellow marrow** is inactive; here, hematopoietic cells are outnumbered by lipid-laden fat cells.
3. In newborns, virtually all of the marrow is red. Yellow marrow gradually replaces most of the red marrow so that, by the pubertal period, most of the marrow is yellow.
4. In adults, red marrow persists in the skull bones, clavicle, vertebrae, sternum, and pelvic bones.

B. MICROSCOPIC ANATOMY

1. Bone marrow is a complex tissue within the cavity of a bone; it is found in the interstices between bony spicules.
2. These marrow compartments are lined by an **osteogenic endosteum** and are filled with an anastomosing network of **vascular channels**.
3. Blood vessels must penetrate a bone to reach the marrow compartment. After arteries enter marrow, they branch rapidly into small vessels that supply complex **vascular sinuses.** These sinuses then connect to the venous drainage of marrow. The circulation in bone marrow has been shown to be closed.
4. In the compartment between the endosteum and the vascular sinuses there is a **hematopoietic compartment** filled with irregularly shaped cords of either hematopoietic cells or fat cells. (A smear of human bone marrow is shown in Figure 9-1.)
   a. In red marrow, the hematopoietic cells outnumber the fat cells.
   b. In yellow marrow, the fat cells outnumber the hematopoietic cells.
5. The vascular sinuses are lined by a continuous endothelium that has a typical basement membrane and surrounding **reticular cells** and **reticular fibers**. These reticular elements serve, respectively, as adventitial cells for the vascular sinuses and as a stromal framework for the hematopoietic compartment. The reticular cells are capable of being converted to fat cells and may be phagocytic.
6. When newly formed blood cells enter the circulation from the hematopoietic compartment, they must pass through the endothelium of the vascular sinuses. Although it seems likely that newly formed blood cells might pass **between** endothelial cells, it is believed that blood cells pass **through** the cytoplasm of the endothelial cells by way of transient **transcellular pores**.

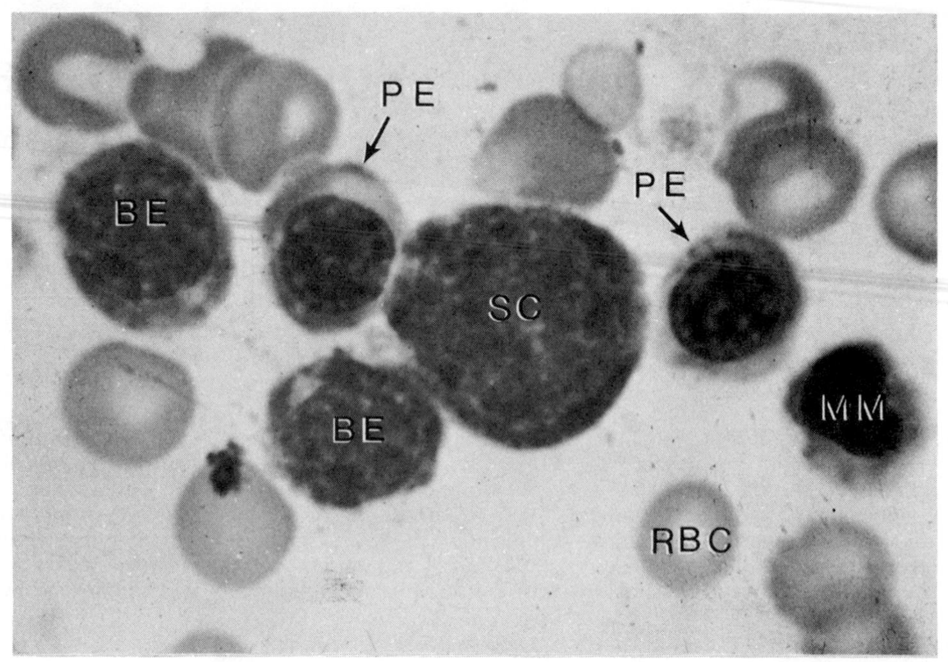

Figure 9-1. Light micrograph of a human bone marrow smear. The primitive stem cell (*SC*) probably is an erythrocyte stem cell. Two basophilic erythroblasts (*BE*), two polychromatic erythroblasts (*PE*), and numerous erythrocytes (*RBC*) can be seen along with a single neutrophilic metamyelocyte (*MM*). (Reprinted with permission from Johnson KE: *Histology: Microscopic Anatomy and Embryology*. New York, John Wiley, 1982, p 118.)

III. HEMATOPOIESIS IN THE EMBRYO

A. EARLY EVENTS

1. Hematopoiesis begins as early as the second week gestation, when **angiogenic cell clusters** first appear in the yolk sac.
2. Soon after, blood vessels and blood cells begin forming in the chorion as well as within the embryo proper.
3. Only large, nucleated **primitive erythrocytes** are formed initially.

B. LATER EVENTS

1. At six weeks gestation, hematopoiesis begins in the **liver** as nests of cells with hematopoietic potential appear among hepatic parenchymal cells. Figure 9-2 shows an example of hepatic hematopoiesis in a human embryo.
2. As the liver contributes enucleated definitive erythrocytes to the circulation, the ratio of primitive to definitive erythrocytes decreases rapidly.
3. For a time, some hematopoiesis also occurs in the **spleen**.
4. Hematopoiesis in **bone marrow** begins during the second month gestation and persists throughout the rest of fetal life. In later fetal life and during the neonatal period, most hematopoiesis occurs in bone marrow.
5. The **thymus** also plays a key role during fetal life in **lymphopoiesis**. Bone marrow stem cells differentiate into **lymphoblasts**, which make their way to the thymus as early as the second month gestation. Differentiation into T lymphocytes then occurs, followed by thymic seeding of the spleen, lymph nodes, and tonsils.

C. THEORIES OF HEMATOPOIESIS

1. Historically, histologists have disagreed about the origin of blood cells.
   - a. Some have argued that blood cells are **monophyletic**, arising from a single stem cell.
   - b. Others have argued that they are **polyphyletic**, arising from several different stem cells (i.e., a specific stem cell for each type of blood cell).

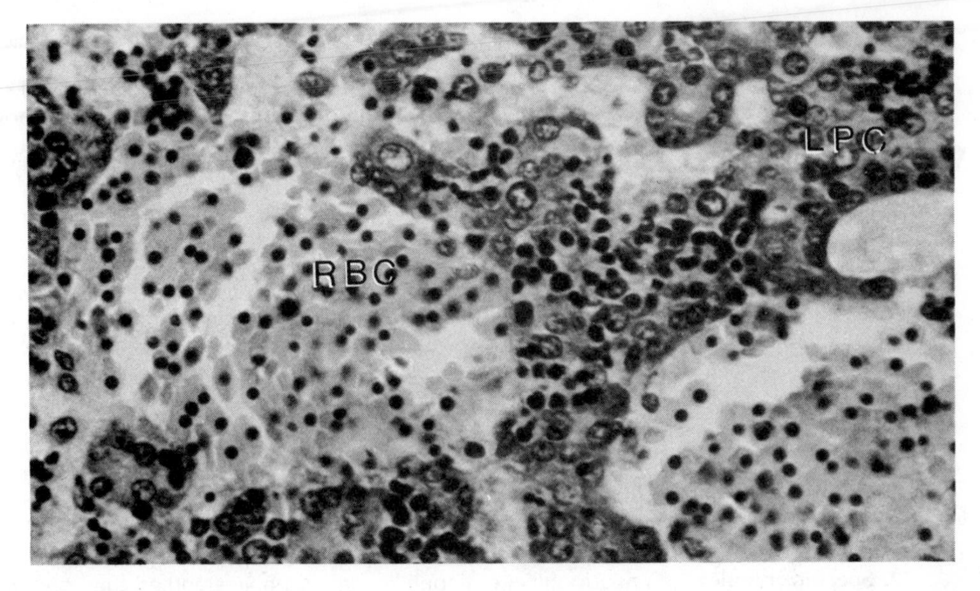

Figure 9-2. Light micrograph of a human fetal liver, taken during active erythropoiesis. Many nucleate erythrocytes (*RBC*) are visible in the sinusoids between liver parenchymal cells (*LPC*). (Courtesy of Dr. Frank Allar Department of Anatomy, George Washington University.)

2. Recent experimental evidence strongly suggests that a specific lymphocyte-like ce represents an authentic multipotential stem cell that is capable of differentiating, under ap propriate stimulation, into all other blood cell types.

IV. ERYTHROPOIESIS

A. EARLY EVENTS

1. Undifferentiated bone marrow stem cells become committed to differentiation under the in fluence of a glycoprotein known as **erythropoietin**.
2. Stem cells differentiate into **erythroblasts**.

B. LATER EVENTS

1. The first recognizable cell in the erythrocyte series is the **basophilic erythroblast**.
   a. This large, free cell is 15 μm in diameter and has a euchromatic nucleus.
   b. The cytoplasmic polyribosomes are involved in hemoglobin biosynthesis and give th cytoplasm its basophilia.
   c. As hemoglobin accumulates, the cytoplasm becomes less basophilic and mor acidophilic.
2. The **polychromatic erythroblast** is found in intermediate stages of erythrocyte development and it soon replaces most of the cytoplasmic polyribosomes with newly made hemoglobin
3. By the time the cytoplasm becomes dominated by hemoglobin, the cell has reached the **orthochromatic erythroblast** stage.
   a. While the hemoglobin is accumulating in the cytoplasm, the nucleus is undergoing heterochromatic involution.
   b. The entire cell shrinks in size as the nuclear chromatin becomes condensed and cytoplasm is shed.
4. Eventually, the nucleus and most mitochondria and polyribosomes are shed to form **definitive erythrocyte** with a 7–8 μm diameter.
5. Some erythrocytes contain variable amounts of residual polyribosomes which can be demon strated by staining with dyes such as cresyl violet. These cells are known as **reticulocytes**.
6. Experimental studies have shown that erythrocytes remain in the peripheral circulation fo 120 days, whereupon they are removed from the circulation and destroyed in the spleen liver, and bone marrow.

a. The hemoglobin is destroyed and degraded to bilirubin and other material, which eventually are excreted in bile.
b. The iron in the hemoglobin is rescued, attached to a serum glycoprotein called **transferrin**, transported to bone marrow, and then reused in the synthesis of new hemoglobin.

V. GRANULOPOIESIS

A. EARLY EVENTS

1. Granulocytes form from undifferentiated bone marrow stem cells.
2. Their first recognizable precursor is the **myeloblast**. This small cell is 10 μm in diameter and has a large euchromatic nucleus with several nucleoi and no granules in its basophilic cytoplasm.
3. Myeloblasts differentiate and begin to produce modified lysosomes called **azurophilic granules**.

B. LATER EVENTS

1. A few azurophilic or **nonspecific** granules accumulate in the cytoplasm, and the nuclear chromatin begins a subtle condensation and heterochromatinization along with a slight indentation. The cell thus formed is called a **promyelocyte**.
2. **Specific granules** (i.e., neutrophilic, eosinophilic, or basophilic granules) then accumulate in the cytoplasm, and the nucleus continues to condense and lobulate. Once specific granules accumulate, cells are called **neutrophilic**, **eosinophilic**, or **basophilic myelocytes**.
3. When numerous specific granules are present in a cell, but the nucleus has not completed the full course of condensation and lobulation, the cell is described as a **metamyelocyte**.
   a. For example, a **neutrophilic metamyelocyte** has a full complement of small neutrophilic specific granules but a relatively uncondensed nucleus with a single indentation rather than the heterochromatic nucleus with 3–5 distinct lobes seen commonly in the definitive neutrophil.
   b. For the neutrophil, the time between the committed myeloblast stage and the definitive neutrophil is approximately two weeks.
4. In the bloodstream, the number of circulating granulocytes is much less than the number of bone marrow myelocytes. Also, metamyelocytes in bone marrow vastly outnumber the mature granulocytes that circulate in blood.

VI. AGRANULOPOIESIS

A. MONOPOIESIS

1. **Monocytes** develop from the same pluripotent bone marrow stem cell as granulocytes, but monocytes go through a different developmental series, with recognizable **monoblast** and **promonocyte** stages.
2. Definitive monocytes briefly pass through the peripheral blood and soon become free **macrophages**.

B. LYMPHOPOIESIS

1. **Lymphocytes** develop from bone marrow stem cells.
2. Two classes of lymphocytes are formed.
   a. **B cells** develop in bone marrow and seed the spleen and lymph nodes.
   b. **T cells** develop in the thymus and seed the spleen and lymph nodes.

VII. THROMBOCYTOPOIESIS

A. MEGAKARYOCYTES

1. **Megakaryocytes** are platelet precursors found in bone marrow. This large cell has a diameter of at least 100 μm and a complex, multilobulated, polyploid nucleus.
2. The megakaryocyte cytoplasm has a perinuclear zone typical for a cell involved in cytoplasmic membrane and granule synthesis.
   a. The cytoplasm contains abundant rough endoplasmic reticulum and a well-developed Golgi apparatus.

b. There also is a prominent peripheral cytoplasmic component rich in granules and filled with a complex network of smooth endoplasmic reticulum.

B. PLATELETS

1. In bone marrow, megakaryocytes lie along vascular channels and slowly release small cytoplasmic fragments called **platelets** into the bloodstream.
2. Megakaryocyte fragmentation results from the fusion of the cell surface membrane with smooth endoplasmic reticular membranes, which allows small bits of megakaryocyte cytoplasm to enter the blood as platelets.
3. Platelets remain in the circulation for 7–10 days and are being replaced constantly by new platelets from bone marrow.
4. There is no nucleus in a platelet.

STUDY QUESTIONS

Directions: Each question below contains four suggested answers of which **one or more** is correct. Choose the answer

A if **1, 2, and 3** are correct
B if **1 and 3** are correct
C if **2 and 4** are correct
D if **4** is correct
E if **1, 2, 3, and 4** are correct

1. Characteristics of megakaryocytes include which of the following?

(1) Their fragmentation results in the release of platelets
(2) They are found in human bone marrow
(3) They have a well-developed Golgi apparatus
(4) They have fewer nuclei per cell than do erythroblasts

2. True statements concerning red marrow include which of the following?

(1) It contains fat cells
(2) Its production decreases following puberty
(3) It is produced from undifferentiated mesenchyme
(4) It can be converted reversibly to yellow marrow

Directions: The group of questions below consists of lettered choices followed by several numbered items. For each numbered item select the **one** lettered choice with which it is **most** closely associated. Each lettered choice may be used once, more than once, or not at all.

Questions 3–8

For each description of blood cell production, choose the appropriate process or processes.

(A) Erythropoiesis
(B) Granulopoiesis
(C) Both
(D) Neither

3. Stimulated by a glycoprotein hormone from the kidneys

4. Active in fetal bone marrow

5. Initiated in the yolk sac during the first month of fetal life

6. Active in the spleen of a normal adult

7. Associated with a decrease in cytoplasmic RNA content and nuclear extrusion

8. Associated with accumulation of modified lysosomes in the cytoplasm

ANSWERS AND EXPLANATIONS

1. The answer is A (1, 2, 3). (*VII A, B*) Megakaryocytes are multinuclear cells with well-developed Golgi apparatus; erythroblasts are mononuclear cells. In bone marrow, megakaryocytes lie along vascular channels and slowly release small cytoplasmic fragments called platelets into the bloodstream.

2. The answer is E (all). (*II B*) Red marrow is marrow involved in active hematopoiesis. It contains some fat but not as much as yellow marrow does. When bones stop growing at puberty, the amount of red marrow also decreases. Red marrow is derived from mesenchyme that surrounds developing bones. Red marrow is converted to yellow marrow, but this conversion is reversible.

3–8. The answers are: 3-A, 4-C, 5-A, 6-D, 7-A, 8-B. (*III B; IV B; V B*) Erythropoietin is a glycoprotein hormone produced by the renal juxtaglomerular apparatus. Both erythropoiesis and granulopoiesis occur in fetal bone marrow. Erythropoiesis begins in the first month gestation, in the yolk sac. The spleen is not active in blood cell formation in the normal adult. Nuclear extrusion does not occur in granulopoiesis. If it did, granulocytes would lack a nucleus, like definitive erythrocytes. The specific granules of granulocytes are modified lysosomes.

10
Immune System

I. INTRODUCTION

A. COMPONENTS

1. **Mononuclear Phagocyte System.** This widely distributed collection of phagocytes of the im mune system includes
   a. Phagocytic cells in bone marrow, lymph nodes, and the spleen
   b. Macrophages
   c. Microglia

2. **Lymphocytes.**
   a. B cells
   b. T cells

3. **Lymphoid Organs.**
   a. Lymph nodes
   b. Spleen
   c. Thymus

B. FUNCTION

1. The human body has evolved an elaborate system to protect itself against invasion by pathc genic organisms and malignant transformation of its own cells. The body also can guard itse from inadvertent introduction of foreign substances.

2. This immune system is highly complex both structurally and functionally, and intens research is underway to unravel these complexities. Biomedical scientists are making im pressive advances in contributing to the understanding of the immune system's functiona aspects.

3. In the future, all physicians probably will have various immunological reagents at the disposal to combat viral disease, malignancy, autoimmune disease, and possibly aging.

II. MONONUCLEAR PHAGOCYTE SYSTEM (MPS)

A. CELLS OF THE MPS.

Throughout the body is found a diverse population of actively **phagocyti cells** derived from **bone marrow stem cells**. These cells are large and have a palely stained, ofte lobulated nucleus and a ruffled cell border that is a consequence of their active migrator behavior.

1. The **monocytes** in the blood, which represent 2–10 percent of the peripheral blood leukocyt population, are an integral part of the MPS.

2. When monocytes leave the peripheral blood, they can move throughout the connectiv tissue domains of the body and differentiate into **macrophages**.
   a. Macrophages can cross epithelial barriers such as the peritoneal or the pleur mesothelium, where they can become actively phagocytic and migratory in either th pleural or peritoneal cavity or within the pulmonary alveoli.
   b. There also are freely mobile macrophages in the spleen, lymph nodes, and bone marrov Fixed macrophages are found in the connective tissues, spleen, lymph nodes, and bon marrow.

3. The phagocytic **Kupffer cells**, lining the hepatic sinusoids, also are part of the MPS.

4. **Microglia** are central nervous system (CNS) phagocytes.

B. MPS FEATURES

1. The basic feature of cells in the MPS is that their phagocytic activity is mediated by immunoglobulins or serum complement factors. This feature distinguishes the MPS from other phagocytic cells in the body (e.g., the polymorphonuclear leukocytes, eosinophils, and pigmented retinal cells).

2. The MPS also is characterized by its ability to meet the body's widespread need for protection against foreign intruders.

III. MACROPHAGES

A. STRUCTURE (Fig. 10-1)

1. With the electron microscope, these cells reveal a large lobulated nucleus with modest amounts of heterochromatin.

2. The cell surface usually is quite irregular and has numerous finger-like or folded projections which function in cell spreading, movement, and phagocytosis.

3. The cytoplasm of this cell has a well-developed rough endoplasmic reticulum and a prominent Golgi apparatus, which is typical for any cell engaged in large amounts of protein synthesis.

4. In addition, these cells have a large number of lysosomes and, at times, a variety of engulfed substances in their cytoplasm.

5. Macrophages have an abundance of microtubules and microfilaments which are involved in the movement of the cell and its organelles.

B. FUNCTION

1. These cells are actively involved in the engulfment and destruction of various substances that enter the body.

2. They also are highly migratory and have the ability to insinuate themselves into the small nooks and crannies of the extracellular matrix of connective tissue.

3. Macrophages ingest some large particulate substances by **phagocytosis**, a process that involves specific attachment and ingestion. Also, they ingest dissolved solutes from their fluid environment by **pinocytosis**.
   a. Once a macrophage has adhered specifically to a microorganism, it can ingest and destroy it. In certain instances, cells such as lymphocytes will adhere specifically to macrophages without being ingested.
   b. Some macrophages, particularly those in the pulmonary alveoli, specialize in the **nonspecific phagocytosis** of inspired particulate matter such as dust, pollen, and cigarette smoke.
   c. **Specific phagocytosis** involves coating microorganisms with specific immunoglobulins or complement factors—a process called **opsonization**—and then **adhesive recognition** of bound immunoglobulins or complement factors by the cell surface proteins of macrophages.
   d. Ingested bacteria are destroyed by a mechanism that probably is related to low lysosomal pH and the presence of hydrolytic enzymes such as **lysozyme**.

4. Macrophages not only specifically ingest opsonized microorganisms, they also can potentiate the lymphocyte antibody production by binding certain antigens to their surface and then interacting with lymphocytes to stimulate antibody production.

5. Macrophages have the ability to migrate in a directed fashion up a concentration gradient of dissolved components of bacterial cell walls. This **chemotactic behavior** probably is responsible for recruiting large numbers of macrophages to areas of tissue destruction or infection.

C. DEVELOPMENT

1. Macrophages originate from undifferentiated bone marrow stem cells. Once differentiated, these stem cells are released into the peripheral circulation as **monocytes**.

2. After several days in the peripheral vasculature, monocytes leave the closed circulation and enter such regions as the connective tissues, peritoneal cavity, and pulmonary alveoli, where they persist for several months as macrophages.

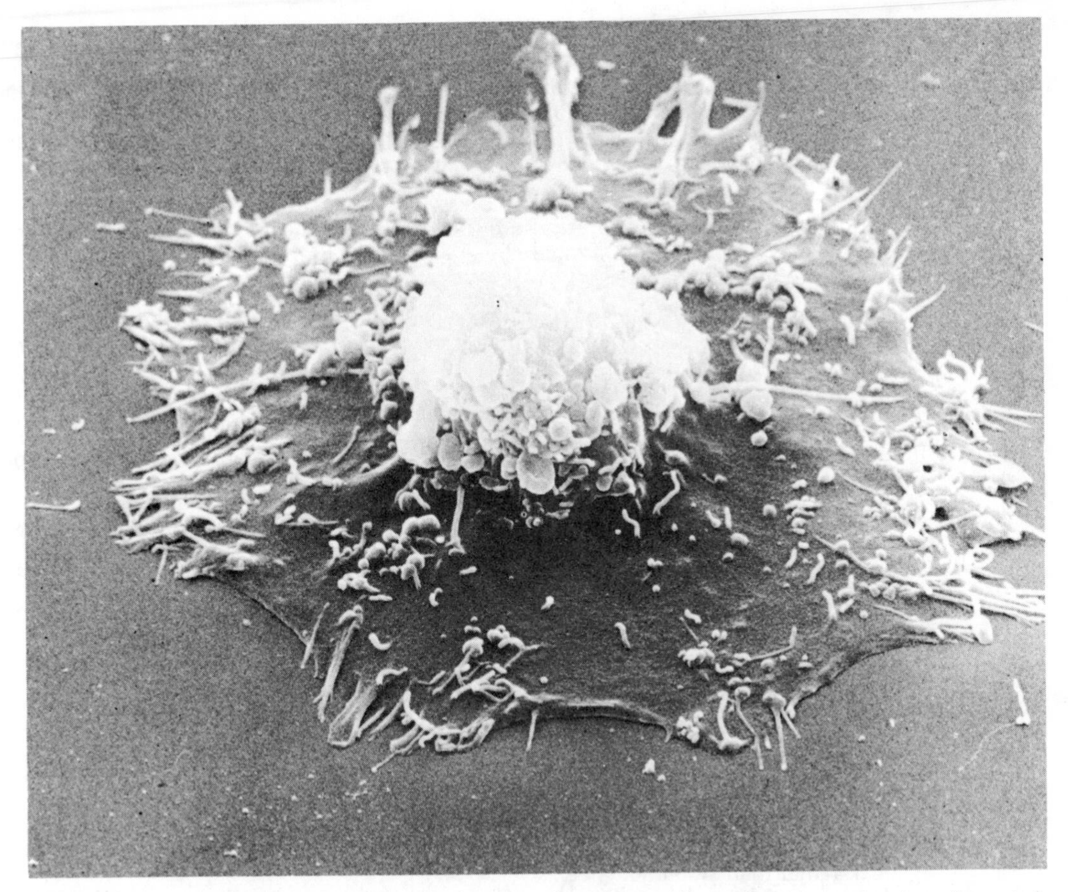

Figure 10-1. A scanning electron micrograph of a macrophage fixed while moving. Notice that the cell margin has numerous flattened lamellipodia and long thin filopodia radiating away from it. The nucleus is conspicuous in the center of the cell along with other surface irregularities, some of which probably represent engulfed debris. (Courtesy of Dr. Ernest N. Albert, Department of Anatomy, George Washington University.)

IV. LYMPHOCYTES

A. HETEROGENEITY

1. The **lymphocyte** is a common leukocyte whose heterogeneous and complex nature was long misunderstood by histologists. Recent studies in cellular immunology, however, have revealed two broad functional classes of lymphocytes: **B lymphocytes** and **T lymphocytes**.

2. Both lymphocyte types originate from undifferentiated bone marrow stem cells.
   a. B lymphocytes leave bone marrow and diffuse throughout the body where they differentiate under poorly understood inductive influences.
      (1) In birds, B lymphocytes differentiate in an organ known as the **bursa of Fabricius** and thus are called **B cells**.
      (2) Although no structure equivalent to the bursa of Fabricius has been clearly identified in any mammal, even man, the name B cell has stuck.
   b. T lymphocytes or **T cells** leave bone marrow and travel to the thymus where they differentiate.

B. B-CELL FUNCTION AND CHARACTERISTICS

1. B cells comprise approximately 35 percent of the circulating lymphocytes and primarily are responsible for humoral immunity (i.e., the production of specific serum immunoglobulins directed against various environmental antigens).

2. B cells function in antibody production in two distinct ways.
   a. In the first and clearly most simple case, antigens bind directly to the B cell's surface; B

cells then undergo a clonal proliferation followed by terminal differentiation into **plasma cells**. Plasma cells are highly specialized for the secretion of immunoglobulins.

b. In the second and more complex case, antigens are bound to the surface immunoglobulins of **helper T cells**. Next, these antigen-antibody complexes are released from helper T cells and bound to **macrophages**. Finally, the B cells interact specifically with stimulated macrophages and subsequently undergo a clonal proliferation and plasma cell differentiation.

3. Certain B cells persist after initial exposure to an antigen and exist in the form of **memory B cells**. These cells can undergo a rapid clonal expansion if a person is exposed again, even years later, to the same antigen.

4. B cells are most heavily concentrated in the germinal centers of aggregates of lymphoid tissue. For example, B cells are found in large numbers in the cortical aggregates in lymph nodes and in the white pulp of the spleen.

C. T-CELL FUNCTION AND CHARACTERISTICS

1. T cells comprise about 65 percent of the circulating lymphocytes and are responsible for a complex phenomenon known as **cellular immunity**.

2. Unlike B cells, a T cell has small amounts of surface immunoglobulins and can bind antigens specifically to its surface. This ability of T cells possibly is due to the fact that their cell surfaces contain gene products from **immune response** genes.

3. Upon antigenic stimulation, T cells undergo clonal proliferation and then differentiate to produce
   a. **Lymphokines**, such as **macrophage inhibitory factor** (MIF), chemotactic factors for other formed elements of peripheral blood, and other nonspecific cytotoxic substances
   b. **Killer T cells**, a T-cell subclass that can contact and specifically kill abnormal cells
   c. **Supressor T cells** and **effector T cells**, two subclasses that modulate B-cell response in the production of humoral immunity

4. In effect, many T-cell functions are stimulated by specific agents, but the response of the T cell per se is localized and relatively nonspecific.
   a. For example, when certain T cells are stimulated to secrete MIF, this substance has a strong **local**, rather than **systemic**, effect.
   b. Macrophages in the immediate vicinity of T cells secreting MIF are immobolized and remain in their location for longer periods of time, thus having a greater opportunity to engulf and destroy the stimulus that initially evoked the MIF production by T cells.

5. T cells are most heavily concentrated in the spleen, specifically in the periarterial lymphatic sheaths, and in the tertiary cortex of lymph nodes (i.e., paracortical regions).

V. LYMPH NODES

A. GENERAL CONSIDERATIONS

1. **Lymph Node Distribution.** These highly organized, encapsulated aggregates of lymphocytes and supporting cells are distributed widely throughout the body. They are particularly abundant in the axillary regions, neck, mediastinum, and retroperitoneal areas in the abdomen. Lymph nodes lie in the path of lymphatic vessels and usually are supplied with several cortical afferent vessels and a smaller number of hilar efferent vessels.

2. **Lymph Circulation.**
   a. Lymph originates when a whole blood filtrate is expressed from capillaries in the periphery of the body; then, lymph is collected into blind-ending lymphatic capillaries and poured into larger lymphatic vessels.
   b. From lymphatic vessels, lymph is drained through lymph nodes, where lymphocytes have a chance to examine the contents in the lymph.
   c. Finally, lymph enters the systemic circulation again by draining into the thoracic duct.

B. STRUCTURE

1. **Microscopic Anatomy** (Fig. 10-2).
   a. Each lymph node is surrounded by a **capsule** made of a dense layer of collagenous connective tissue with some smooth muscle cells in it. Trabeculae penetrate from the capsule into the central portion of the lymph node.
   b. A lymph node usually has one **convex surface** served by several afferent lymphatics. One

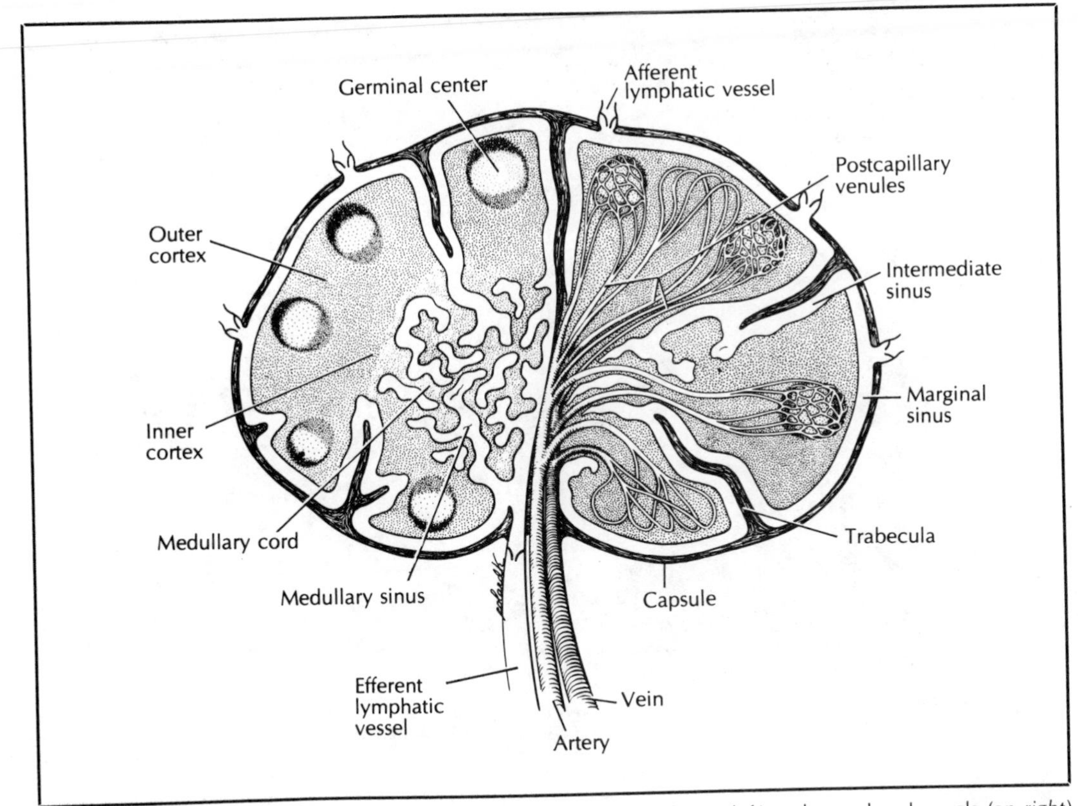

Figure 10-2. Diagram of a lymph node, showing lymphatic channels (*on left*) and vascular channels (*on right*). (Reprinted with permission from Bloom W, Fawcett D: *A Textbook of Histology*. Philadelphia, WB Saunders, 1975, p 473.)

or two efferent lymphatics drain the organ at its indented **hilus**, where blood vessels also enter and leave.

c. In the spaces between the capsule and trabeculae is found a complex **reticulum** of cells, extracellular fibers, and free cells of the immune system such as lymphocytes and macrophages.

(1) The reticulum serves essentially as a scaffolding for lymphocytes and for a complex system of phagocytes.

(2) It is composed of stellate **reticular cells** and argyrophilic (i.e., silver-staining), periodic acid-Schiff (PAS)-positive collagenous **reticular fibers**. These fibers presumably are secreted by the reticular cells.

2. Lymphocyte Distribution.

a. Each lymph node has a cortex densely packed with lymphocytes and a medulla with looser, cord-like arrangements of lymphocytes and sinuses.

b. Within the nodes are found dense, lymphocyte aggregates called **nodules**.

(1) Primary nodules are homogeneous collections of densely packed small lymphocytes. (Figure 10-3 shows an example of a primary nodule.)

(2) Secondary nodules have dense peripheral regions filled with small lymphocytes and **germinal centers** occupied by larger lymphocytes and some macrophages.

c. Between the cortical nodules are found internodular cortical lymphocytes; the **tertiary cortex** is found between the cortical nodules and the medulla.

d. Immature B cells are found most commonly in primary nodules, whereas differentiating B cells are found in the germinal centers along with macrophages and certain types of T cells.

e. The lymphocytes of the tertiary cortex primarily are T cells.

3. Lymphatic Channels.

a. Lymph drains into the convex surface of the node through afferent lymphatics that communicate with **cortical sinuses**.

b. Cortical sinuses pass between cortical nodules and through the tertiary cortex to connect with **medullary sinuses**. These sinuses follow trabeculae to the central portion of the node,

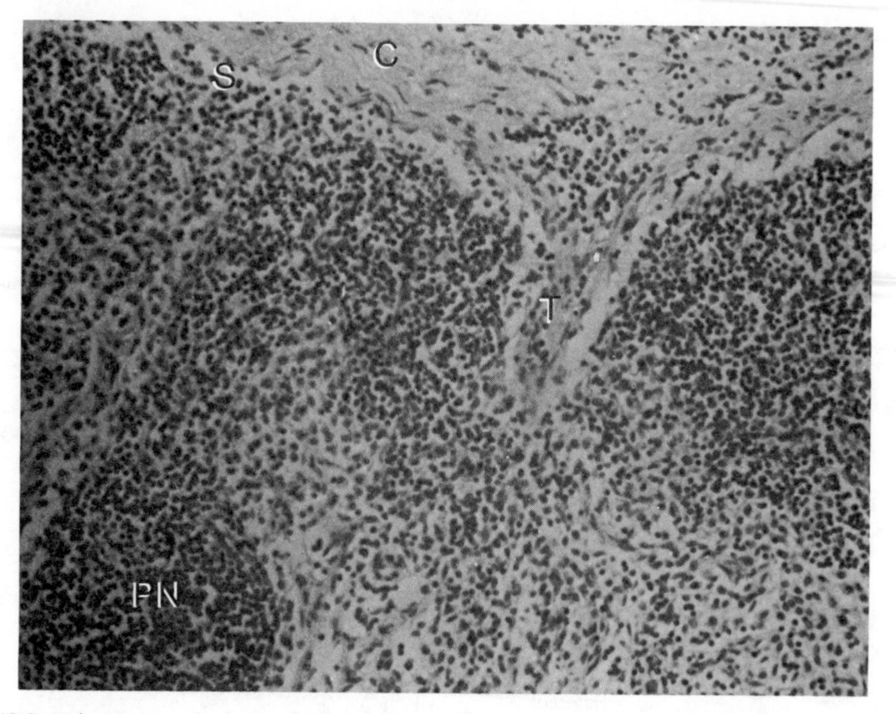

Figure 10-3. Light micrograph of a lymph node. A connective tissue trabecula (*T*) extends from the capsule (*C*) into the cortex of the node. Subcapsular lymph sinuses (*S*) are visible along with a primary nodule (*PN*). (Reprinted with permission from Johnson KE: *Histology: Microscopic Anatomy and Embryology*. New York, John Wiley, 1982, p 132.)

where they form an anastomosing network of channels that drain eventually into hilar efferent vessels.

c. Blood vessels enter and leave the lymph node at the hilus. Blood vessels within lymph nodes are unremarkable except for the post-capillary venules in and around the tertiary cortex. Here are found cuboidal or columnar endothelial cells that pass both T cells and B cells **between** their lateral boundaries.

C. **FUNCTION.** Lymph nodes have three important functions in the immune system.

1. They serve as a **filter** for both particulate material and soluble antigens. Lymph percolates slowly through the tortuous sinuses, thereby maximizing the time available for phagocytic activity and immune surveillance.

2. Lymph nodes represent sites for the **clonal expansion** of specific populations of B cells and T cells.

3. Macrophage-lymphocyte interaction in **immune responsiveness** is intense in lymph nodes.

VI. **THE SPLEEN** is a highly complex organ involved in hematopoiesis, removal of effete erythrocytes, and immune surveillance. In a healthy adult, the spleen weighs more than 100 g and, in spite of its complex function, is not essential to life. It develops in the dorsal mesogastrium and is wholly mesodermal in origin.

A. **MICROSCOPIC ANATOMY**

1. The spleen has a dense connective tissue capsule from which numerous trabeculae penetrate into the splenic parenchyma.

2. The compartments between trabeculae are filled with red and white pulp.
   a. **Red pulp** is rich in erythrocytes and tortuous vascular channels.
   b. In contrast, **white pulp** is composed of lymphocyte collections similar to those of the primary and secondary nodules found in lymph nodes. White pulp exists in two forms.
      (1) Dense aggregates of small lymphocytes form cylindrical clusters around central arteries. These clusters collectively are named the **periarterial lymphatic sheath** (PALS).

(2) In addition, **secondary nodules** with well-formed germinal centers contribute to the white pulp.

c. Where white pulp and red pulp meet, a **marginal zone** is formed.

3. In the spleen, the PALS and marginal zone surrounding secondary nodules contain T cells, whereas the secondary nodules are rich in B cells.

B. BLOOD CIRCULATION

1. Splenic arteries enter the spleen at the hilus and soon after branch into **trabecular arteries**.
2. Trabecular arteries branch into **central arteries**, the vessels that pass through the PALS.
3. Central arteries divide into **follicular arteries**. These supply secondary nodules in the white pulp and also feed into a series of smaller and smaller arterioles that empty in one of two ways.
   a. Arterioles that empty directly into the reticular meshwork of irregularly arranged sinusoids complete the **open splenic circulation**.
   b. Arterioles that empty into splenic sinuses lined by elongated cells complete the **closed splenic circulation** (Fig. 10-4).
4. From the open reticular meshwork and from splenic sinuses, blood drains into **veins of the pulp** and then into **trabecular veins**. At the hilus, trabecular veins feed into **splenic veins** that carry the blood away from the spleen.

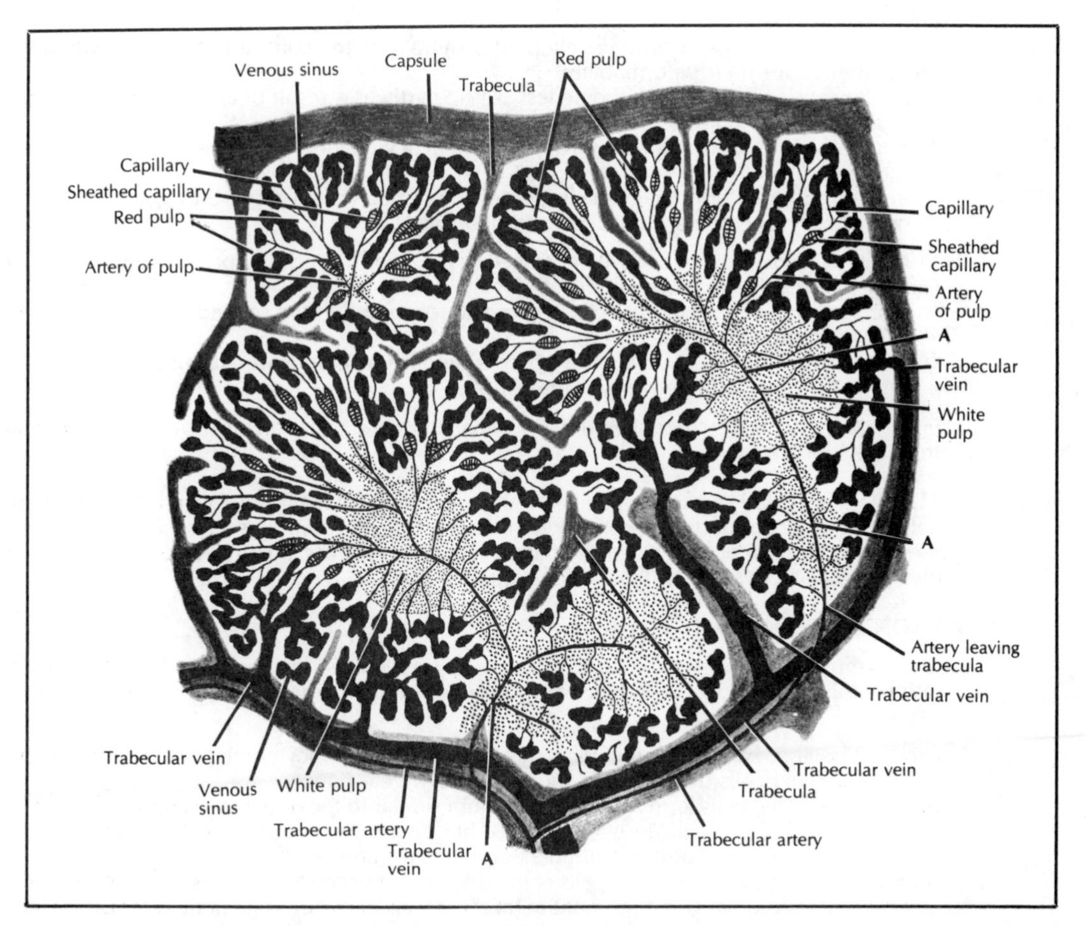

Figure 10-4. Diagram of the microscopic anatomy of the spleen, illustrating closed splenic circulation. Trabecular arteries branch into central arteries (*A*) which, in turn, branch into follicular or pulp arteries. These branch into capillaries that empty directly into venous sinuses. (Reprinted with permission from Bloom W, Fawcett D: *A Textbook of Histology*. Philadelphia, WB Saunders, 1975, p 493.)

C. RETICULAR ELEMENTS

1. Throughout much of the parenchyma of the spleen are found reticular cells and reticular fibers similar to those found in lymph nodes.

2. These cells and fibers form a scaffolding that supports both the white pulp and the red pulp.

3. Reticular cells and associated reticular fibers create a support system for the PALS, surrounding central arteries with several concentric layers rather than a spongy meshwork.

D. SINUSES AND CORDS.
A complex mixture of **splenic sinuses** and a reticular meshwork of **splenic cords** are found in red pulp.

1. Splenic sinuses are peculiar vascular channels that measure approximately 40 μm in diameter and are lined by tapered cells shaped like barrel staves.
   a. These cells are elongated parallel to the long axis of the sinus, and there are no attachment specializations between them.
   b. They also are loaded with contractile microfilaments whose suggested role is to contract and open the gaps between the "staves".

2. Splenic sinuses are surrounded by a fenestrated basement membrane and are contacted frequently by adventitial reticular cells that are part of the splenic cords.

3. Blood cells migrate from the splenic cords to the walls of splenic sinuses, where they pass between the lining cells and into the lumina of the sinuses. From there blood cells eventually exit the spleen through splenic veins.

E. FUNCTION

1. The spleen functions as a complex filter. For example, in the cords and sinuses, erythrocytes are examined for their deformability.
   a. As they age, erythrocytes become less plastic partly as a result of age-dependent changes in the chemical composition of their plasma membranes.
   b. Young, healthy cells can pass through the phagocyte-laden cords into sinuses and out of the spleen.
   c. In contrast, aged or damaged cells are less deformable; as a result they get caught in the cords and eventually are engulfed and destroyed.

2. In addition to destroying aged or damaged erythrocytes, white pulp functions in immune surveillance much like lymph nodes do.
   a. After entering white pulp, B cells become concentrated in the secondary nodules while T cells become concentrated in the PALS.
   b. The spleen functions to trap and process **blood-borne** antigens. (Lymph nodes serve a similar function with antigens in the lymph.)

3. Finally, the splenic macrophages and phagocytic reticular cells can clear particulate material from the blood.

VII. THYMUS.
This irregularly shaped organ lies in the neck, close to the thyroid gland. It has a unique organization for an organ of the immune system. Superficially it resembles a lymph node in that it contains numerous lymphocytes. When examined in detail, however, the thymus clearly is different from a lymph node both structurally and functionally.

A. MICROSCOPIC ANATOMY (Fig. 10-5).

1. The thymus has a connective tissue capsule with trabeculae radiating away from the capsule into the lobes of the organ. (In this way the thymus is different from lymph nodes, which lack lobes.)

2. Each lobe is approximately 1 mm wide and contains a lymphocyte-rich cortex and a medulla rich in epithelial cells derived from the third pharyngeal pouch.
   a. These epithelial cells form a cellular reticulum similar to the one formed by mesenchymally derived reticular cells in lymph nodes, bone marrow, and the spleen; in the thymus, however, there are not large numbers of reticular fibers.
   b. Instead the epithelial reticular cells of the thymus are linked at their tips by desmosomes in a way that presumably is a remnant of their early epithelial arrangement in the pharyngeal pouches.
   c. These reticular epithelial cells form a meshwork that is infiltrated by numerous lymphocytes and macrophages.

3. The thymus also contains a conspicuous medulla that connects with the medullary portions of the lobes and is composed of cells derived from the pharyngeal pouches.

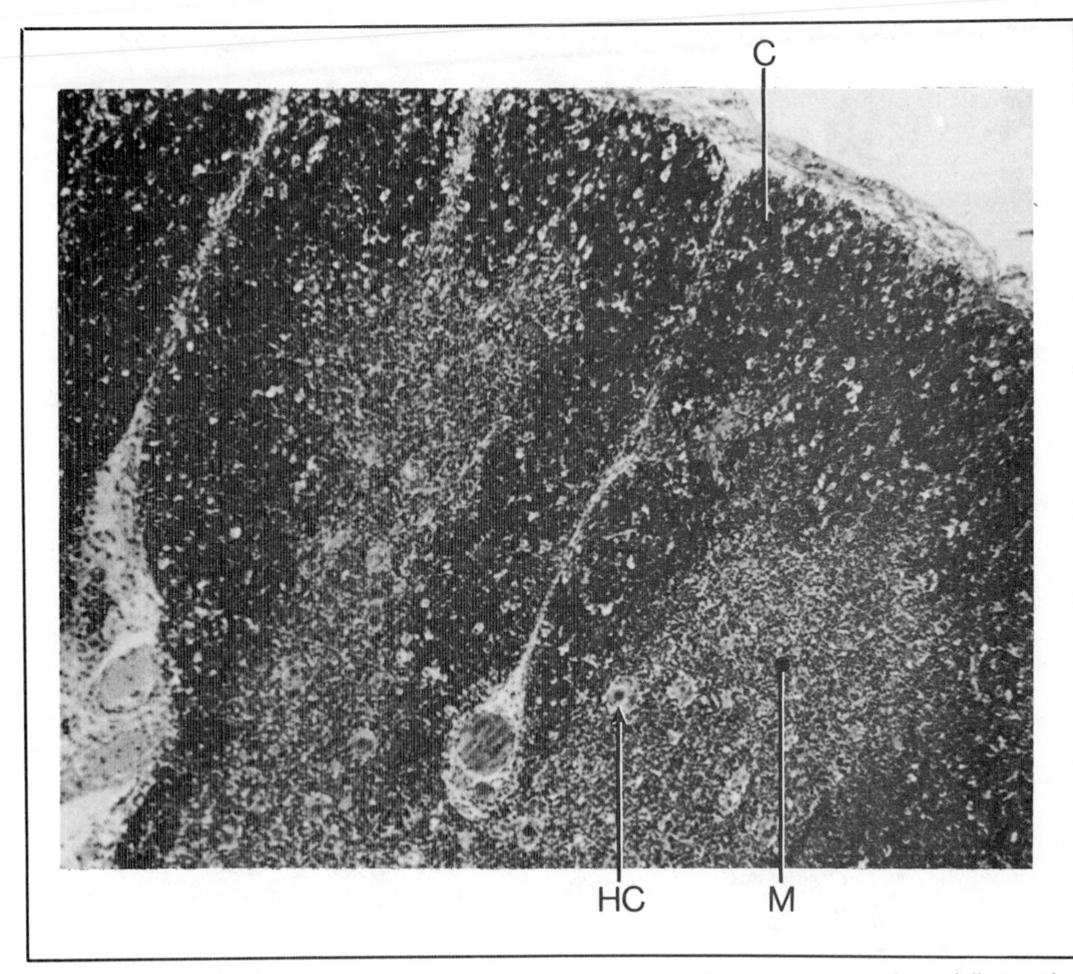

Figure 10-5. Light micrograph of a lobe of the thymus. Each lobe is divided into a cortex (*C*) and a medulla (*M*). The medullary portions of each lobe are particularly rich in epithelial reticular cells and Hassall's corpuscles (*HC*). (Reprinted with permission from Johnson KE: *Histology: Microscopic Anatomy and Embryology*. New York, John Wiley, 1982, p 141.)

4. The medullary portions of the thymus contain many concentrically arranged groups of degenerating epithelial cells known as **Hassall's** or **thymic corpuscles**.

B. FUNCTION

1. The thymus is a key organ in the immune system. It represents the site where bone marrow–derived lymphocytes proliferate and differentiate into T cells.
2. This differentiation is mediated by a **lymphokine** known as **thymosin**, which is secreted by the reticular epithelial cells.
3. Neonatal thymectomy in animals results in a complete lack of the ability to reject heterologous tissue grafts, and this deficiency can be restored by grafting thymic tissue.

C. DEVELOPMENT

1. The thymus is relatively large in a newborn but does not grow as fast as other tissues in the body.
2. At puberty it reaches its largest absolute size but is relatively small compared to the rest of the body's tissues.
3. In later life, the thymus undergoes a slow involution, and thymic reticular cells sometimes are replaced by fatty, degenerated-looking cells.

STUDY QUESTIONS

Directions: Each question below contains five suggested answers. Choose the **one best** response to each question.

1. All of the following cells are components of the mononuclear phagocyte system EXCEPT

(A) histiocytes
(B) neutrophils
(C) monocytes in peripheral blood
(D) splenic macrophages
(E) lymphoid macrophages

2. All of the following statements concerning B cells are true EXCEPT

(A) they can differentiate into plasma cells
(B) they must interact with macrophages before they can produce antibody
(C) they are derived from bone marrow
(D) they are abundant in secondary nodules in lymph nodes
(E) they are less common than T cells in the periarterial lymphatic sheaths (PALS) in the spleen

Directions: The question below contains four suggested answers of which **one or more** is correct. Choose the answer

A if **1, 2, and 3** are correct
B if **1 and 3** are correct
C if **2 and 4** are correct
D if **4** is correct
E if **1, 2, 3, and 4** are correct

3. True statements concerning the thymus include which of the following?

(1) Its reticular epithelial cells are derived from the third pharyngeal pouch
(2) Its Hassall's corpuscles are clusters of degenerated T cells
(3) Its T cells decrease in number after birth
(4) It represents the body's major source of plasma cells

Directions: The group of questions below consists of lettered choices followed by several numbered items. For each numbered item select the **one** lettered choice with which it is **most** closely associated. Each lettered choice may be used once, more than once, or not at all.

Questions 4–8

For each description of sites or structures within a lobe of the thymus, choose the appropriate lettered component shown in the micrograph below.

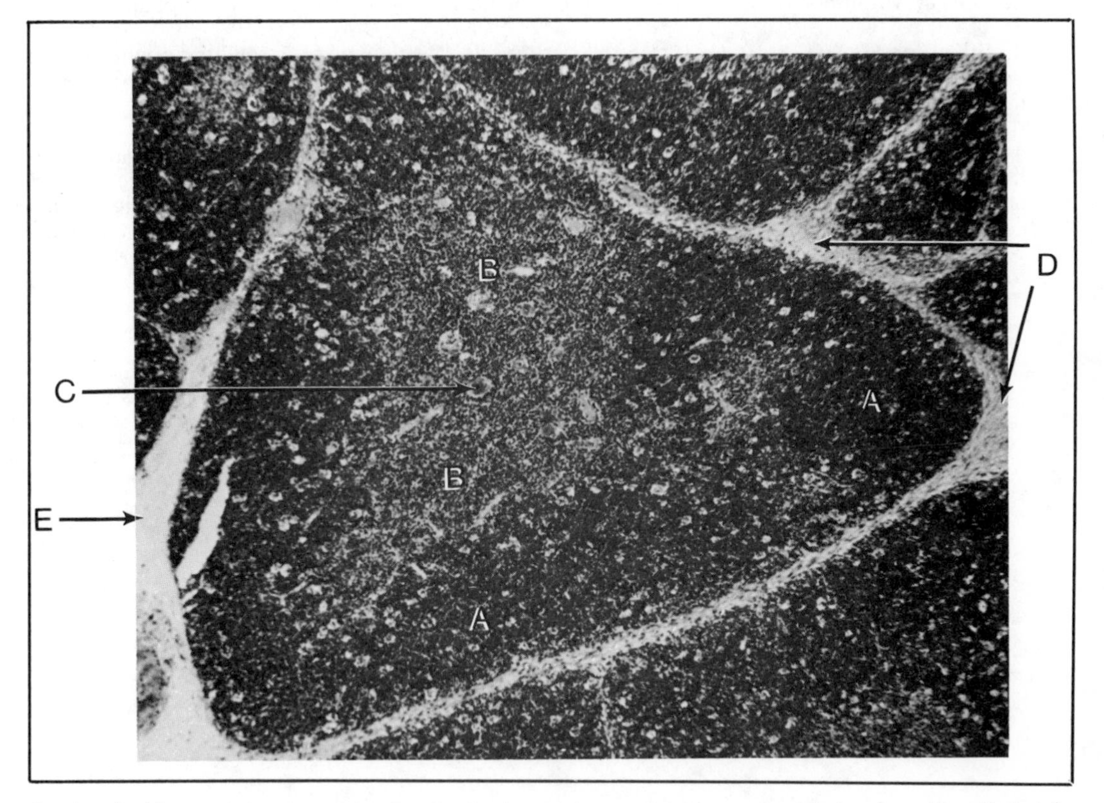

Reprinted with permission from Johnson KE: *Histology: Microscopic Anatomy and Embryology*. New York, John Wiley, 1982, p 137.

4. Cluster of degenerated thymic reticular epithelial cells

5. Area rich in reticular epithelial cells

6. Area rich in T cells

7. Area rich in cells derived from bone marrow

8. Area rich in cells that are derived from third pharyngeal pouch and are joined by desmosomes

ANSWERS AND EXPLANATIONS

1. The answer is B. (*II A*) The exclusion of neutrophils is a somewhat arbitrary one, but most histologists do not treat them as part of the mononuclear phagocyte system (MPS). Neutrophils are phagocytic cells and have a similar phagocytic function to other elements of the MPS.

2. The answer is B. (*IV B*) Memory B cells can differentiate into plasma cells without mediation by macrophages. B cells are derived from bone marrow stem cells and commonly differentiate into plasma cells with the help of helper T cells and macrophages. B cells are abundant in secondary nodules in lymph nodes and are less common than T cells in the periarterial lymphatic sheaths.

3. The answer is B (1, 3). (*VII A C*) Hassall's corpuscles are derived from reticular epithelial cells, not degenerated T cells. Plasma cells are derived from B cells, which exist in limited numbers in the thymus. B cells are derived predominantly from the lymph nodes and the spleen.

4–8. The answers are: 4-C, 5-B, 6-A, 7-A, 8-B. (*VII A*) Hassall's corpuscles (C) are clusters of degenerated thymic reticular epithelial cells. The medulla (B) of thymic lobules is rich in reticular epithelial cells, which are derived from the third pharyngeal pouch and are joined by desmosomes. T cells are abundant in the cortex (A) of thymic lobules. T cells are derived from bone marrow.

11
Cardiovascular System

I. INTRODUCTION

A. CARDIOVASCULAR SYSTEM

1. The **heart** is a large, four-chambered organ designed to propel the fluid and cellular components of blood through the entire body.
   a. The right atrium of the heart collects blood from the systemic circulation via the inferior and superior venae cavae and pumps deoxygenated blood to the lungs via the pulmonary arteries.
   b. Oxygenated blood returns from the lungs to the left atrium of the heart; from there it passes to the left ventricle and is pumped out into the systemic circulation via large **elastic arteries**.
2. These elastic arteries soon branch into smaller, **muscular arteries** that supply various organs and regions of the body.
   a. The muscular arteries communicate with **arterioles**, which in turn empty into a vast network of **capillaries**. There are many types of capillaries to serve the various essential functions in the body.
   b. The capillary bed drains into post-capillary **venules**, which in turn drain into veins of ever-increasing diameter.
3. The largest **veins** then drain back into the right atrium of the heart.
   a. The blood pressure in the veins is relatively low, and often the flow is sluggish.
   b. Consequently, many large veins are equipped with their own system of **valves** to prevent reflux and venous stasis.

B. LYMPHATIC SYSTEM

1. The lymphatic system begins in the most distal parts of the body as a series of blind-ended **lymphatic capillaries**.
2. These capillaries drain lymph, a blood filtrate, into small **lymphatic vessels** that drain different parts of the body, often pass through lymph nodes, and eventually empty into the **thoracic duct**.
3. This terminal vessel empties into the venous circulation at the junction of the left internal jugular and left subclavian veins.

II. BASIC HISTOLOGY OF THE CARDIOVASCULAR SYSTEM (CVS)

A. ENDOTHELIUM

1. The entire CVS is lined by a simple squamous epithelium known as **endothelium.**
2. The structure of the endothelium varies with its location.
   a. In the heart, the endothelium is a continuous smooth lining that practically assures maximum blood flow through the heart with minimal resistance.
   b. In contrast, the endothelial lining of the capillaries is specialized in several ways to facilitate gas and nutrient exchange or to allow rapid exit of cells through the blood vessels into the surrounding connective tissue.

B. BLOOD CIRCULATION

1. Blood is propelled through the CVS by strong muscular contractions of the heart.
2. During **systole**, blood is ejected from the left ventricle under considerable pressure. This pressure immediately expands the walls of the large elastic arteries (aorta, pulmonary, common carotid, and other arteries).

3. During **diastole**, the stretched elastic fibers in the walls of the elastic arteries contract passively. In this manner, some of the propulsive energy of the systolic contractions is stored in the wall of the CVS and used immediately thereafter to smooth out and maintain blood flow.

4. Blood then is distributed to the muscular arteries—vessels that are composed of much smooth muscle and are innervated by the autonomic nervous system.
   a. The blood supply to different regions in the body is controlled by regulating the diameter of the lumina of muscular arteries.
   b. As blood is propelled away from the heart, the hydrostatic pressure decreases and the flow becomes more even and less pulsatile.

5. Arterioles are equipped with smooth muscle that can contract upon sympathetic stimulation.
   a. They also may respond to local chemical conditions to regulate blood to a particular region.
   b. Arterioles further reduce the hydrostatic pressure in the blood until it is delivered to the capillary beds at a considerable rate but relatively low hydrostatic pressure.

6. Gas, nutrient, and waste exchange occurs in the thin-walled capillaries; this occurs also in the post-capillary venules. These vessels drain into veins where the walls are relatively thin.

7. Venous drainage of the extremities is aided in part by muscle contraction, and many larger veins are equipped with valves.

8. The volume capacity of veins is potentially larger than the volume capacity of arteries. Consequently, veins represent a storage system within the CVS.

C. LAYERS OF THE CVS

1. The **tunica intima** is the layer found nearest to the lumen. It is composed of endothelium and a basement membrane.

2. The **tunica media** is the next layer outward from the lumen. It is composed primarily of smooth muscle cells and extracellular fibrils.

3. The **tunica adventitia** is the outermost layer, and it is usually composed of a thin layer of fibroblasts and collagen fibers.

D. FUNCTIONAL VARIATION OF LAYERS

1. Relative importance of each layer varies tremendously in different areas. In fact, not all layers exist in all areas.
   a. The tunica media is very thick in elastic arteries and composed of many layers of smooth muscle cells, collagen fibers, and **elastic laminae**.
   b. There is no tunica media, however, in a capillary.

2. In the heart, all three layers are present but are named differently.
   a. Tunica intima is called **endocardium**.
   b. Tunica media is called **myocardium**.
   c. Tunica adventitia is called **epicardium**.

III. HEART

A. MICROSCOPIC ANATOMY

1. **Endocardium.**
   a. The endocardium is continuous with the tunica intima of the great vessels entering and leaving the heart; however, the endocardium on the left side of the heart is not directly continuous with the one on the right. Instead, they are separated by septa of cardiac muscle.
   b. This layer consists of the cardiac endothelium and a thin but continuous basement membrane.
   c. A subendothelial connective tissue exists and contains collagen, elastic fibers, smooth muscle fibers (cells), small blood vessels, and nerves.
   d. The **Purkinje fibers** of the conducting system also are found in the ventricles along the endocardial surface.
   e. The subendothelial connective tissue is continuous with the connective tissue of the myocardium.

2. **Cardiac valves** all are similar and represent a connective tissue extension of the subendothelial tissue. These valves lack blood vessels and nerves.

3. **Myocardium.**
   a. This layer is composed primarily of cardiac muscle fibers and fibroblasts.
   b. The myocardium has an impressive blood supply of its own delivered by the coronary arteries and their tributaries. Consequently, histologic sections reveal many capillaries in the myocardium.
   c. In the atria, the myocardium is relatively thin and is rich in collagen and elastic fibers. (In contrast, the myocardium in the ventricles is relatively thick.) Atrial cells have many more gap junctions connecting them, and the conduction velocity of impulses initiating contractions is higher here.
   d. The cardiac muscle fibers deep in the ventricles are arranged in broad sweeping spirals around the ventricles, and individual muscle cells found here are larger and have a more elaborate T-tubule system than the cardiac cells of the atria.
   e. The connective tissue elements, primarily fibroblasts, proliferate in regions where the cardiac muscle fibers have been damaged. Unfortunately, damaged cardiac muscle cells have no regenerative capacity.
   f. Individual muscle cells are joined together by **intercalated discs**.

4. **Epicardium.**
   a. This outermost cardiac layer has an external coat of mesothelium, on which rests the visceral layer of the **pericardium**.
   b. Beneath the mesothelium are found collagen fibers, fat, and nerve fibers.
   c. In obese patients, there may be considerable amounts of adipose tissue in the epicardium.

5. **Cardiac Skeleton**. The cardiac skeleton is a connective tissue mass in the myocardium, which serves as an insertion site for the fibers in valves as well as for the myocardial cells themselves.
   a. The bulk of the cardiac skeleton is made up of **annuli fibrosi** that surround both atrioventricular canals, a **trigona fibrosa**, and the **septum membranaceum**, which makes up the upper portion of the interventricular septum.
   b. In old age, parts of the cardiac skeleton may undergo either fatty degeneration or calcification.

B. CONDUCTING SYSTEM.

This modified system of cardiac muscle fibers usually is separated from the rest of the cardiac muscle by connective tissue.

1. The **sinoatrial (SA) node** lies in the medial wall of the right atrium near the entrance of the superior vena cava. It conducts an impulse quickly to another mass of conducting cells, the **atrioventricular (AV) node**, located on the right side of the **interatrial septum**.

2. The impulse spreads rapidly down the interventricular septum in the **AV bundle**.

3. Ramifications of the AV bundle, called **Purkinje fibers**, course through the subendothelial connective tissue and contact cardiac muscle fibers.

4. Like cells in the AV and SA nodes, Purkinje fibers are modified cardiac muscle fibers that have few myofibrils per cell and may have an abundance of glycogen granules.

IV. ELASTIC ARTERIES

ELASTIC ARTERIES include the aorta, pulmonary arteries, and common carotids.

A. MICROSCOPIC ANATOMY

1. The tunica intima is unremarkable, with an endothelium and its basement membrane attached directly to a subendothelial layer of collagen, elastic fibers, and some smooth muscle cells. The boundary between the intima and the media is marked by the **elastica interna.** (The elastica interna is bridged here and there by **myoendothelial junctions**.)

2. The media is made up of about 50 concentric sheets of elastic fibers. These sheets are fenestrated, and concentric layers are connected to one another (Fig. 11-1). Interspersed between the fenestrated elastic laminae are numerous smooth muscle cells attached by collagen fibers.

3. The elastica externa separates the media from the adventitia, an unremarkable layer of fibroblasts and collagen fibers. Blood vessels from the adventitia, the **vasa vasorum**, nourish a portion of the media.

4. Some arteries may have longitudinal smooth muscle fibers in either the intima, media, or adventitia or some combination of these. In the aorta and pulmonary arteries, cardiac muscle fibers may spill over from the heart.

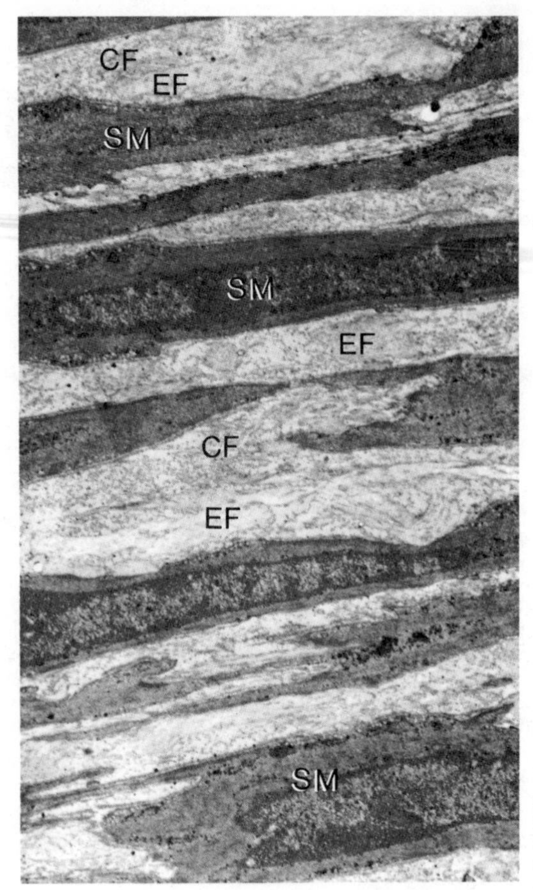

Figure 11-1. Electron micrograph through the tunica media of a large elastic artery. This layer consists of many layers of concentrically arranged smooth muscle cells (*SM*), collagen fibers (*CF*), and elastic fibers (*EF*). (Courtesy of Dr. Ernest N. Albert, Department of Anatomy, George Washington University.)

B. FUNCTION

1. When the left ventricle contracts, the blood pressure rises suddenly in the elastic arteries near the heart.
2. These elastic arteries expand, absorbing and storing the contractile energy of the left ventricle.
3. As a result, the pulsatile flow of blood is smoothed out somewhat.

V. MUSCULAR ARTERIES include most of the smaller, named arteries that are responsible for distribution of blood to all the organs. For example, the radial artery is a muscular artery (Fig. 11-2).

A. MICROSCOPIC ANATOMY

1. The tunica intima of a muscular artery is much like that found in an elastic artery. Myoendothelial junctions also are present, although their function is not understood.
2. The difference is in the media, where there are many smooth muscle fibers and few other types of cells. The elastic fibers are less prominent and exist in discontinuous layers interspersed between the 3–40 layers of smooth muscle cells, all of which are arranged helically or concentrically around the lumen. The smallest muscular arteries have a prominent **elastica interna** but little other elastic tissue in the media.
3. The adventitia is indistinctly demarcated from the media because the **elastica externa** is not prominent. The adventitia of muscular arteries is somewhat more robust than that seen in elastic arteries but otherwise contains the usual fibroblasts, collagen fibers, and vasa vasorum.
4. Certain branches of the abdominal aorta have an inner layer of elastic tissue in the media.

B. FUNCTION

1. The contraction and relaxation of smooth muscle cells in muscular arteries regulate blood pressure.

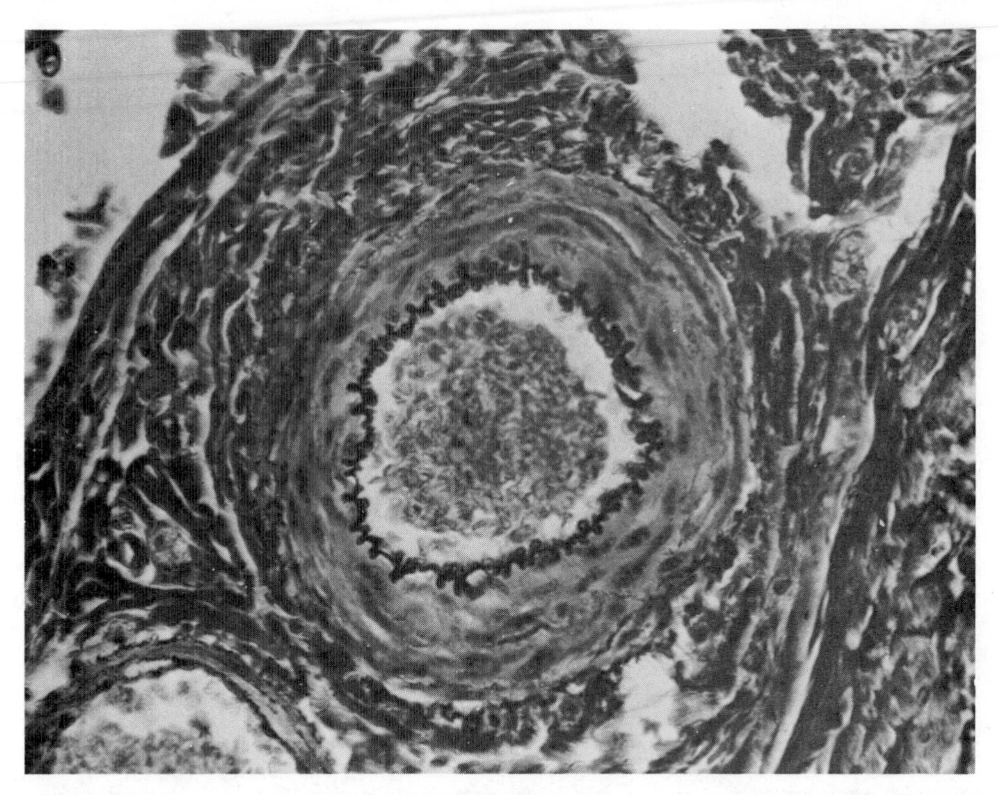

Figure 11-2. Light micrograph of a small muscular artery. This blood vessel is clogged with clumped erythrocytes. The internal elastic lamina shows very clearly because the elastic fibers in it have been specially stained. Smooth muscle cells make up most of the tunica media, and a mixture of collagenous and elastic fibers occur in the adventitia. (Reprinted with permission from Johnson KE: *Histology: Microscopic Anatomy and Embryology.* New York, John Wiley, 1982, p 159.)

2. Muscular arteries also participate, to some extent, in regulating the perfusion of different parts of the body under different physiologic conditions.

VI. ARTERIOLES

A. MICROSCOPIC ANATOMY

1. In these smallest arterial branches, the intima is wholly composed of endothelium.
2. Endothelial cells have projections passing through the fenestrated elastica interna in the form of myoendothelial junctions.
3. The media is usually two to four cells thick, and the adventitia is reduced to a smattering of collagen fibers.

B. FUNCTION

1. In some locations, just proximal to a capillary bed, the arteriolar lumen becomes reduced considerably and the smooth muscle becomes more prominent.
2. These muscular structures are precapillary sphincters and are thought to dilate and constrict rhythmically at all times.

VII. CAPILLARIES

A. MICROSCOPIC ANATOMY

1. These are the smallest branches of the CVS. The diameter of a typical capillary is 5–10 μm.
2. They are composed of an endothelium, a basement membrane, and possibly a small number of **pericytes** that are thought to be modified smooth muscle cells and may represent remnants of arteriolar media.

3. The adventitia consists of only a few strands of collagen and merges into the surrounding connective tissue.

B. DISTRIBUTION

1. The relative abundance of capillaries varies greatly with body function.
   a. In the myocardium or lung parenchyma there are thousands of capillaries in each mm^2, but only scores per mm^2 are found in most connective tissue.
   b. The hyaline and elastic cartilage and any type of epithelium are completely devoid of capillaries.
2. Only a small portion of the blood is found in capillaries at any given instant, but the surface area of all capillaries in both the systemic and pulmonary circulations is about 100 m^2.
3. Small capillaries are quite common in the lungs. Red blood cells must squeeze through these small capillaries, and as they do they are deformed and pressed against the wall of the capillary. In this manner, they move slowly and have a greater opportunity for gas exchange.
4. Several types of capillaries vary in structure and function. Variations in capillary structure include the contact relations between the endothelial cells comprising the vessel and the character of the capillary wall.

C. VARIETIES

1. **Continuous Capillaries.**
   a. These are the most common type, being predominant in all forms of muscle, the central nervous system, the lungs, and in bony **haversian systems**.
   b. Individual endothelial cells are joined together by adherent junctions where membranes come into close apposition.
   c. There is also an occluding junction at the capillary lumen.
   d. Pinocytotic vesicles are numerous in continuous capillaries. Their likely function is transport of materials across the endothelium, probably in both luminal and abluminal directions.
   e. Continuous capillaries have a substantial basement membrane and a few collagen fibrils in the pericapillary spaces. The basement membrane is split by pericytes that enclose the membrane in a sheath of collagen.
   f. Although the walls of continuous capillaries are thicker than those of other types of capillaries, they are still extremely thin (about 0.2–0.4 μm.)
2. **Fenestrated Capillaries.**
   a. These are common in the renal glomeruli, gastrointestinal tract, choroid plexus, ciliary body, and endocrine glands. Fenestrated capillaries at these interfaces are highly modified for rapid and controlled fluid or solute transport across the endothelium.
   b. Fenestrated capillaries have numerous transcellular pores in their walls. These openings have a diameter of approximately 70 nm.
   c. The entire fenestration is lined by the plasma membrane of endothelial cells, and the fenestrations represent aqueous channels **through** the vessel walls.
   d. Fenestrations may or may not have a diaphragm across them. The exact functional significance of the fenestrations and diaphragms is unclear, but they are probably involved in regulation of solute transport across the endothelium. Fenestrated capillaries in the renal glomeruli have no diaphragms but most others do.
   e. These capillaries come equipped with a continuous basement membrane which is usually fragile. The basement membrane of the fenestrated renal glomerular capillaries, however, is robust and serves as part of the protein filtration mechanism that prevents some blood proteins from entering the forming urine.
   f. The walls of the fenestrated capillaries are only 0.1 μm thick.
3. **Discontinuous Capillaries.**
   a. An abundance of these are found lining sinusoids in the reticuloendothelial system, especially in the liver, spleen, and bone marrow. In the liver, large gaps separate individual endothelial cells.
   b. The large intercellular gaps allow massive protein transport from hepatic parenchymal cells into the blood or cellular transport from the splenic pulp or bone marrow into the closed circulation.
   c. Some endothelial cells may be specialized for phagocytosis (**Kupffer cells**).
   d. The basement membrane is thin and discontinuous at best, and it may be absent entirely in some instances.

VIII. VEINS AND VENULES

A. VEINS

1. Microscopic Anatomy.

a. In veins, the divisions between intima, media, and adventitia are less distinct than the divisions in arteries of corresponding size.

b. Also, there are great structural differences in veins of the same diameter, which depend on whether they are above or below the heart and whether or not they are in a fixed or mobile position in the body.

c. Some veins change drastically in structure over relatively short portions of their course. For example, in parts of the venae cavae, the media is extremely thin and there is an abundance of longitudinally arranged smooth muscle in the adventitia. Nearby, cardiac muscle fibers may spill over into the media of the venae cavae.

d. In spite of these variations, all veins
   (1) Have walls that usually are thinner than those of companion arteries
   (2) Usually have less muscular and elastic tissue in their media than do companion arteries
   (3) Have a relatively thin intima with a continuous endothelium and basement membrane
   (4) Have a relatively thick adventitia with either a few (medium veins) or an abundance (large veins) of smooth muscle cells

e. Most medium and large veins have a robust vasa vasorum originating in the adventitia and sometimes penetrating almost to the intima.

2. Functions.

a. Veins serve as a blood reservoir.

b. They also are equipped with valves to prevent retrograde blood flow.

B. VENULES

1. **Microscopic Anatomy.** The smallest of these vessels is only slightly larger than a capillary (10–30 μm in diameter) and may be associated with a few pericytes (sometimes called **pericytic venules**) or two to four layers of smooth muscle cells (**muscular venules**).

2. **Function.** Venules are found in the body wherever there are rapid changes in the perfusion of an area for homeostatic reasons (e.g., temperature regulation in the skin).

IX. EMBRYOLOGY

A. EARLY DEVELOPMENT OF THE CVS

1. The entire CVS is derived from mesoderm via a differentiation of mesenchymal cells.

2. Beginning in the third week gestation, blood vessels arise spontaneously in the embryo proper and in the chorion and yolk sac in **angiogenic cell clusters**.

3. These blood vessels rapidly form anastomoses and communicate with the developing heart.

4. The mesenchymal cells in the embryo differentiate into vascular endothelium and **hemocytoblasts** that initially give rise to primitive nucleated erythrocytes.

B. EARLY DEVELOPMENT OF THE HEART

1. The heart arises from angiogenic cell clusters formed in the splanchnic mesoderm, anterior and lateral to the developing neural plate.

2. As the head fold of the amnion grows, the forming heart tube is brought under the head and into the midline of the body.

3. The **extraembryonic coelom** extends around the head region and is tucked under the head during flexion to give rise to the pericardial cavity.

4. At the earliest stages, the human heart is represented by a simple tube with a pair of **aortic roots** emerging from the **bulbus cordis** at the cranial end. These roots communicate directly with the **primitive ventricle**, which has a **left** and **right sinus horn** draining into the **sinus venosus**.

5. The heart tube lengthens within the pericardial cavity and consequently undergoes a **looping** where the cranial and caudal portions of the heart change relative positions. Thus, the ventricle comes to lie caudal to the atria.

6. During looping, the bulbus cordis divides into a **truncus arteriosus**, which feeds the **aortic arches**, and a **conus cordis**, which communicates directly with the primitive right ventricle.

7. In early stages, both the atria and the ventricles are in direct communication (i.e., the atrial and ventricular septa are not formed yet).

C. **ATRIAL SEPTATION.** The atrial septation process itself is complex. This essential barrier between the right and left atria is not complete until birth.

1. Initially, the right and left atria are not separated.
2. A curtain of tissue, the **septum primum**, grows down from the top of the heart, between the atria, toward the **endocardial cushions** in the center of the heart. Before the septum primum reaches the endocardial cushions, a hole, the **ostium primum**, allows blood flow from right to left.
3. As the septum primum grows and obliterates the ostium primum, a new hole, the **ostium secundum**, forms in the septum primum, maintaining communication between right and left atria.
4. Now a **septum secundum** grows downward toward the endocardial cushions. It does not grow down completely but leaves another channel, **the foramen ovale**.
5. Septation of the atria is completed after birth, when there is a decrease in the amount of blood entering the heart on the right side due to cessation of flow from the umbilical vein together with a sudden inflation and perfusion of the lung parenchyma.
   a. The increased perfusion of the lungs increases the pressure in the blood returning to the heart via the pulmonary veins. A subsequently increased pressure in the left atrium forces the septum primum against the septum secundum, obliterating the ostium secundum and foramen ovale and completing septation.
   b. At birth, the smooth muscle cells in the wall of the **ductus arteriosus** constrict and suddenly reduce the amount of blood being shunted past the pulmonary arteries. This also increases blood flow to the lungs.

D. **VENTRICULAR SEPTATION** results from two processes.

1. A mass of muscular tissue grows from the apex of the heart toward the endocardial cushions. This mass later becomes the **muscular portion** of the interventricular septum.
2. Also, a downgrowth from the inferior endocardial cushions completes the septation of the ventricles. This downward growth later contributes to the **membranous portion** of the interventricular septum.

E. **DIVISION OF VENTRICULAR OUTFLOW**

1. The truncus arteriosus and the conus cordis are subdivided by the fusion of a series of swellings from their walls.
2. In this manner, a division is created in the ventricular outflow of the heart, leading eventually to a separation of the systemic and pulmonary ventricular output.

F. **SINUS VENOSUS**

1. The primitive unseptated atrium is fed by reasonably symmetrical branches of a left and right sinus venosus.
2. **Anterior** and **posterior cardinal veins** drain into a **common cardinal vein**, which in turn empties into the sinus venosus along with the **vitelline** (yolk sac) and **umbilical** veins (placenta). This is duplicated on both the left and right sides of the primitive heart.
3. The **left** sinus venosus undergoes considerable degeneration and eventually forms part of the coronary sinus and venous drainage of the heart.
4. The **right** sinus venosus becomes incorporated into the wall of the right atrium and also contributes to the superior and inferior venae cavae.

STUDY QUESTIONS

Directions: Each question below contains four suggested answers of which **one or more** is correct. Choose the answer

A if **1, 2, and 3** are correct
B if **1 and 3** are correct
C if **2 and 4** are correct
D if **4** is correct
E if **1, 2, 3, and 4** are correct

1. Features of continuous capillaries include

(1) luminal tight (occluding) junctions
(2) abundant pinocytotic vesicles
(3) a continuous basal lamina
(4) prominent luminal microvilli

2. True statements concerning arterioles include which of the following?

(1) Smooth muscle fibers are common in the media
(2) Fenestrated elastic membranes are prominent in the media
(3) The endothelium is continuous
(4) The adventitia contains many smooth muscle cells

Directions: The group of questions below consists of lettered choices followed by several numbered items. For each numbered item select the **one** lettered choice with which it is **most** closely associated. Each lettered choice may be used once, more than once, or not at all.

Questions 3–7

For each description, select the most appropriate capillary.

(A) Continuous capillary
(B) Fenestrated capillary
(C) Discontinuous capillary
(D) Lymphatic capillary

3. Abundant in skeletal muscle

4. Predominant vessel in areas where such blood filtrates as urine are produced

5. Lines the bone marrow sinusoids

6. Present in the choroid plexus

7. Responsible for the blood-brain barrier

ANSWERS AND EXPLANATIONS

1. The answer is A (1, 2, 3). (*VII C*) Continuous capillaries are closed barriers of epithelium. They have luminal tight junctions, abundant pinocytotic vesicles, a continuous basal lamina, but no microvilli at the lumen.

2. The answer is B (1, 3). (*VI*) Arterioles have very little elastic tissue in their media. The adventitia also is quite sparse and does not contain smooth muscle cells.

3–7. The answers are: 3-A, 4-B, 5-C, 6-B, 7-A. (*VII C*) Continuous capillaries are a feature of skeletal muscle, the lungs, and the brain. They provide a continuous epithelial barrier that restricts diffusion of materials from the blood into the tissues. Fenestrated capillaries are present in renal glomerular capillaries. The fenestrations are holes through the walls of the capillary endothelial cells. Discontinuous capillaries are present in the liver, bone marrow, and spleen. There are gaps between capillary endothelial cells, which are large enough to pass cells. For example, in the bone marrow, mature erythrocytes move from the hematopoietic compartment in the bone marrow into the blood through the intercellular gaps in the discontinuous capillaries. The choroid plexus is the site of cerebrospinal fluid (CSF) production. The CSF is a blood filtrate containing some of the proteins found in whole blood. The fenestrated capillaries of the choroid plexus allow a selected subset of blood proteins to enter the CSF. Continuous capillaries in the brain are responsible for the blood-brain barrier. They restrict flow of various substances from the blood to the brain parenchyma.

12
Respiratory System

I. INTRODUCTION

A. COMPONENTS

1. Conducting Airways.
- **a.** Nasal cavity
- **b.** Nasopharynx
- **c.** Larynx
- **d.** Trachea
- **e.** Bronchi

2. Epithelium for Gas Exchange.
- **a.** Type I (squamous) cells
- **b.** Type II (surfactant-secreting) cells

B. FUNCTIONS

1. Oxygenates blood by way of inspired air

2. Releases CO_2 from blood by way of expired air

3. Contains organ for olfaction

C. LUNG DEVELOPMENT

D. LUNG DEFENSES

1. Filters

2. Mucus and cilia

3. Lymphoid infiltration of lamina propria

4. Macrophages

II. UPPER RESPIRATORY TRACT AND OLFACTORY EPITHELIUM

A. AIR CONDITIONING IN UPPER RESPIRATORY TRACT

1. Air entering the nasal cavity is warmed, moistened, and often cleared of significant amounts of water-soluble air pollutants. These varied functions are accomplished by the **respiratory mucosa** of the nasal cavity.
- **a.** The nasal cavity has a pseudostratified ciliated columnar (PCC) epithelium. Glands that produce mucous and serous (watery) secretions are plentiful in the respiratory mucosa.
- **b.** An extensive vascular bed assures that inspired air is warmed.

2. The structure of the nasal cavity causes inspired air to flow with considerable turbulence.

B. OLFACTORY MUCOSA

1. The olfactory mucosa also has a PCC epithelium, which is quite tall and modified for olfaction.

2. There are three cell types in the olfactory epithelium.

a. Olfactory Cells.

(1) Olfactory cells have very long, nonmotile cilia which are the site of chemoreceptors. Serous (Bowman's) glands in the mucosa keep these cilia moist so that chemicals in inspired air can be dissolved.

(2) Humans have a poorly developed sense of smell; their olfactory cells have relatively short cilia in comparison to the olfactory cells of other animals such as cats.

(3) The mechanism of action of chemoreceptors is poorly understood. Most likely the receptor cell membrane has receptor proteins that recognize structural differences in stimulants. After interacting with the stimulant molecules, the receptor proteins then cause depolarization of the olfactory cell membrane and, subsequently, generation of an action potential.

(4) The olfactory cell is a neuron with a dendrite and an axon. It synapses directly with neurons in the olfactory bulb. The axons of many epithelial cells are gathered together into **fila olfactoria**. These carry the action potential generated by the stimulating compound to the olfactory lobes of the central nervous system.

b. **Sustentacular cells** have numerous microvilli and a well-developed apical Golgi apparatus. In general, they have the appearance of secretory cells. Sustentacular cells contact the olfactory cells.

c. **Basal cells** are undifferentiated and retain the capacity to divide and differentiate into either olfactory or sustentacular cells. (One curious feature of many sensory cells is that the sensory portion of the cell is frequently a modified cilium. This is true for the cells that are sensitive to light in the retina, for example cones and rods, and those that are sensitive to mechanical stimulation in the cochlea and the vestibular apparatus.)

C. NASOPHARYNX AND LARYNX

1. The nasopharynx has a PCC epithelium, which is occasionally interrupted by stratified squamous epithelium of the oropharynx. The mucosa of the nasopharynx is thin and lacks glands.

2. The larynx connects the pharynx and trachea. It is a hollow, bilaterally symmetrical structure made of cartilage, muscle, and ligaments that prevent food from entering the respiratory system during swallowing. Also, the larynx contains the vocal cords, which are ligaments for speech.

III. TRACHEA

A. MICROSCOPIC ANATOMY (Fig. 12-1)

1. The trachea is a tubular structure which begins below the larynx and ends at the bifurcation, where the bronchi begin.

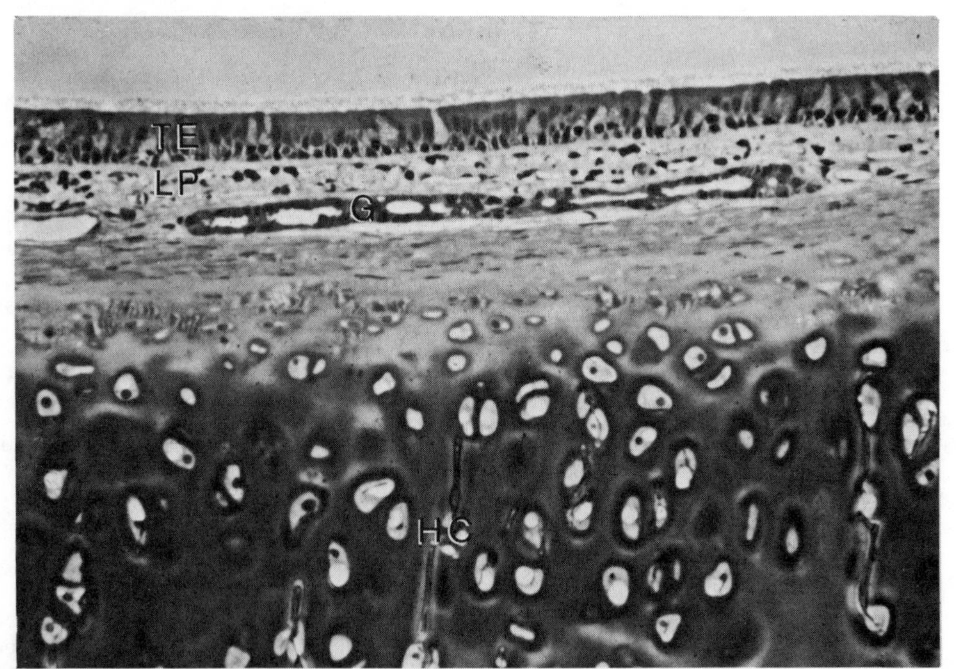

Figure 12-1. Light micrograph of a section of the trachea. The tracheal epithelium (*TE*) rests on a lamina propria (*LP*). Parts of glands (*G*) are evident in the lamina propria. Beneath the lamina propria, a large piece of hyaline cartilage (*HC*) is shown.

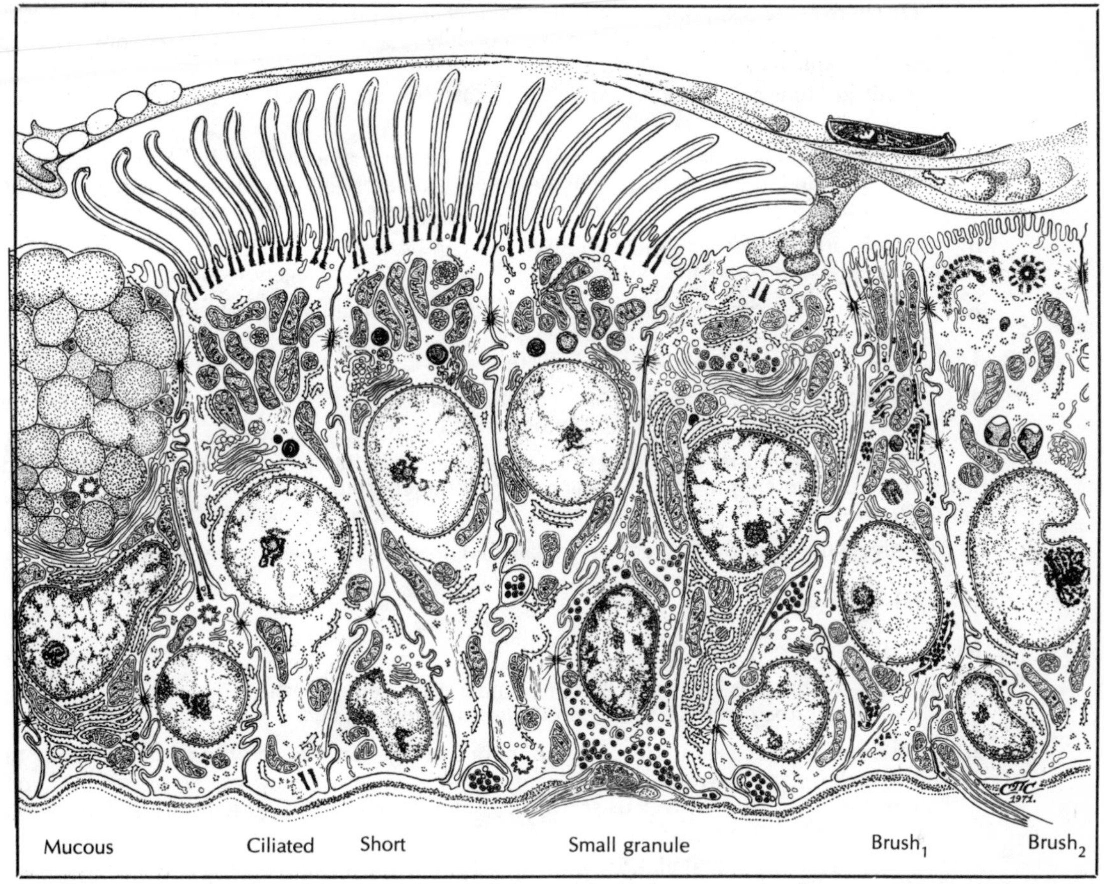

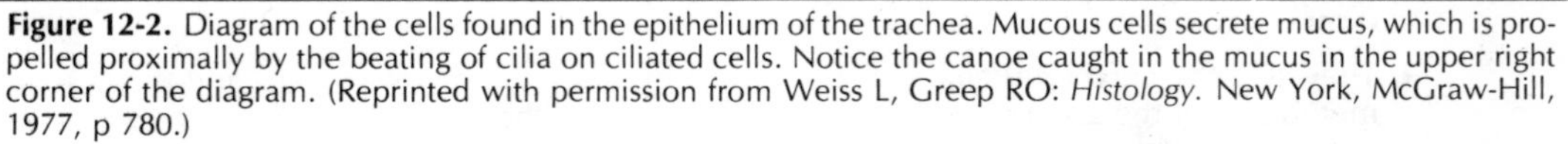

Figure 12-2. Diagram of the cells found in the epithelium of the trachea. Mucous cells secrete mucus, which is propelled proximally by the beating of cilia on ciliated cells. Notice the canoe caught in the mucus in the upper right corner of the diagram. (Reprinted with permission from Weiss L, Greep RO: *Histology.* New York, McGraw-Hill, 1977, p 780.)

2. The trachea has a **mucosa**, **submucosa**, poorly developed **muscularis**, and an **adventitia**.

3. The diverse tasks of this PCC epithelium are reflected by the fact that it has six cell types (Fig. 12-2).

a. Mucous (Goblet) Cells.

(1) These cells produce large numbers of mucous droplets which coalesce during secretion to form larger drops. Parts of the mucous cell are lost during this **apocrine** secretion.

(2) The mucous secretion forms a continuous layer on the tracheal epithelium which traps inspired debris, including dirt, pollen, and microorganisms.

b. Ciliated Cells.

(1) These cells are abundant in the epithelium. Each cell has one cilium which contains nine peripheral doublet microtubules. The cilium is anchored in the apical cytoplasm of the cell by a **basal body**.

(2) Adenosine triphosphate (ATP) is supplied for ciliary motility by a profusion of mitochondria in the apices of the ciliated cells.

(3) The cilia beat toward the pharynx and propel mucus and entrapped debris away from the lungs, up the trachea, and through the pharynx into the gastrointestinal tract, where they are eventually eliminated from the body.

c. Short Cells.

(1) These undifferentiated cells also are abundant.

(2) Short cells rest on the basement membrane of the epithelium but do not reach the lumen.

d. Brush Cells.

(1) Brush cells are so-named because they have a microvillous apical border (**brush border**).

(2) There are two kinds of brush cells—some that are probably basal cells differentiating into ciliated cells and others that are probably mucous cells that have recently released their mucus.

e. **Small granule cells** contain granules of catecholamines and are thought to influence mucus secretion.

4. There is a prominent basement membrane below the tracheal epithelium, and occasional glands are found in the connective tissue below the basement membrane. The demarcation between the lamina propria and the submucosa is not sharp in the trachea or bronchi.
5. Mast cells and polymorphonuclear leukocytes can be seen wandering through the matrix of capillaries, small lymphatics, fibroblasts, and extracellular collagenous fibers.
6. The connective tissue of the submucosa blends into the perichondrium of the cartilage.

B. TRACHEAL CARTILAGES AND ADVENTITIA

1. C-shaped cartilaginous rings help keep the trachea open during inspiration. Dorsally, where the cartilage is absent, there is a prominent band of smooth muscle.
2. Flattened **perichondrial cells** form several layers around the cartilage. These cells fatten and differentiate into **chondroblasts**, which secrete the constituents of the cartilage matrix and wall themselves off from the neighboring cells, thus becoming **chondrocytes**.
3. The adventitial layer of the trachea contains adipose tissue, blood vessels, and nerves.

IV. BRONCHIAL TREE (Fig. 12-3)

A. BRONCHI

1. The trachea bifurcates into bronchi immediately before entering the lungs. Bronchi continue to branch, approximately 20 times, until they are quite small.
2. As the bronchi decrease in size, the amount of cartilage reduces and the amount of smooth muscle in the space between the basal lamina and the cartilage increases. Also, the PCC epithelium becomes smaller. Mucus-secreting goblet cells and ciliated cells are still common here.
3. Bronchi contain **bronchial glands**, which produce mucous and serous secretions (from two distinct cell types).
4. They also have **myoepithelial cells**, which lie between the secretory cells and the basement membrane of the epithelium. These cells have many characteristics of smooth muscle cells and are thought to aid secretion by their contraction.

B. BRONCHIOLES

1. There are no mucus-secreting goblet cells in bronchioles. Ciliated cells are common, as are **bronchiolar cells** (Clara cells) which produce serous secretions.
2. Distally, bronchioles become smaller, and their walls contain scattered **alveoli**. At their alveoli, bronchioles have the dual function of providing a conducting airway to alveolar ducts and a surface for gas exchange.
3. These **respiratory bronchioles** branch several times and lead to **alveolar ducts**, which communicate with numerous alveoli at the end of the bronchial tree.
4. Respiratory bronchioles contain ciliated and bronchiolar cells where there are no alveoli. Smooth muscle and connective tissue also are prominent here.

C. LOBES

1. The left lung has an upper and a lower lobe. The right lung has an upper, middle, and lower lobe.
2. Each lobe is supplied by a variable number of **segmental bronchi**, which represent third- or fourth-order branches of the bronchi.
3. Segmental bronchi then branch many times until they become **terminal bronchioles**, the smallest bronchioles, which are not involved in gas exchange.
4. Blood vessels follow the segmental bronchi. Segments of the lobes have independent blood supplies and are demarcated by connective tissue septa.

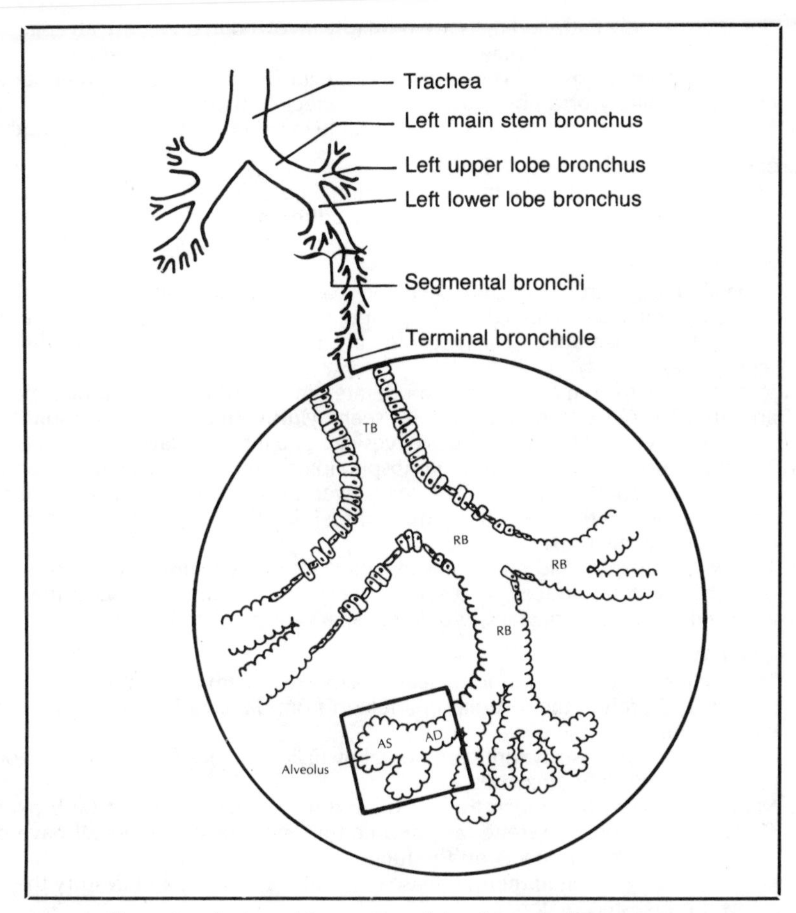

Figure 12-3. Diagram illustrating the branching pattern of the airway in the respiratory system. In this illustration, an acinus (*circled area*) is sometimes defined as all units of the lung distal to the terminal bronchioles. An acinus is sometimes called a secondary lobule. A primary lobule (*boxed area*) is defined as all units distal to an alveolar duct. *RB* = respiratory bronchiole; *TB* = terminal bronchiole; *AS* = alveolar sac; *AD* = alveolar duct. (Reprinted with permission from Johnson KE: *Histology: Microscopic Anatomy and Embryology*. New York, John Wiley, 1982, p 176.)

5. This segmentation pattern is regular and allows removal of seriously diseased portions of the lung without the loss of structure or function of other healthy segments.

V. ALVEOLI

A. MICROSCOPIC ANATOMY

1. Alveoli are blind sacs; they constitute the major respiratory surface of the lung, with a collective area about half the size of a tennis court.
2. The alveolar epithelium is exceedingly thin, frequently much less than 1 μm, and rests on a typical basement membrane. Capillary endothelial cells and connective tissue fibroblasts are intimately associated with the alveolar epithelium.
3. Gases from the alveolar space must diffuse twice across the plasma membrane of the alveolar cell, through the basement membrane, and twice through the plasma membrane of the capillary endothelial cells.
4. The gas then dissolves in the blood plasma and finally diffuses across the erythrocyte membrane and interacts with the hemoglobin-carrier molecule.
5. This journey is easily accomplished in a diffusion path of 1 μm or less.

B. ALVEOLAR EPITHELIUM AND MACROPHAGES

1. **Squamous Alveolar Epithelial Cells (Type I Cells).**

a. These extremely flattened cells are primarily involved in lining the alveolar spaces and are specialized for gas exchange.
b. They have a nucleus that is difficult to distinguish from a nucleus in either an endothelial cell in a capillary or a fibroblast cell in connective tissue.
c. The cytoplasm of type I cells is unremarkable except that it has many **pinocytotic vesicles.**

2. **Great Alveolar Cells (Type II Cells).**
a. These rounded cells are joined by junctional complexes (i.e., **zonula occludens**, or tight junction; **zonula adherens**; and **macula adherens**, or desmosome) to the type I cells, forming a continuous epithelium which is joined to the bronchial and bronchiolar epithelium, the gut, and the skin.
b. Type II cells protrude into the alveolar spaces and have many of the characteristics of secretory cells. With the light microscope, they are conspicuous because of their empty, vacuolated appearance. The vacuoles are remnants of extracted **multilamellar bodies (cytosomes)**.
c. With the electron microscope, a moderate amount of rough endoplasmic reticulum and an extensive Golgi apparatus can be seen. **Multivesicular bodies**, multilamellar bodies, and structures containing mixtures of vesicles and intermediate lamellae also are common.
d. The multilamellar bodies contain phospholipids, mucopolysaccharides, and proteins; this secretory product of type II cells spreads over the surface of the alveoli. One predominant phospholipid in this secretion, **dipalmitoyl lecithin**, is an active component of the pulmonary **surfactant**.
e. The polar phospholipid molecules in surfactant reduce surface tension of the water layer over the alveolar surfaces. This surface tension resists an increase in the alveolar surface area and, without surfactant, would cause alveolar collapse.

3. **Alveolar Macrophages.**
a. These cells, part of the mononuclear phagocyte system, constantly cleanse the epithelial surface and protect the alveolar epithelium from damage by microorganisms or other inhaled irritants.
b. They engulf and destroy foreign material by fusing the membranes of phagocytic vacuoles to sacs of hydrolytic enzymes.
c. Many substances that reach the alveoli and macrophages cannot be broken down.
(1) Coal dust, soot, cigarette tars, and cotton and asbestos fibers all have disastrous and potentially lethal effects on the lungs.
(2) Macrophages engulf them, release lysosomal enzymes, and destory the alveoli instead of the ingested materials.
(3) In addition, many of these ingested materials are carcinogenic and, when concentrated, can lead to malignant transformation of cells in the lung.

VI. LUNG DEVELOPMENT

A. EARLY DEVELOPMENT

1. At three weeks gestation, a **respiratory diverticulum** forms as a ventral bud from the foregut. Soon after its formation, the respiratory diverticulum is separated from the rest of the gastrointestinal tract by the growth of an **esophagotracheal septum**.

2. By four weeks gestation, this septum has nearly completed its separation, and the respiratory diverticulum begins to branch into a left and a right lung bud. The epithelium lining the entire respiratory system develops from this diverticulum and is therefore of **endodermal origin**.

B. GROWTH OF RESPIRATORY DIVERTICULUM

1. The lung buds grow and branch caudally and laterally. On the right side, three main buds form while only two from on the left side. Eventually these buds form lobes—three lobes on the right and two on the left.

2. As the lung buds grow, they become associated with splanchnic mesoderm. The mesoderm contributes to the cartilages, smooth muscle, blood vessels, and connective tissues of the respiratory system.

3. The lung buds and associated splanchnic mesoderm undergo a reciprocal inductive interaction that is essential for normal pulmonary development.

4. The lateral growth of the lungs causes a progressive obliteration of the **pericardioperitoneal canals**. In the primitive condition, the pericardioperitoneal canals are part of the extraembryonic coelom. They are lined with the same mesothelium that lines the pericardial and peritoneal cavities.

5. The developing lungs become coated with a **visceral pleura**, which is reflected upon a **parietal pleura** at the mediastinum.

C. DEVELOPMENT OF FUNCTIONAL LUNGS

1. The growth and branching of the lung buds continues until the embryonic lungs assume a glandular appearance.
2. Between six and seven months gestation, the epithelium lining the many branches of the lung buds differentiates into flattened cells and surfactant-secreting type II cells.
3. Once type II cells have formed, a fetus is viable. Premature infants without type II cells have great difficulty inflating their lungs, due to the lack of surfactant-producing cells. Premature infants are at high risk for developing **respiratory distress syndrome**.

VII. LUNG DEFENSES

A. DELETERIOUS MATERIAL. Air contains oxygen for respiration but also may contain bacteria, pollen, dust, rubber particles, cigarette smoke, and other contaminants. Everything in air but oxygen and nitrogen is either extraneous or frankly harmful.

B. FILTERS

1. The nares contain hairs that filter the air as soon as it enters the body.
2. The nasal turbinates and other structures in the nasal cavity provide a tortuous pathway for inspired air. Dust and other particulate matter travel in a straight line once inspired and commonly collide with a mucus-moistened surface equipped with cilia. The dust is trapped and eventually expelled from the body via the gastrointestinal tract.
3. The nasal cavity is moist and well-supplied with capillaries, which can either warm or cool inspired air and increase the relative humidity of dry air.
4. The nasal and tracheal epithelia also are equipped with nerve endings that, when stimulated, can set off a reflex expulsion of air (a sneeze) from the airways.

C. TRACHEAL EPITHELIUM

1. The tracheal epithelium is coated with a network of strands of mucus secreted by the goblet cells.
2. The cilia on this epithelium contribute to a ''mucociliary escalator'' that constantly propels unwanted substances proximally for expulsion from the gastrointestinal tract.
3. In addition, apical tight junctions and a robust basement membrane prevent bacteria from crossing the respiratory epithelium.

D. LYMPHOCYTES AND MACROPHAGES

1. Bacteria crossing the respiratory epithelium immediately encounter a lamina propria rich in lymphocytes and neutrophils.
2. The lymphatic drainage of the entire respiratory system is extensive, as is the aggregate of mediastinal lymph nodes.
3. In the alveoli are many alveolar macrophages, which also serve to protect the lungs.
4. All of these defensive features protect the gossamer epithelium of the alveoli.

VIII. CIGARETTES AND THE LUNGS

A. THE SMOKING PATIENT

1. Many patients are nicotine addicts. They damage their cardiovascular systems by smoking, but they damage their lungs also in a fundamental way.
2. The smoking habits of patients should be determined so that these patients can be advised accordingly. Patients who smoke excessively have a serious medical problem.

B. HARMFUL CONSTITUENTS OF CIGARETTE SMOKE

1. Cigarette smoke contains a large collection of carcinogenic materials. It contains various residues of burned tobacco leaves, insecticides, asbestos fibers from the manufacturing plant,

and—from the fertilizer used in tobacco farming—polonium 210 (a naturally occurring isotope which upon decay releases potent ionizing radiations).

2. These materials are carried deep into the lungs where they accumulate at branch points of the airway or in the macrophages. The macrophages become overloaded and self-destruct, releasing lysosomal hydrolases that will destroy the lung parenchyma, leading to **emphysema**.

C. CARCINOGENESIS AND SMOKING

1. Carcinogens are deposited in the connective tissues and lymph nodes around the airways and remain there, thus providing long-term exposure to carcinogens.
2. The polonium 210 in fertilizers is a radioisotope that enters the lungs in smoke and irradiates epithelial cells. This exposure probably causes the malignant transformation of these cells.
3. In addition to the harmful agents mentioned, some unknown component of the smoke immobilizes and destroys the cilia.
4. Many smokers do not have a PCC epithelium in much of their respiratory system. Instead, they have a squamous metaplasia and, thus, lose a good deal of their defense system.

STUDY QUESTIONS

Directions: Each question below contains five suggested answers. Choose the **one best** response to each question.

1. All of the following statements concerning the olfactory epithelium are true EXCEPT

(A) Bowman's glands secrete a serous product
(B) sustentacular cells contact the olfactory cells
(C) olfactory cells have long motile cilia
(D) basal cells rest on a basement membrane
(E) axonal projections from the olfactory epithelium enter the central nervous system

2. All of the following statements concerning the alveolar epithelium are true EXCEPT

(A) type I and type II cells are joined by junctional complexes
(B) there is a well-developed basement membrane under this epithelium
(C) alveolar epithelial cells are closely associated with capillary endothelium
(D) ciliated cells are found in alveoli
(E) type II cells contain multilamellar bodies of surfactant

Directions: The group of questions below consists of lettered choices followed by several numbered items. For each numbered item select the **one** lettered choice with which it is **most** closely associated. Each lettered choice may be used once, more than once, or not at all.

Questions 3–7

For each of the following descriptions of an adult respiratory structure, select the appropriate embryonic rudiment.

(A) Respiratory diverticulum
(B) Splanchnic mesoderm
(C) Both
(D) Neither

3. Produces smooth muscle, cartilage, and blood vessels in the trachea

4. Produces ciliated and goblet cells in bronchi

5. Produces type I and type II cells

6. Produces both alveolar macrophages and cartilage

7. Produces capillaries and connective tissue in the distal lungs

ANSWERS AND EXPLANATIONS

1. The answer is C. (*II B*) Olfactory cells are neurons; each has an axon and a dendrite. The axons of many olfactory epithelial cells are gathered into fila olfactoria, which carry the action potential to the central nervous system. Olfactory cells have long cilia but they are not motile. Serous secretions from Bowman's glands in the mucosa keep these cilia moist.

2. The answer is D. (*V A, B*) The alveolar epithelium is very thin and rests on a typical basement membrane. Capillary endothelial cells and connective tissue fibroblasts are intimately associated with the alveolar epithelium. This epithelium contains type I and type II cells; these cell types are joined by junctional complexes to form a continuous epithelium. There are no ciliated cells in the alveolar epithelium.

3–7. The answers are: 3-B, 4-A, 5-A, 6-C, 7-B. (*VI A, B*) All epithelia lining the respiratory system, from the larynx all the way distally to the alveoli, are derived from the respiratory diverticulum. Everything else comes from splanchnic mesoderm, except for macrophages. These phagocytes come from bone marrow.

13 Upper Gastrointestinal Tract and Development of the Face

I. INTRODUCTION. The digestive process begins when food enters the mouth. The lips, tongue, and teeth start the digestive processing of food by crushing or tearing it into manageable pieces and mixing it with saliva so the food may be swallowed and transported to the stomach. The sight, smell, and taste of food initiate salivation (as well as peristalsis in the more distal portions of the digestive system). Besides its role in initiating digestion, the tongue also is important for forming the sounds in human speech.

The following subjects are discussed in this chapter.

A. ORAL CAVITY

B. TEETH

C. TONGUE

D. SALIVARY GLANDS

E. TONSILS AND PHARYNX

F. DEVELOPMENT OF FACE, PALATE, AND PHARYNGEAL ARCHES

II. MICROSCOPIC ANATOMY OF THE ORAL CAVITY

A. LININGS

1. The oral cavity is lined everywhere by a **stratified squamous epithelium**. This layer is nonkeratinized in most locations but is keratinized in areas subject to abrasion during ingestion and chewing of food (e.g., the gums near the teeth, the hard palate, and the dorsal surface of the tongue). Whether or not it is keratinized, the stratified squamous epithelium of the oral cavity is 10–50 cell layers thick and often rests on high papillae of the underlying connective tissue lamina propria.
2. There is little smooth muscle in the oral cavity wall, even when the entire structure is quite thick, and there is little submucosal connective tissue. All layers that are characteristic of the skin, from stratum basale to stratum corneum, also are seen in the oral cavity. Likewise, the epithelial cells are joined by a plethora of desmosomes.

B. MINOR SALIVARY GLANDS

1. A wide variety of minor salivary glands are scattered throughout the mouth. These glands are small invaginations of the surface epithelium. (The major salivary glands have a similar but more elaborate structure.)
2. Some minor salivary glands are **mixed-type** glands (labial glands), producing mucous and serous secretions; others are purely **mucous glands** (palatine glands).

III. TEETH are complex structures formed from invaginations of the surface epithelium in the oral cavity. Cells from oral ectoderm and head mesenchyme interact in the production of a unique set of 20 **primary** (deciduous) teeth, lost during childhood, and 32 **secondary** teeth, meant to last the duration of one's life.

A. DEVELOPMENT

1. In a seven-week-old fetus, a thickening in the oral epithelium covers both the developing mandible and the developing maxilla. By proliferation, this **dental lamina** penetrates the underlying mesenchyme and gives rise to a cup-shaped **enamel organ**.
2. A separate enamel organ forms for each of the primary teeth and grows larger and becomes deeply invaginated. A mesenchymal condensation called a **dental papilla** forms deep within this invagination. It consists of a dense collection of fibroblastic cells that will eventually form most of the tooth proper, including the pulp cavity and dentin.
3. On the surface of the enamel organ, ectodermally derived **ameloblasts** form **enamel** and mesenchymally (from neural crest) derived **odontoblasts** form **dentin.** Tooth formation involves an epithelial-mesenchymal interaction between the ectodermally derived enamel organ and the mesenchymally derived dental papilla (Fig. 13-1).

B. DENTIN AND ENAMEL

1. Initially, odontoblasts secrete a collagenous extracellular matrix known as **predentin**. This becomes calcified into a hard, bony material called **dentin**. Dentin, like bone, contains hydroxyapatite crystals.
2. The formation of dentin seems to induce ameloblasts to produce a crown consisting of a second calcified extracellular matrix called **enamel**. The crown forms first on the cusps of the developing teeth and then grows toward the area where the roots of the teeth will form.
3. Roots, lacking enamel, are formed next by odontoblasts. The formation of the root seems to induce certain mesenchymal cells to differentiate into **cementoblasts**.
4. These cells then secrete a third kind of calcified extracellular matrix, the **cementum**, which helps affix the teeth in the bony sockets in the maxilla and mandible.

C. AMELOBLASTS

1. Ameloblasts form an epithelial layer highly specialized for the **basal** secretion of enamel.

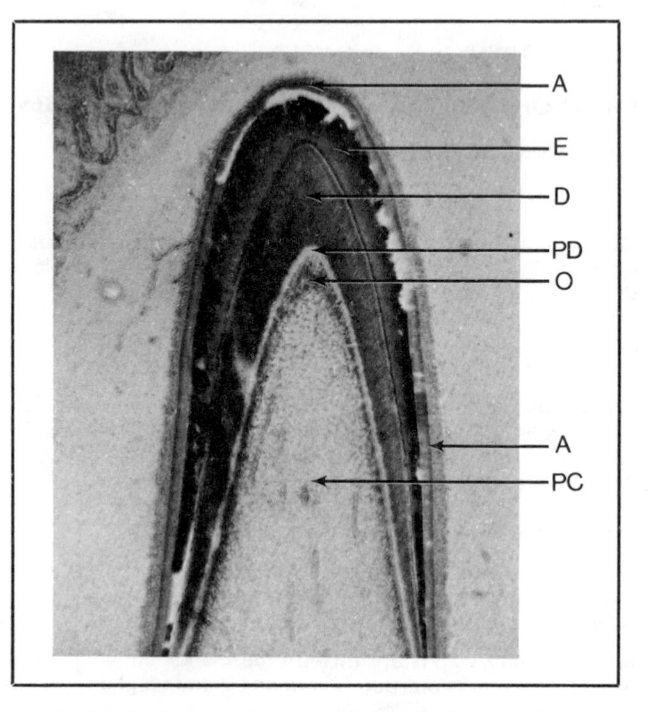

Figure 13-1. Light micrograph of a histologic section through a developing tooth in the mandible of a pig embryo. The enamel (*E*) and dentin (*D*) form around a mesenchymal condensation known as a pulp cavity (*PC*). A layer of odontoblasts (*O*) secrete predentin (*PD*), which rapidly calcifies into dentin. A layer of ameloblasts (*A*) secrete the enamel that is cracked and chipped extensively as a sectioning artifact in this slide. (Reprinted with permission from Johnson KE: *Histology: Microscopic Anatomy and Embryology.* New York, John Wiley, 1982, p 187.)

2. These cells are derived from an invagination of the oral epithelium, and for a time they retain its basement membrane. As predentin and dentin accumulate adjacent to the basement membrane of the ameloblasts, and as ameloblasts secrete enamel, this basement membrane disappears.

3. Ameloblasts are exceedingly tall and have an apical collection of mitochondria, an apical nucleus, and a well-developed endoplasmic reticulum (ER) in the basal compartment of the cell near the nucleus.

4. Their rough endoplasmic reticulum (RER) and Golgi apparatus migrate basally in preparation for their secretion of the extracellular matrix of enamel.

5. Secretory granules of the extracellular organic matrix of enamel accumulate in the basal portion of the cells and are released into the forming enamel region. Almost immediately this organic material calcifies.

6. **Enamel Chemistry.**
   a. The **organic** portion of enamel contains proteins and glycoproteins, but these are different from collagen and keratin.
   b. The **inorganic** portion is composed predominately of hydroxyapatite that has a different crystalline arrangement from the hydroxyapatite of bone.
   c. In addition, there are traces of other calcium salts in enamel.

7. Mature enamel differs from bone in being completely devoid of any cellular residents. Even the processes of ameloblasts are excluded from enamel. Perhaps this exclusion of cellular processes accounts for the extreme hardness of enamel.

D. ODONTOBLASTS

1. Odontoblasts differentiate from mesenchymal cells of the dental papilla that is immediately adjacent to the basement membrane of the enamel organ. They become extremely elongated and form an epithelioid layer.

2. As odontoblasts form predentin and dentin, they recede centripetally, resulting in a progressive decrease in the size of the dental papilla and leaving behind **odontoblast processes** similar to the processes of osteocytes.

3. These processes (often called Tome's fibers) become surrounded by dentin and are contained in **dentinal tubules**, which are much like the canaliculi in bone.

4. The ultrastructure of the odontoblast is unusual for a polarized cell that secretes protein masses from one end.
   a. The nucleus is located in the basal portion of the cell, that is, nearest to the pulp cavity.
   b. The supranuclear zone is packed with RER and mitochondria, with the cisternae of the RER and the mitochondria elongated and parallel to the long axis of the cell.
   c. In the middle of the cell, between the supranuclear ER zone and the apex, there are numerous flattened Golgi profiles as well as membrane-bound vesicles that contain procollagen and proteoglycan, both predentin precursors.
   d. In the apical zone, similar to the supranuclear region, is found a second mass of RER and mitochondria.

5. Odontoblasts are joined into a cohesive sheet by apical junctional complexes.

6. The odontoblast processes contain numerous filamentous structures but have no membrane-bound organelles.

7. **Dentin Chemistry.**
   a. Predentin is a mixture of collagen, glycoproteins, and several types of proteoglycans.
   b. Following secretion by the odontoblasts, predentin is calcified into dentin by a type of hydroxyapatite.
   c. Dentin is like bone in many respects, but it is much denser and harder than bone.

IV. TONGUE

A. MICROSCOPIC ANATOMY

1. The tongue has great masses of **skeletal muscle** running in several directions and allowing complex voluntary movements during food intake and speech.

2. The tongue is covered by a **stratified squamous epithelium**, which is keratinized on the top (dorsal surface) but unkeratinized on the underside (ventral surface).

B. LINGUAL PAPILLAE

1. The dorsal surface is covered by three types of papillae.
   a. Most of the papillae are the simple **filiform** type. These occur in rows all over the dorsal surface of the tongue.
   b. Scattered throughout the filiform papillae are **fungiform** papillae. These are especially numerous at the tip of the tongue.
   c. Finally, there is a V-shaped group of about 10 **circumvallate** papillae at the root of the tongue.
2. The fungiform and circumvallate papillae are associated with **taste buds**.
3. The four taste sensations of the tongue are
   a. Salty
   b. Acid
   c. Bitter
   d. Sweet
4. Different fungiform papillae have different sensitivities to these four primary tastes.
5. In addition to taste, the odor of food is involved in the complex flavor sensations experienced while eating.

V. MAJOR SALIVARY GLANDS

A. SALIVA COMPOSITION

1. The **parotid**, **submandibular**, and **sublingual** salivary glands are paired structures that produce most of the saliva.
2. Saliva is hypotonic and is produced at the rate of about 1 L per day.
3. It contains water, mucins and other glycoproteins, ions, and ptyalin, an enzyme involved in the initial hydrolysis of carbohydrates.
4. Also, saliva contains considerable antibacterial activity due to its substantial secretory immunoglobulin A (IgA) and lactoperoxidase content.

B. MICROSCOPIC ANATOMY—COMMON FEATURES

1. Major salivary glands all are drained by a complex system of large **excretory ducts** which carry the secretions into the oral cavity. These glands are controlled by the autonomic nervous system.
2. The secreting cells are arranged into masses of acinar structures at the distal termination of the ducts.
3. All acini are equipped with intraepithelial (i.e., between the acinar cells proper and basement membrane) smooth muscle-like cells called **myoepithelial cells**.
4. Presumably, these cells function by expressing secretions from the acini into the ducts that drain the gland.
5. Acinar cells are arranged in pyramidal clusters with their apical portions draining into a minute lumen. They have a basal nucleus, an abundance of RER, a prominent Golgi apparatus, and apical granules of either serous or mucous material.

C. VARIATIONS IN ACINAR STRUCTURE

1. The acini of the different glands vary tremendously.
   a. The majority of the acini in the **parotid** glands are **serous** in character (i.e., they produce a thin proteinaceous secretion).
   b. In the **submandibular** glands, acini may be either **serous** or **mixed** (consisting of mucous acini with serous demilunes).
   c. In the **sublingual** glands, most acini are **mucous**, but **mixed** acini may occur here also.
2. The acini in the parotid and submandibular glands are drained by small **intercalated ducts** which are lined by a low cuboidal or squamous epithelium. Intercalated ducts then drain into **striated** (secretory) **ducts**.
   a. Each striated duct has a tall columnar epithelial lining with deep basal infolding of the plasma membrane. Many elongated mitochondria are found between the infolded basal membrane clefts.

b. This arrangement suggests that striated ducts exhibit ion pumping and that they probably change the ionic composition of the serous secretions formed in the parotid and submandibular glands.

3. The sublingual glands do not have this complex system of ducts. Instead, mucous acini drain into simpler secretory ducts that have, at most, a poorly developed system for concentrating their secretion.

VI. TONSILS AND PHARYNX

A. TONSILS

1. Near the base of the tongue are paired structures called the **palatine** and **lingual tonsils**. There are slight differences in the anatomy of the tonsils, but both are essentially deep cryptic infoldings of the stratified squamous epithelium in this region, which along with surrounding connective tissue become massively infiltrated with lymphocytes.
2. This lymphocytic infiltration can be pronounced enough to obscure completely the epithelium that faces the oral cavity.
3. Also, there is considerable lymphoid infiltration of the epithelium, mucosa, and submucosa.
4. The palatine tonsils reach their maximum development in childhood and are thought to be involved in preventing bacterial invasion of the more distal portions of the gastrointestinal tract.

B. PHARYNX

1. The pharynx is the space between the mouth and nares and the esophagus.
   a. The upper portion (above the soft palate) is the **nasopharynx**; it has a pseudostratified ciliated columnar epithelium with goblet cells and it communicates with the nasal passages.
   b. The lower portion consists of two sections.
      (1) The **oropharynx**, between the soft palate and the epiglottis, has a stratified squamous epithelium.
      (2) The **laryngopharynx**, with a pseudostratified ciliated columnar epithelium, communicates with the larynx and esophagus.
2. A thick layer of elastic fibers occupies the space normally filled with muscularis mucosae in the rest of the gastrointestinal tract.
3. The lamina propria of the pharynx contains many lymphocytes.
4. Several layers of muscular tissue make up the rest of the wall of the pharynx.

VII. DEVELOPMENT OF FACE, PALATE, AND PHARYNGEAL ARCHES

A. FACE (Fig. 13-2)

1. In a four-week-old embryo, the face is not recognizable as human. The anterior neuropore is still patent, and the face, such as it is, consists of a **frontal prominence**, two lateral **maxillary swellings**, and two **mandibular swellings** that are fused at the ventral midline.
2. Soon, two widely spaced **nasal placodes** form and become invaginated. Meanwhile, the maxillary swellings grow toward the midline and carry the laterally placed nasal placodes medially where their medial portions meet and fuse.
3. The **maxillary processes** grow from the maxillary swellings; these also meet and fuse in the midline, below the developing nasal structures, to form the upper portion of the jaw and the lip.
4. As the maxillary processes grow medially, they appear to carry the forming eyes in toward the nose.
5. The mechanics of these movements are not well understood. When the eye rudiments first form they are projecting laterally, similar to the eyes in a fish. The eyes move to a forward-projecting position as the face develops and flattens.

B. PALATE

1. At five weeks gestation, the nasal and oral cavities are in direct communication with the

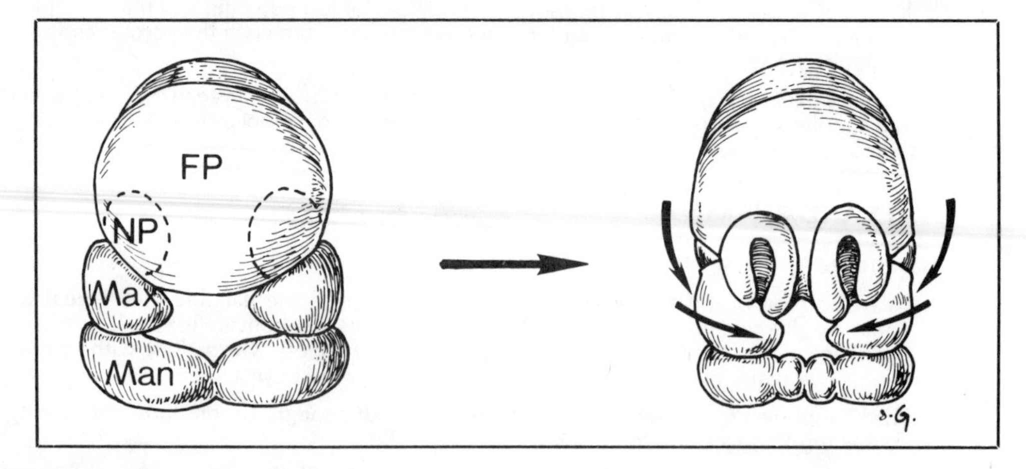

Figure 13-2. Diagram illustrating the development of the anterior aspect of the human face. The frontal prominence (*FP*) and nasal placodes (*NP*) are labeled along with the maxilla (*Max*) and mandible (*Man*). The arrows indicate the general direction of the convergence of these structures. Notice that the eye is brought from a lateral projection into the frontal plane of the face during development. (Reprinted with permission from Johnson KE: *Histology: Microscopic Anatomy and Embryology.* New York, John Wiley, 1982, p 192.)

stomodeum. The separation of these two cavities results mostly from the development of the **primary** and **secondary palate**.

2. The primary palate forms near the point where the nasal swellings fuse. It grows away from the nasal swellings along the ventral surface of the oral cavity. Meanwhile, a vertical **nasal septum** appears in the frontal prominence and grows downward toward the primary palate.
3. Also, two laterally placed **lateral palatine processes** begin to grow ventromedially.
4. The developing **tongue** is elevated at this stage and interposed between the lateral palatine processes (Fig. 13-3). Next, the tongue is depressed and the lateral palatine processes elevate dorsally while continuing to grow toward the midline.
5. At this point, the nasal septum, primary palate, and two lateral palatine processes meet and fuse, resulting in a final separation of two nasal cavities from the oral cavity. The hard palate is formed by the primary palate and lateral palatine processes.
6. At first, the walls of the nasal cavities are relatively smooth. Soon, a series of convolutions develop in the wall, eventually giving rise to the **nasal conchae.**

C. PHARYNGEAL ARCHES

1. While the oral cavity is undergoing this complex series of events, equally dynamic changes are occurring in the foregut. The ornate gross anatomy of the mature head and neck is foreshadowed by diverse developmental events.
2. A series of paired, segmentally arranged structures form on the side of the future neck region of the embryo. These structures are the so-called **pharyngeal arches**, which are condensations of mesenchymal tissue.
3. They are coated by the surface epithelium of the embryo on the lateral side and by the epithelium that lines the developing foregut on the medial side. The lateral epithelium invaginates into a series of paired **pharyngeal clefts**, and the medial epithelium evaginates to form a series of **pharyngeal pouches**.
4. The first pharyngeal arch forms just caudal to the site of the buccopharyngeal membrane. The structures that are derived from the pharyngeal pouches (also called pharyngeal grooves) are endodermal derivatives.
5. Four pharyngeal arches are formed.
   a. The **first pharyngeal arch** gives rise to **Meckel's cartilage** (which later becomes surrounded by mesenchymally derived cells that form the **mandible**), the **malleus** and **incus**, and the **muscles of mastication**.
   b. The **second pharyngeal arch** forms **Reichert's cartilage**, which later forms the stapes, the

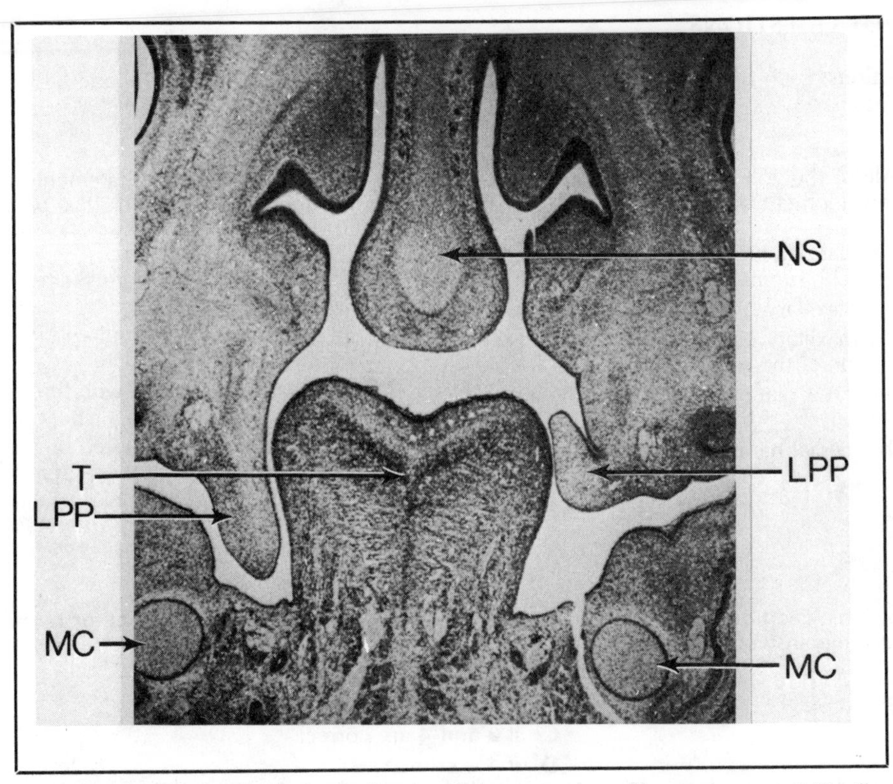

Figure 13-3. Section through the developing oral cavity of a human embryo. The nasal septum (*NS*) will meet with the lateral palatine processes (*LPP*) once the tongue (*T*) is depressed. Sections through Meckel's cartilage (*MC*), a derivative of the first pharyngeal arch and precursor of the mandible, also are evident.

styloid process of the temporal bone, the stylohyoid ligament, and the lesser horn of the hyoid bone. Also, the second arch gives rise to the **muscles of facial expression**.

c. The **third pharyngeal arch** forms the remainder of the hyoid bone and some minor pharyngeal muscles.

d. The **fourth pharyngeal arch** gives rise to the cartilages of the larynx and many pharyngeal and laryngeal muscles.

D. PHARYNGEAL CLEFTS AND POUCHES

1. The **first** pharyngeal **cleft** forms the **external auditory meatus**, and the **first** pharyngeal **pouch** forms the **auditory tube**.

2. The second, third, and fourth pharyngeal clefts are obliterated by a downgrowth of tissue from the second arch. However, these clefts contribute to a **cervical sinus**, a transient embryonic structure that normally disappears but can give rise to cysts.

3. The **second** pharyngeal **pouches** grow laterally and eventually invaginate into a complex series of convolutions. Next, they become infiltrated with lymphocytes, thus forming the **palatine tonsils**.

4. The **third** pharyngeal **pouches** form the **inferior parathyroids** and part of the thymic epithelium.

5. The **fourth** pharyngeal **pouches** form the **superior parathyroids** and perhaps part of the thymic epithelium. The fate of the epithelial cells that form from the fourth pouch is not well-known.

6. Recent experimental studies, using nuclear markers and transplantation, suggest that the parafollicular (C) cells of the thyroid actually are derived from the neural crest.

STUDY QUESTIONS

Directions: Each question below contains five suggested answers. Choose the **one best** response to each question.

1. All of the following statements concerning the development of the face are true EXCEPT

(A) the mandibular processes form caudad to the frontal prominence
(B) the maxillary processes fuse in the midline
(C) the maxillary processes contribute to the formation of the upper lip
(D) the nasal placodes form medially and move laterally
(E) the medial nasal swellings fuse in the midline

2. All of the following statements concerning the development of the definitive palate are true EXCEPT

(A) the primary palate forms after the lateral palatine processes fuse
(B) the majority of the adult palate is derived from the secondary palate
(C) the nasal septum fuses with the palate
(D) during development, the lateral palatine processes are widely separated by the tongue
(E) the primary palate and secondary palate both contribute to the adult hard palate

Directions: Each question below contains four suggested answers of which **one or more** is correct. Choose the answer

A if **1, 2, and 3** are correct
B if **1 and 3** are correct
C if **2 and 4** are correct
D if **4** is correct
E if **1, 2, 3, and 4** are correct

3. True statements concerning the pharynx include which of the following?

(1) The oropharynx is lined by a nonkeratinized stratified squamous epithelium
(2) The nasopharynx is lined by a pseudostratified ciliated columnar epithelium with goblet cells
(3) Lymphoid infiltration of the pharyngeal lamina propria is extensive
(4) The pharynx contains a prominent muscularis mucosa

4. True statements concerning the development of the teeth include which of the following?

(1) Ameloblasts form enamel
(2) Odontoblasts have a well-developed rough endoplasmic reticulum
(3) Ameloblasts have a well-developed Golgi apparatus
(4) Odontoblast processes contain membrane-bound organelles

Directions: The group of questions below consists of lettered choices followed by several numbered items. For each numbered item select the **one** lettered choice with which it is **most** closely associated. Each lettered choice may be used once, more than once, or not at all.

Questions 5–9

For each description of components of a submaxillary saliva gland, select the appropriate lettered structure shown in the micrograph below.

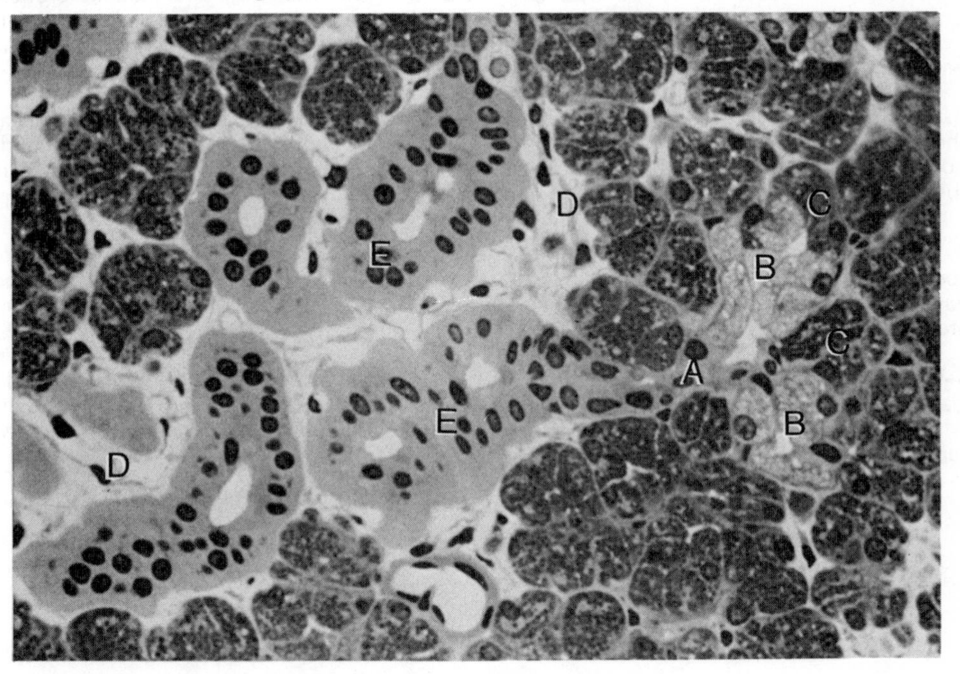

5. Most distal portion of ducts
6. Connective tissue separating lobules
7. Secretes mucus
8. Proximal to intercalated duct
9. Secretes proteinaceous product

ANSWERS AND EXPLANATIONS

1. The answer is D. (*VII A*) The developing face is formed by the frontal prominence, nasal swellings, and maxillary and mandibular processes. The nasal placodes form laterally and move medially during development of the face.

2. The answer is A. (*VII B*) The definitive palate is composed of an anterior hard palate and a posterior soft palate. For the development of the palate, the primary palate forms at about the same time as the lateral palatine processes. Next, the primary palate fuses with the two lateral palatine processes. The primary palate becomes the anterior and medial portion of the hard palate. The lateral palatine processes form the posterior portions of the hard palate and all of the soft palate. The nasal septum fuses with the palate.

3. The answer is A (1, 2, 3). (*VI B*) The pharynx is connected to the nasal cavity and the oral cavity and therefore is lined by both nonkeratinized stratified squamous epithelium and pseudostratified ciliated columnar epithelium with goblet cells. There are many lymphocytes in the pharynx; they exist in the tonsils as well as in more diffuse collections. There is no muscularis mucosa to speak of in the pharynx.

4. The answer is A (1, 2, 3). (*III C, D*) Both ameloblasts (enamel) and odontoblasts (dentin) are highly specialized for secretion of extracellular matrix substances. Thus, both cell types would be expected to have a prominent rough endoplasmic reticulum and Golgi apparatus. The processes of odontoblasts are devoid of organelles.

5–9. The answers are: 5-A, 6-D, 7-B, 8-E, 9-C. (*V C*) This is a submaxillary (submandibular) salivary gland. It has mucous acini (B) with serous demilumes (C). The acini drain into intercalated ducts (A), which drain into the more proximal interlobular ducts (E). Lobules are separated by connective tissue septa (D).

14
Esophagus and Stomach

I. INTRODUCTION

A. ESOPHAGUS

1. After food has been thoroughly chewed and mixed with saliva, small portions of it are swallowed a bit at a time. The food **bolus** formed during swallowing then is conveyed, by a reflex action, down the esophagus to the stomach.
2. Little digestion occurs in the esophagus other than the action of salivary amylase.
3. Contractions of the muscles in the esophageal wall are stimulated by the bolus and aid in passing it into the stomach.
4. A sphincter at the gastroesophageal junction controls passage of food into the stomach. It also helps to prevent reflux of the acid contents of the stomach back into the esophagus once swallowing is completed.

B. STOMACH

1. The stomach is a muscular sac lined by an epithelium that secretes large amounts of acid, pepsinogen, and mucus.
2. The digestive process occurs to a limited extent in the stomach.
3. The partially digested contents of the stomach are known as **chyme**.
4. When the chyme reaches an appropriate consistency, the pyloric sphincter opens and admits the chyme into the duodenum.

II. COMMON ANATOMIC FEATURES OF TUBULAR VISCERA (Fig. 14-1)

A. MUCOSA

1. This is the layer nearest the lumen of the organ.
2. It is composed of the following layers (from inside to outside)
   a. Epithelium
   b. Basement membrane
   c. Lamina propria
   d. Muscularis mucosae
3. The composition of these layers varies tremendously depending on function.
   a. The esophagus has a **stratified squamous** epithelium (to resist abrasion), while the stomach has a simple **columnar** epithelium, with cells that secrete mucus, hydrochloric acid (HCl), or pepsinogen.
   b. The lamina propria may be infiltrated with a only few lymphocytes (as in the duodenum) or may be massively infiltrated with lymphocytes (as in Peyer's patches in the ileum).

B. SUBMUCOSA

1. This layer extends from the muscularis mucosae away from the lumen to the muscularis externa.
2. It may be a nondescript mass of loose connective tissue (as in the esophagus) or may have an abundance of glands (as in the duodenum).

C. MUSCULARIS EXTERNA

1. This layer usually is composed of several layers of muscular tissue.

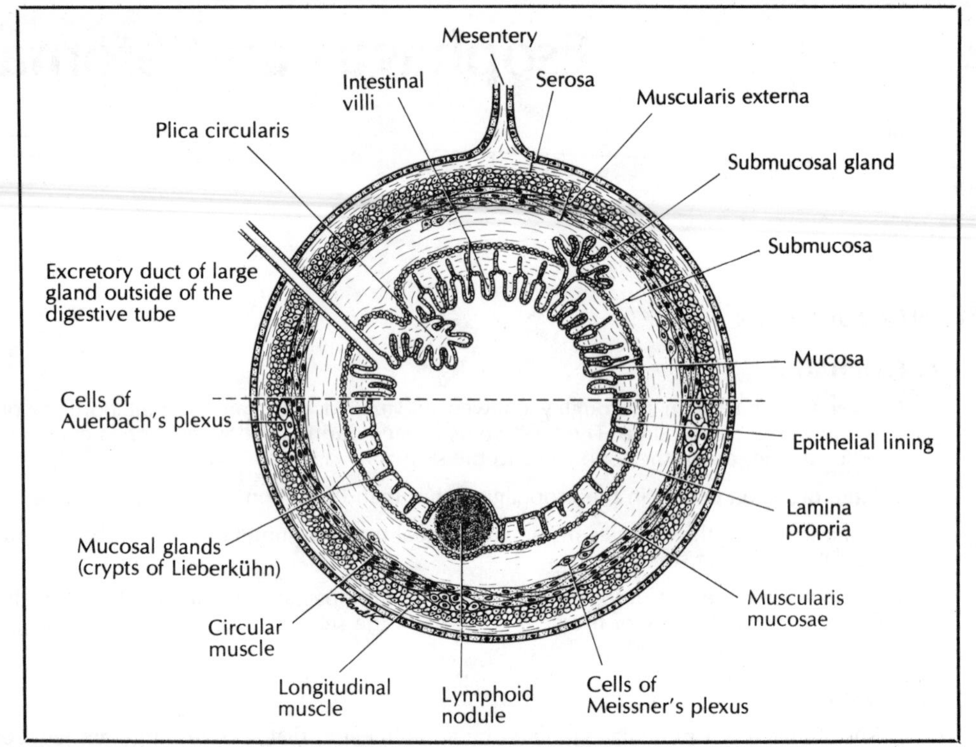

Figure 14-1. Diagram illustrating the histologic organization of the gastrointestinal tract. This diagram is not really typical of any particular portion of the gastrointestinal tract but shows all of the structural variants in one picture. (Reprinted with permission from Bloom W, Fawcett DW: *A Textbook of Histology.* Philadelphia, WB Saunders, 1975.)

2. Most frequently, the muscularis externa contains smooth muscle controlled by the autonomic nervous system.
3. This layer mixes contents or moves them from one place to another.

B. ADVENTITIA

1. This is an external layer of connective tissue.
2. It may be coated with a **mesothelium**. When coated with a serous mesothelium, the adventitia usually is called a **serosa**.
3. The adventitia may lack a mesothelium on the retroperitoneal portions of the gut tube.

III. ESOPHAGUS

A. MICROSCOPIC ANATOMY

1. The **mucosa** of the esophagus is covered by an unusually thick stratified squamous epithelium which usually is unkeratinized. The epithelium can become keratinized if subjected to continuous ingestion of abrasive foods.
   a. In some areas, especially near the top of the esophagus and at the gastroesophageal junction, there are deep glandular invaginations of the surface epithelium. Some of these glands are restricted to the mucosa, but others are found in the submucosa. The acini are strictly mucous type.
   b. The esophageal epithelium rests on a well-developed basement membrane.
   c. The lamina propria is not as cellular as it is in the distal portions of the gastrointestinal tract. It lacks lymphoid infiltration except around the orifices of mucous glands.
   d. The muscularis mucosae is quite pronounced. It begins at about the level of the cricoid cartilage where it replaces the pharyngeal elastic fibers. The fibers here run somewhat longitudinally.
2. The **submucosa** is rather hypocellular, contains some mucous glands, and has an abundant

blood supply. In the submucosal glands, the acini are connected to the surface epithelium by ducts, the largest of which are lined by a stratified squamous or cuboidal epithelium.

3. The **muscularis externa** has some anatomic variations from the usual pattern found elsewhere in the gastrointestinal tract.
   a. In the upper quarter of the esophagus, the muscularis externa contains an **inner circular** or obliquely spiraling layer surrounded by an **outer longitudinal** layer of **skeletal** fibers.
   b. In the second quarter, **skeletal** and **smooth** muscle fibers are intermingled with smooth fibers that increase in number distally.
   c. In the lower half, the muscle fibers all are **smooth** but still have the inner circular and outer longitudinal arrangement in their layers.

4. The **adventitia** is a thin layer of collagenous and elastic fibers.
   a. Until the esophagus passes through the diaphragm, the adventitia lacks a mesothelium.
   b. Below the diaphragm, the short pregastric segment of the adventitia is covered by a thin reflection of the peritoneal mesothelium.

B. FUNCTION

1. The esophagus is approximately 25 cm long in the adult human. It serves as a simple conduit for food passing from the oropharynx to the stomach.

2. It is quite distensible, due to a series of longitudinal folds formed by the mucosa and part of the submucosa.

3. As a bolus passes down through the esophagus, the folds are temporarily flattened out but reappear when the bolus has passed.

IV. STOMACH

A. MICROSCOPIC ANATOMY (Fig. 14-2)

1. **Gastric Regions.** The stomach is divided into a cardiac **antrum**, a **corpus** and **fundus**, and a **pyloric portion** that connects it to the duodenum. At the gastroesophageal junction there is a sharp transition from the stratified epithelium of the esophagus to the simple columnar epithelium of the cardiac stomach.

2. **Rugae.**
   a. In its undistended state, the gross surface of the fundic and corpic stomach is covered by many folds that run more or less parallel to the longitudinal axis of the organ.
   b. These are folds in the mucosa and submucosa known as **rugae.** When the stomach is distended by a meal, the rugae flatten somewhat due to stretching of the stomach.

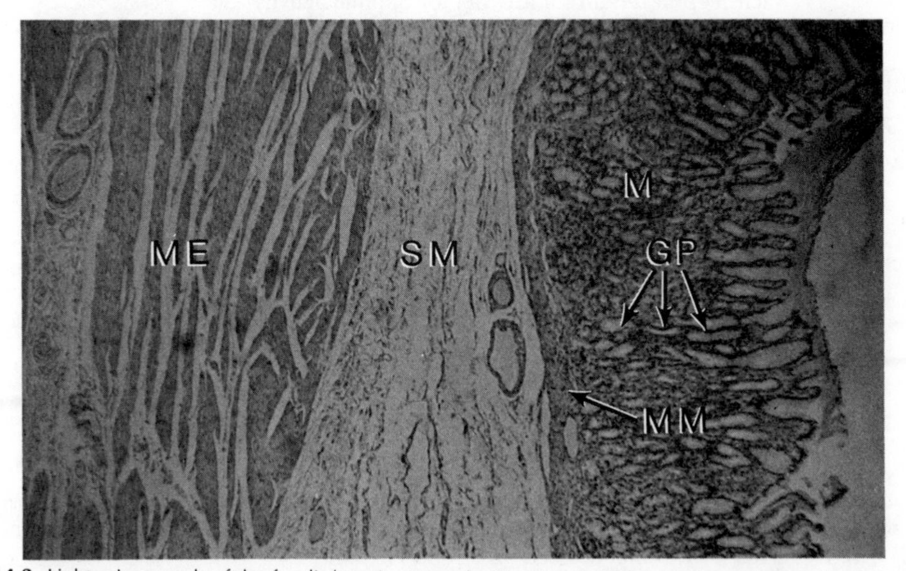

Figure 14-2. Light micrograph of the fundic/corpic stomach. Gastric pits (*GP*) contain parietal and chief cells. The muscularis mucosae (*MM*) divides the mucosa (*M*) from the submucosa (*SM*). Smooth muscle cells are arranged in irregular bundles in the muscularis externa (*ME*). (Reprinted with permission from Johnson KE: *Histology: Microscopic Anatomy and Embryology*. New York, John Wiley, 1982, p 208.)

3. The surface is further subdivided into **gastric areas** (1–6 mm in diameter) by a series of shallow grooves. Each gastric area is deeply indented by numerous deep surface invaginations known as **gastric pits**.

4. **Gastric pits** in the cardiac and pyloric portions of the stomach lead to glands that are particularly coiled. These glands are lined mostly by mucous cells, whose apical cytoplasm is dominated by a palely stained mucus that pushes the nucleus of the cell basally.

B. GASTRIC GLANDS are found in the corpus and fundus. These glands are different from those in the cardiac and pyloric stomach in several ways.

1. They are relatively straight and not as coiled.

2. The mucus-producing cells are limited to the upper portions of the glands. Mucous surface cells produce mucus that is chemically distinct from that produced by the mucous neck cells.

3. The glands contain **parietal** (oxyntic) **cells** and **chief cells**.
   a. **Parietal Cells.**
      (1) These are large acidophilic cells which are remarkabley adapted for the secretion of **HCl.** They pump hydrogen ions into the gastric lumen, with resulting passive diffusion of chloride ions and a pH of 2.0 in the stomach. (This hydrogen ion concentration, thus, is 100,000 times greater than it is in the blood.)
      (2) Also, parietal cells are believed responsible for the secretion of a glycoprotein called **intrinsic factor**. This material binds to vitamin B_{12} and facilitates its intestinal absorption. Patients with pernicious anemia often have serious atrophy of the parietal cells.
      (3) With the electron microscope, parietal cells reveal two unique features (Fig. 14-3).
         (a) Their apical surfaces are deeply invaginated by a cone-shaped depression of the surface. These invaginations are lined further by an interdigitating second folding of microvilli, resulting in a great increase in the overall surface of these cells.
         (b) There is a tremendous number of large mitochondria that completely surround the cone-shaped apical surface invaginations.
      (4) Parietal cells are most common in the upper portions of the gastric pits.
   b. **Chief Cells.**
      (1) These have the typical ultrastructure of cells that secrete large amounts of zymogen granules.
      (2) They have large nuclei with prominent nucleoli and a strong cytoplasmic basophilia.
      (3) The electron microscope reveals that the cytoplasmic basophilia is due to the large amount of rough endoplasmic reticulum in the basal portions of these cells.
      (4) In addition, chief cells have a well-developed apical Golgi apparatus and numerous zymogen granules that are membrane-delimited sacs of **pepsinogen**. The pepsinogen is secreted and transformed to an active form, **pepsin**, in the acid milieu of the stomach. Pepsin requires a low pH for optimal activity.
      (5) **Rennin** is another proteolytic enzyme secreted by chief cells. It is involved in the digestion of milk proteins.
   c. The gastric mucosa also contains cells that are part of the enterochromaffin system.

V. GASTROINTESTINAL HORMONES AND ENTEROCHROMAFFIN CELLS

A. FUNCTIONAL CONSIDERATIONS

1. A group of complex endocrine cells are widely distributed throughout the stomach and intestines.

2. At one time they were lumped together as one type of cell known as the **enterochromaffin system** or argentaffin cells. Now it is known that there are many different cell types, each secreting a different hormone. These cells sometimes are called **APUD cells**, an acronym that stands for **a**mine **p**recursor **u**ptake and **d**ecarboxylation cells.

3. Enterochromaffin cells are intraepithelial cells that secrete low molecular weight polypeptide hormones involved in the regulation of gut motility and secretion. Nearly a score of gastrointestinal hormones have been identified and are partially understood.

4. These endocrine cells may be either in contact with or isolated from the luminal contents.

5. With the electron microscope, these cells reveal large nuclei, moderate amounts of cytoplasmic mitochondria and rough endoplasmic reticulum, and prominent, **basal**, membrane-delimited vesicles filled with polypeptide hormones.

B. CELLS OF THE ENTEROCHROMAFFIN SYSTEM

1. **Gastrin** or **G cells** are found in the pyloric antrum.

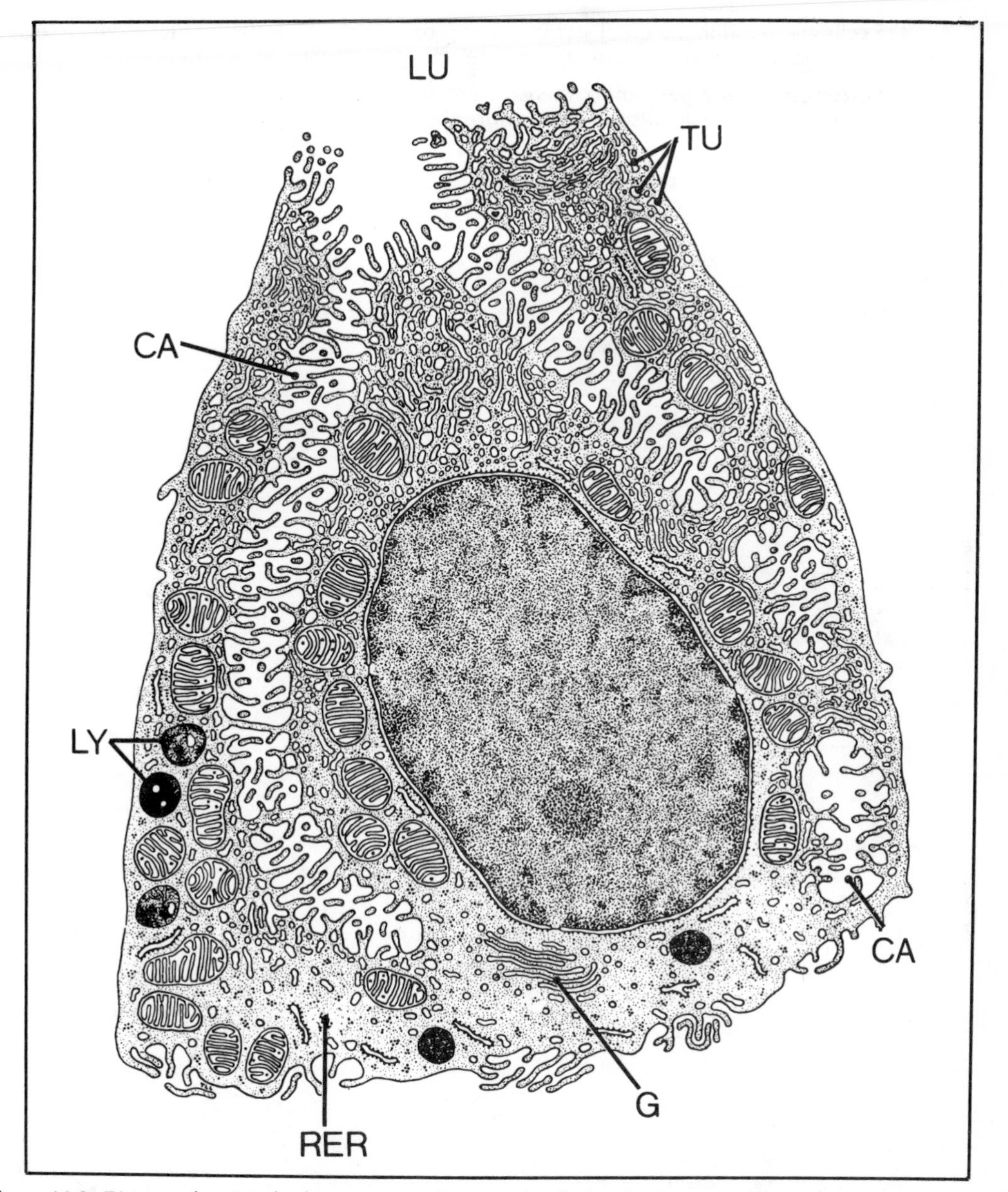

Figure 14-3. Diagram showing the fine structure of a parietal cell. Canaliculi (*CA*) are deep invaginations of the cell surface at the lumen (*LU*) of the stomach in close association with tubules (*TU*) and vesicles in the cell cytoplasm. This cell also has lysosomes (*LY*), a rough endoplasmic reticulum (*RER*), and a Golgi apparatus (*G*). (Reprinted with permission from Lentz TL: *Cell Fine Structure.* Philadelphia, WB Saunders, 1971.)

- **a.** They secrete a group of polypeptide hormones known as **gastrin**, the most potent of which contains 17 amino acids.
- **b.** Gastrin promotes parietal cell secretion of HCl and also is mitogenic for parietal cells.
- **c.** Also, gastrin stimulates antral motor activity.
- **d.** When foods find their way to the pyloric antrum, microvilli of G cells in contact with the lumen are stimulated and release gastrin.
- **e.** Gastrin release is apparently inhibited by acidity.

2. I cells are found in much of the stomach and all of the intestines.
- **a.** I cells secrete a polypeptide known as **cholecystokinin (pancreozymin)**, which is 28 amino acids long.
- **b.** It stimulates the release of protein-rich secretions from the pancreatic acinar cells and the release of the stored contents of the gallbladder.

3. **S cells** are found in the distal stomach and proximal small intestine. They produce **secretin**, a 27-amino acid polypeptide that stimulates copious alkaline secretion from the pancreas.
4. **EC cells** are found throughout the gastrointestinal mucosa, from the stomach to the rectum. These cells secrete **serotonin** and **endorphins**. Serotonin promotes smooth muscle contraction.
5. **A cells** are scattered throughout the duodenum and jejunum. They secrete **glucagon**, like A cells of the pancreas.

STUDY QUESTIONS

Directions: The question below contains five suggested answers. Choose the **one best** response to the question.

1. All of the following statements concerning parietal cells are true EXCEPT

(A) they have a deeply invaginated apical surface
(B) they are acidophilic
(C) they have many mitochondria
(D) they secrete gastric intrinsic factor
(E) they secrete pancreozymin

Directions: Each question below contains four suggested answers of which **one or more** is correct. Choose the answer

A if **1, 2, and 3** are correct
B if **1 and 3** are correct
C if **2 and 4** are correct
D if **4** is correct
E if **1, 2, 3, and 4** are correct

2. True statements concerning the microanatomy of the muscularis externa in the esophagus include which of the following?

(1) Skeletal muscle fibers are predominant in the upper quarter
(2) Smooth muscle fibers are most abundant in the second quarter
(3) The lower half of the esophagus contains only smooth muscle fibers
(4) In the upper quarter, the muscularis externa contains only one layer

Directions: The group of questions below consists of lettered choices followed by several numbered items. For each numbered item select the **one** lettered choice with which it is **most** closely associated. Each lettered choice may be used once, more than once, or not at all.

Questions 3–7

For each description of a structure or site in the esophageal mucosa, choose the appropriate lettered component shown in the micrograph below.

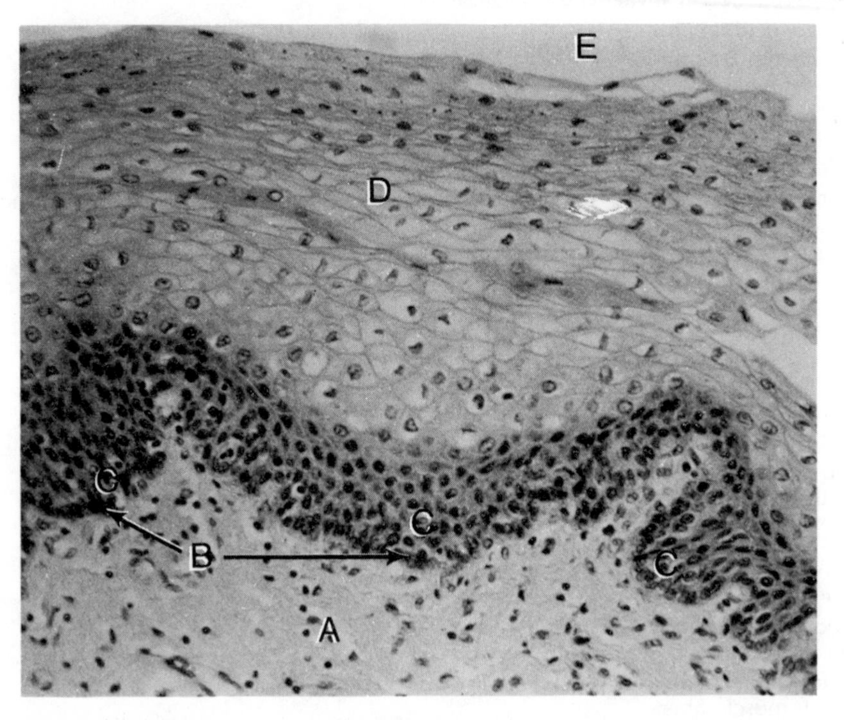

Reprinted with permission from Johnson KE: *Histology: Microscopic Anatomy and Embryology*. New York, John Wiley, 1982, p 207.

3. Cells sloughed from the epithelium as a food bolus passes

4. Basement membrane between the esophageal epithelium and the lamina propria

5. Glands and numerous lymphocytes found here

6. Actively dividing layer of cells

7. A loose areolar connective tissue of the mucosa

ANSWERS AND EXPLANATIONS

1. The answer is E. (*IV B 3*) Parietal cells are responsible for secreting HCl into the stomach. To accomplish this function, these cells have surface invaginations and many mitochondria, which give their cytoplasm a distinct acidophilia. Parietal cells also secrete gastric intrinsic factor. Pancreozymin is a secretion product of enterochromaffin cells.

2. The answer is B (1, 3). (*III A 3*) The esophageal muscularis externa contains skeletal muscle fibers in its upper portion and smooth muscle fibers in its lower portion. Smooth muscle fibers are least abundant in the second quarter of the muscularis externa of the esophagus. All portions of the muscularis externa contain two layers of muscular tissue.

3–7. The answers are: 3-D, 4-B, 5-A, 6-C, 7-A. (*III A 1*) This is a micrograph of the esophageal mucosa. The esophagus has a stratified squamous epithelium in it. This epithelium rests on a basement membrane (B). The basal cells of the epithelium proliferate (C), and the apical cells are sloughed off (D). The lamina propria (A) is a loose areolar connective tissue which may have glands and lymphocytes in it.

15
Intestines

I. INTRODUCTION

A. DIGESTION

1. Ingested food is thoroughly fragmented, mixed, and partially digested in the oral cavity and the stomach. After appropriate gastric digestion, acidic chyme empties into the duodenum.
2. Here, the chyme is neutralized quickly by copious mucous and bicarbonate secretions of the submucosal **Brunner's glands**.
3. In the proximal portion of the duodenum, the **common bile duct** joins the **main pancreatic duct**; the **accessory pancreatic duct** joins the gastrointestinal tract here at a second opening. These ducts empty **bile**, mixtures of pancreatic hydrolytic **enzymes**, or both into the duodenal lumen.

B. ABSORPTION

1. Most of the hydrolysis of foodstuffs as well as the absorption of breakdown products occurs in the **small intestine**, and the residue of digestion is passed into the **large intestine**.
2. Selective absorption, mainly of water, and mucous secretions from numerous colonic goblet cells concentrate the undigested residue and aid in forming feces. These are expelled from the body as waste material, thereby completing the process of digestion.

II. SMALL INTESTINE

A. OVERVIEW

1. **Gross Anatomy.** The small intestine is about four meters long, extends from the pyloric sphincter of the stomach to the colon, and has three divisions: **duodenum**, **jejunum**, and **ileum**. A thickening of the mesentery, the **suspensory ligament** of Trietz, marks the junction between the duodenum and jejunum. All segments of the small intestine are histologically similar but are distinguished by small differences.
2. **Surface Elaboration.**
   - **a.** Because of its absorption function, the surface of the small intestine is highly modified.
     - **(1)** The mucosa and submucosa form a series of circular folds known as the **plicae circulares**.
     - **(2)** These structures are similar to the gastric rugae (in that they include mucosa and submucosa), but the plicae circulares form rings rather than longitudinal folds. Unlike rugae, they are not distensible.
   - **b.** Finger-like projections (**villi**) of the mucosa and deep surface invaginations (**crypts** of Lieberkühn) also increase intestinal surface area.
     - **(1)** The individual epithelial cells that line the villi have numerous **microvilli** on their apical surfaces.
     - **(2)** Each microvillus has a dense **glycocalyx**, which contains several enzymes for the final stages of digestion.

B. COMMON HISTOLOGIC FEATURES

1. **Epithelium.**
   - **a.** The intestinal mucosa consists of simple columnar epithelial cells covered with numerous microvilli and interspersed between scattered goblet cells that produce protective mucus (Fig. 15-1).
   - **b.** Between villi, the surface is invaginated into crypts of Lieberkühn (**intestinal glands**).
   - **c.** At the base of each crypt are **Paneth cells**. These cells have apical granules that look somewhat like zymogen granules, although their exact function is not known.

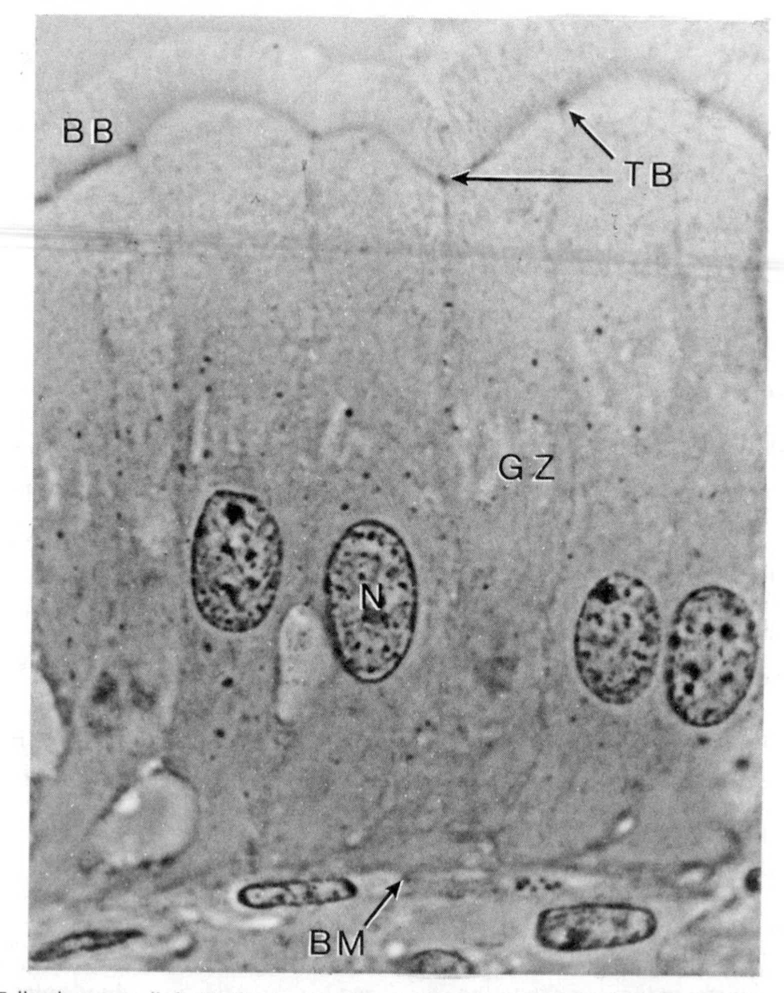

Figure 15-1. Tall columnar cells lining the ileum. These cells rest on a basement membrane (*BM*). At their apical surfaces, they have a brush border (*BB*) made of numerous microvilli. Cells are joined together at junctional complexes, parts of which show up as terminal bars (*TB*). A prominent Golgi zone (*GZ*) also is visible above the nucleus (*N*). (Reprinted with permission from Johnson KE: *Histology: Microscopic Anatomy and Embryology*. New York, John Wiley, 1982, p 231.)

d. Recent studies show that Paneth cell granules contain **lysozyme**. This enzyme has bactericidal properties and may be involved in controlling the bacterial population of the small intestine.

e. The tall columnar epithelial cells are equipped with a **brush border** of **microvilli**. Each microvillus is thought to have independent motile capability due to an internal cytoskeleton of actin-like filaments.

f. The microvilli are coated with a glycocalyx which is periodic acid-schiff (PAS)-positive. This cell coat is composed of carbohydrate-rich glycoproteins, and it contains enzymes for binding and splitting certain disaccharides.

g. In electron micrographs, the glycocalyx has minute filaments that project outward from the unit membrane of the intestinal epithelial cell.

h. Turnover.

(1) The intestinal epithelial cells have a relatively short life span.

(2) New cells are produced in the crypts by **mitosis**, after which they migrate up the lateral portions of the villi to the tips.

(3) Here, old cells are sloughed into the lumen, aided most likely by the intrinsic motility of the villi.

i. Apical Junctional Complexes.

(1) Intestinal epithelial cells are joined to one another by substantial apical **junctional complexes** designed to prevent the contents of the lumen from entering the extracellular compartments between intestinal epithelial cells.

(2) The intestinal lumen contains hydrolytic enzymes that are fully capable of digesting the wall of the intestinal tract.
(3) These enzymes are prevented from digesting their own container by a mucous coat on the apices of the cells, by the apical junctional complexes, and probably by the glycocalyx as well.

j. Other Common Features.
(1) Enterochromaffin cells also are found scattered among some intestinal epithelial cells.
(2) The epithelium rests on a thick basement membrane composed of collagen fibrils and other glycoconjugates.

2. Lamina Propria.

a. Beneath the basement membrane is found a lamina propria composed of loose areolar connective tissue. Lymphocytes, macrophages, granulocytes, and plasma cells abound in this compartment.

b. Each villus has a core of highly cellular lamina propria. Also, each has an afferent arteriole and an efferent venule which are connected by a rich capillary plexus.

c. Each villus is drained by a single, blind-ending lymphatic capillary known as a **lacteal**.

d. In some instances, the lamina propria contains nodules of lymphocytes rather than just scattered lymphocytes. In the ileum, clusters of larger nodules and lymphatic follicles are prominent enough to be macroscopically visible as Peyer's patches.

e. Slips of smooth muscle extend along the length of the villi, and when they contract they probably pump absorbed materials in the lacteals and capillary plexus out of the villi.

f. The boundary between the mucosa and submucosa is the **muscularis mucosae**. It is a relatively prominent band of smooth muscle fibers running along the bases of the villi. The muscularis mucosae also is included in the plicae circulares, along with much submucosal tissue.

3. External Layers.

a. The **muscularis externa** of the small intestine has an inner circular layer and an outer longitudinal layer (Fig. 15-2). Ganglia for the neurons that make up **Auerbach's plexus** are

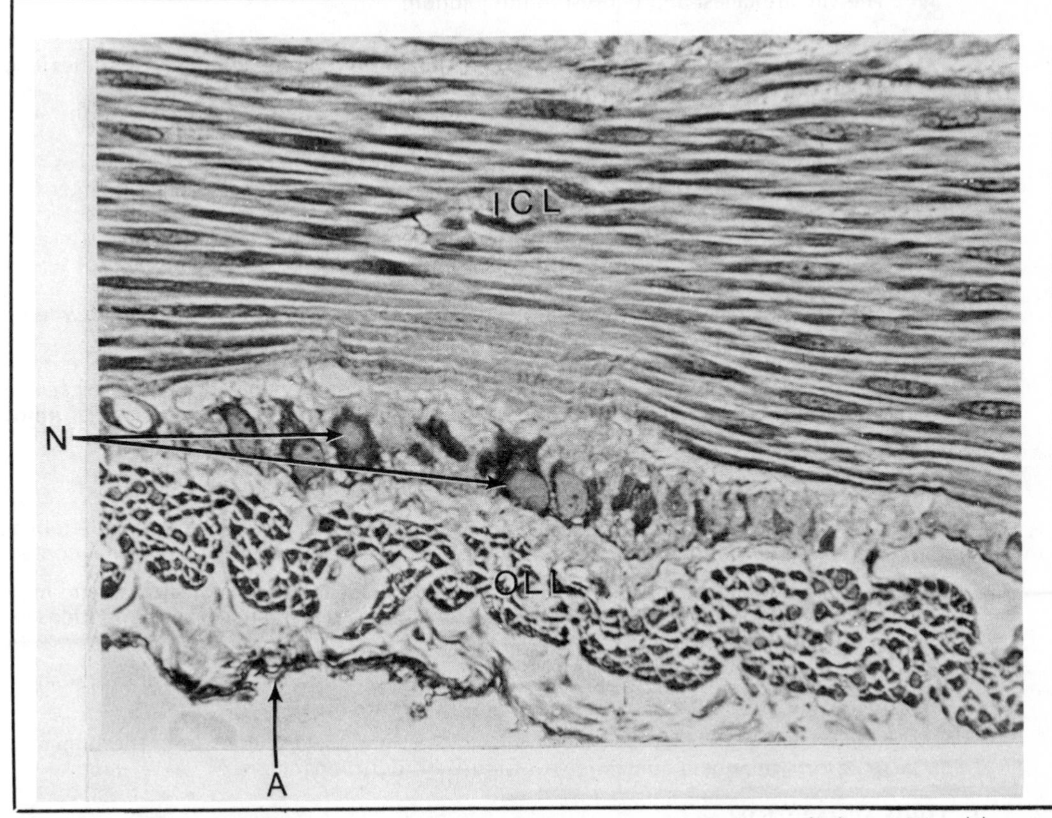

Figure 15-2. The neurons (*N*) of Auerbach's plexus are located between the inner circular layer (*ICL*) and the outer longitudinal layer (*OLL*) of the muscularis externa. Also, the connective tissue layer of the adventitia (*A*) is clearly shown. (Reprinted with permission from Johnson KE: *Histology: Microscopic Anatomy and Embryology*. New York, John Wiley, 1982, p 206.)

quite prominent and lie between the two layers of the muscularis externa. These nerve cells are part of the motor innervation of the small intestine.

b. Also, a less prominent **Meissner's plexus** is found in the submucosa of the small intestine.

c. The **adventitia** or serosa of the small intestine is a thin layer of connective tissue coated in most places by the same simple squamous mesothelium that lines the entire peritoneal cavity.

C. ANATOMIC VARIATIONS

1. Duodenum.

a. The duodenum is the most proximal portion of the small intestine.

b. It has a massive array of submucosal **Brunner's glands**.

(1) These glands are extensions of the intestinal glands; the epithelial cells lining Brunner's glands are continuous with the epithelial cells that line the crypts and villi.

(2) Brunner's glands secrete a mucus that is histochemically different from that produced in the goblet cells. Most likely this granular secretion is protective, as it is quite alkaline and rich in buffering bicarbonate ions.

c. With the electron microscope, the secretory cells reveal a morphology that is intermediate between zymogenic cells and goblet cells. They have a particularly prominent apical Golgi apparatus, presumably involved in the synthesis of the carbohydrate portion of their mucous secretion.

d. In humans, Brunner's glands usually are restricted to the duodenum, but they also may appear in the pyloric part of the stomach.

e. They decrease in size in the distal portions of the duodenum and may occur for a short length in the most proximal portions of the jejunum.

2. Jejunum.

a. The jejunum lacks the submucosal Brunner's glands that are characteristic of the duodenum.

b. The plicae circulares are most prominent here.

c. The villi are tallest and densest in the jejunum.

3. Ileum.

a. In the ileum, the villi are not as tall nor as numerous as in the jejunum. Near the ileocecal valve, the villi almost disappear.

b. The epithelium of the ileum contains more goblet cells than any other part of the small intestine.

c. Also, the ileum contains the greatest density of lymphoid nodules. In some places are seen distinct aggregates of lymphoid follicles known as **Peyer's patches**. These larger follicles are not restricted to the ileum but are so concentrated here that they are grossly visible.

(1) The lymphoid follicles contain large numbers of **plasma cells** that produce a secretory form of immunoglobin A (IgA) called **SIgA**, which is essentially a dimer of serum IgA chains linked together by a **J polypeptide**.

(2) Before being secreted, SIgA molecules are passed through epithelial cells where they have a glycoprotein moiety added to them.

(3) This added glycoprotein **secretory component** makes the SIgA molecules more resistant to proteolysis by luminal proteases. Also, it presumably increases their half-life in the intestinal lumen, where the molecules probably serve as antibacterial **immunoglobulins**.

D. EXTRACELLULAR DIGESTION

1. The digestive capacity of the small intestine is far greater than would be predicted from a simple measurement of the hydrolytic enzymes that are dissolved freely in luminal contents.

2. Preparations of intestinal brush border contain substantial amounts of **sucrase** and **maltase**. Also, these preparations contain a good deal of **alkaline phosphatase**, **aminopeptidases**, and **glyceride synthetase**.

3. All these enzymes, except the glyceride synthetase, are involved in nutrient degradation and probably are enzymatic activities of the glycocalyx itself.

4. Glyceride synthetase may be involved in transporting dietary lipids from the lumen to the lacteals, or it may be a contaminant of brush border preparations.

E. LIPID ABSORPTION

1. The columnar epithelial cells play an important role in the transport of digested fats from the intestinal lumen to the lacteals.

2. Pancreatic lipolytic enzymes that are secreted into the intestinal lumen break triglycerides into fatty acids and monoglycerides.
3. After breakdown, these components are transported into epithelial cells by a poorly understood mechanism that is unlike simple pinocytosis.
   a. Perhaps the triglyceride synthetase activity of the brush border is involved in the uptake and transport of the triglyceride-breakdown products.
   b. At any rate, triglyceride-breakdown products make their way into epithelial cells where they are resynthesized into triglycerides.
4. Small lipid-rich vesicles accumulate in the apical cytoplasm in the cisternae of the smooth and rough endoplasmic reticulum. The triglycerides are combined with cholesterol esters and proteins to make various lipoprotein micelles.
5. Once formed into lipoprotein-containing structures, these materials are expelled from the lateral borders of the columnar intestinal epithelial cells.
6. Following expulsion into the extracellular compartment, these lipoprotein micelles are known as **chylomicrons** and are visible with the light microscope. Chylomicrons then travel through the basement membrane of the epithelium, through the connective tissue of the lamina propria, and into the lymphatic capillaries.
7. From the lacteals, chylomicrons move into the thoracic duct and finally into the systemic circulation.
8. After entering the systemic circulation, dietary lipids are taken to the liver where they are combined with other proteins to form very low-density lipoproteins (VLDLs), a major transport form of dietary triglycerides in the systemic circulation.

III. LARGE INTESTINE

A. FUNCTIONAL CONSIDERATIONS

1. This terminal portion of the gastrointestinal tract extends from the ileocecal valve to the anus. Here, the nutrient-poor remnants of digestion are concentrated by water absorption and mixed with mucus. This final stage aids in formation of feces that can be moved through the large intestine to the rectum and anus without causing mucosal damage.
2. A limited amount of digestion of foodstuffs occurs in the large intestine by the action of left-over pancreatic enzymes and by the ubiquitous bacterial flora of the large intestine.
3. No digestive enzymes have been identified as secretion products of the colonic mucosa.

B. MICROSCOPIC ANATOMY

1. The large intestine is about 160 cm long and quite consistent in histological detail throughout its various gross segments.
2. Grossly, the surface of the large intestine is relatively smooth, lacking the rugae and plicae circulares characteristic of the stomach and small intestine, respectively. The surface of the large intestine is deceptively large, in spite of the fact that there are no villi, because the surface epithelium is deeply invaginated into a series of **pits**.
3. The surface **epithelium** is a mixture of goblet cells and tall columnar cells with a microvillous brush border. Goblet cells are extremely abundant in the pits, and a few enterochromaffin cells commonly are seen at the bases of the pits.
4. The **lamina propria** contains a large amount of lymphoid tissue. Small aggregates of lymphocytes exist in some places, but solitary nodules, occasionally large enough to extend into the submucosa, also can be found.
5. **Muscularis Externa.**
   a. The external longitudinal smooth muscle is distributed in three bands called the **teniae coli**.
   b. The internal circular band is interrupted periodically by the insertion of smooth muscle fibers from the teniae coli.
   c. The independent tonic contractions of the teniae coli in different segments of the large intestine result in the formation of gross sacculations in the wall known as **haustra**.
6. The **adventitia** of the large intestine is not completely invested with a mesothelium in the ascending and descending limbs of the colon.

7. The **serosa** of the large intestine often has an unusually large amount of fatty tissue which may take the form of grossly visible appendages known as **appendices epiploicae**.

8. The **cecum** and **vermiform appendix** are like the rest of the large intestine in histologic appearance except for a lack of teniae coli and a presence of massive mucosal and submucosal lymphoid infiltration in the vermiform appendix.

9. **Rectum.**
   a. The rectum is slightly different from the large intestine in that the mucosal epithelium has an abundance of goblet cells and the external longitudinal smooth muscle layer is continuous.
   b. At the rectoanal junction there is a transition from the columnar epithelium with goblet cells to a stratified squamous epithelium, which is characteristic of the rest of the body's external surface.

IV. DEVELOPMENTAL ANATOMY OF THE GUT

A. ENDODERMAL DERIVATIVES

1. The epithelial lining of the entire gastrointestinal tract, from the pharynx to the rectum, is derived from the endoderm.

2. The boundary between the stomodeum (the primitive oral cavity) and the foregut is called the **buccopharyngeal membrane**. This structure has a covering of ectodermal epithelium on the buccal side and a covering of endodermal epithelium on the pharyngeal side, with no intervening mesoderm.

3. This situation is duplicated at the other end of the gut, at the **cloacal membrane**, which serves as the boundary between the hindgut and the proctodeum.

B. PRIMITIVE GUT

1. **Formation.**
   a. The endodermal lining of the yolk sac is brought into medial fusion by growth of the lateral folds of the amnion.
   b. A tubular, endodermally lined canal is formed along the entire length of the embryo.
   c. This endodermal tube is interrupted at the **vitelline duct** (a narrow connection between the yolk sac and the midgut), becomes wrapped by splanchnic mesoderm, and gives rise to the epithelium of the gut.
   d. The splanchnic mesoderm forms the smooth muscles, submucosa, and lamina propria of the gut as well as the mesenteries and ligaments that suspend and fix the gut to the body wall.

2. **Divisions.**
   a. **Foregut.**
      (1) The foregut, in conjunction with the pharyngeal pouches, develops into the complex structures in the oral cavity.
      (2) The caudal portions of the foregut form the esophagus, stomach, and part of the duodenum.
      (3) Diverticula from the foregut develop into the liver parenchyma, biliary apparatus, and pancreas.
      (4) The **celiac artery** and its branches from the dorsal aorta supply the foregut derivatives.
   b. **Midgut.**
      (1) The midgut forms the remainder of the duodenum as well as the jejunum, ileum, and large intestine up to and including the proximal two-thirds of the transverse colon.
      (2) The superior mesenteric artery supplies midgut derivatives.
   c. **Hindgut.**
      (1) The hindgut forms the rest of the transverse colon, the descending colon, and the rectum.
      (2) The inferior mesenteric artery supplies hindgut derivatives.

C. LATER DEVELOPMENT

1. The **esophagus** initially is short and lined by columnar epithelium that may be ciliated. As the trunk of the embryo elongates, so does the esophagus, and the epithelium becomes stratified squamous.

2. The **stomach** appears first as a fusiform dilation of the foregut just caudal to the esophagus. It is suspended by a **dorsal mesogastrium** and a **ventral mesogastrium**. The latter attaches it to

the **septum transversum**. The **liver** forms from a diverticulum of the foregut that grows into the septum transversum.

3. Next, the stomach begins a complex morphogenesis that alters its position in the peritoneal cavity.
   a. There is a differential growth of the stomach so that the greater curvature becomes larger than the lesser curvature.
   b. The stomach rotates along its longitudinal axis, causing the primitive left side to lie ventrally and the primitive right side to lie dorsally. This concept of gastric rotation is controversial; simple differential growth, according to some authors, may account for the final position of the stomach.
   c. The dorsal mesogastrium grows tremendously and becomes the **greater omentum**, which eventually forms a curtain-like fold that covers the rest of the small and large intestines.
   d. The ventral mesogastrium forms the **lesser omentum**.

D. ROTATION OF MIDGUT LOOPS

1. Next, the remainder of the foregut and most of the midgut lengthen tremendously. For a time, the midgut loops are so voluminous that they undergo a normal herniation into the umbilical cord.
2. Much of the cranial portion of the foregut and the caudal portion of the midgut becomes fixed to the body wall as their mesenteries fuse with the body wall.
3. While herniated into the umbilical cord, the free portions of the gut loop make several important changes.
   a. The gut loop rotates about the axis formed by the vitelline duct. (If facing the ventral surface of the embryo, this is a counterclockwise rotation of about 90 degrees.) The effect of this rotation is to bring the cranial portion of the gut loop to the left and the caudal portion to the right.
   b. The cranial portion of the gut loop grows much more rapidly, tremendously increasing the length of the small intestine.
   c. The vitelline duct becomes much smaller and the yolk sac shrinks.
   d. Eventually, the vitelline duct disappears, leaving the mouth and anus as the only connections (through the body wall) between the outside world and the lumen of the gut.
4. At approximately three months gestation, the abdominal cavity grows considerably larger. Gradually, the gut loops are withdrawn into the abdominal cavity. As they are drawn in, they rotate further from right to left.
5. The cecum and appendix move into the right upper quadrant of the abdominal cavity. From there, the cecum descends to the iliac fossa, its final resting place.
6. After its rotation, the gut affixes to the body wall.
7. The fusion of several mesenteries occurs so that parts of the duodenum, as well as parts of the ascending and descending colon, move into the retroperitoneal space.
8. Also, the pancreas is placed retroperitoneally, with the fusion of the dorsal mesoduodenum to the body wall.

STUDY QUESTIONS

Directions: The question below contains five suggested answers. Choose the **one best** response to the question.

1. Which statement does not describe lipid digestion and transport?

(A) Dietary triglycerides are degraded in the lumen by the pancreatic lipases
(B) Free fatty acids are transported into the intestinal epithelial cells by pinocytosis
(C) Triglycerides are complexed with proteins in the smooth endoplasmic reticulum prior to expulsion from the cells
(D) Chylomicrons contain triglycerides and proteins
(E) Dietary lipids are complexed with proteins in the liver

Directions: Each question below contains four suggested answers of which **one or more** is correct. Choose the answer

A if **1, 2, and 3** are correct
B if **1 and 3** are correct
C if **2 and 4** are correct
D if **4** is correct
E if **1, 2, 3, and 4** are correct

2. The jejunum is a portion of the small intestine where

(1) substantial absorption of digested food occurs
(2) the plicae circulares are most prominent
(3) the villi are densest
(4) enterochromaffin cells are absent

3. True statements concerning the ileum include which of the following?

(1) It has an inner circular and an outer longitudinal layer of smooth muscle cells
(2) Peyer's patches can extend into the submucosa
(3) Villi are shorter here than in the jejunum
(4) Lymphoid nodules are abundant

Directions: The group of questions below consists of lettered choices followed by several numbered items. For each numbered item select the **one** lettered choice with which it is **most** closely associated. Each lettered choice may be used once, more than once, or not at all.

Questions 4–8

For each description, select the appropriate intestinal components.

(A) Brunner's glands
(B) Appendices epiploicae
(C) Haustra
(D) Teniae coli
(E) Plicae circulares

4. Found to a variable extent throughout most of the small intestine

5. Sacculations in the wall of the colon

6. Homologous to rugae and contain submucosa

7. Mucus and bicarbonate secreting glands restricted to the submucosa of the duodenum

8. Longitudinal bands of smooth muscle fibers in the muscularis externa of the colon

ANSWERS AND EXPLANATIONS

1. The answer is B. (*II E*) Lipids are treated in a complex fashion during their digestion and transport from the lumen of the gastrointestinal tract into the blood. After triglycerides are broken down into free fatty acids and monoglycerides in the lumen of the small intestine, they are transported into the epithelial cells by nonpinocytotic, receptor-mediated transport. Here they are resynthesized into triglycerides. Triglycerides are complexed with proteins in the smooth endoplasmic reticulum and are expelled as lipoprotein complexes known as chylomicrons. The liver also plays an important role in modifying serum lipid content.

2. The answer is A (1, 2, 3). (*II C 2*) The jejunum is the middle portion of the small intestine. The bulk of the absorption of digested food occurs here. Thus, the villi are most numerous and most prominent here, and the plicae circulares are well-developed. Enterochromaffin cells are found throughout the gastrointestinal tract.

3. The answer is E (all). (*II C 3*) The ileum is the last portion of the small intestine. The entire small intestine has an inner circular and an outer longitudinal layer of smooth muscle. The ileum has particularly prominent patches of lymphoid tissue called Peyer's patches. These are elaborations of the lymphoid infiltration of the lamina propria which commonly extend into the submucosa of the wall.

4–8. The answers are: 4-E, 5-C, 6-E, 7-A, 8-D. (*II A; III B*) The small intestines have plicae circulares throughout their length. There are Brunner's glands present only in the duodenum. The large intestines have longitudinal bands of smooth muscle called the teniae coli which, upon their contraction, produce sacculations in the wall of the colon. These are known as haustra.

16 Liver, Gallbladder, and Pancreas

I. INTRODUCTION

A. LIVER

1. The liver serves both an **endocrine** and an **exocrine** function as well as being a major component of the reticuloendothelial system.
   a. As its endocrine function, the liver synthesizes serum albumin and other plasma proteins and then secretes them into the blood. Although serum albumin usually is not considered a hormone, it is secreted directly into the circulatory system.
   b. As its exocrine function, the liver secretes **bile**, which aids in digestion.
2. In the embryo during the middle portion of intrauterine life and also in some pathologic states in the adult, the liver serves as a site of extramedullary **hematopoiesis**.
3. The liver is supplied with blood recently drained from the gastrointestinal tract by the portal circulation; it also is supplied by the systemic circulation. The liver processes nutrients absorbed into the circulation by the gastrointestinal tract.
   a. The lacteals in the intestinal villi carry absorbed chylomicrons rich in triglycerides into the thoracic duct and systemic circulation where they eventually make their way to the liver.
   b. Here, the triglycerides are combined with carrier lipoproteins.
4. The liver is a site for the metabolism of steroids and the detoxification of drugs, alcohol, and other poisons. Also, thyroxine (T_4) is converted to its more active form, triiodothyronine (T_3), in the liver.

B. GALLBLADDER

1. The gallbladder stores and concentrates bile, the main exocrine secretion of the liver. Bile drains from the liver via the left and right hepatic ducts into a common hepatic duct.
2. The gallbladder itself is drained by the cystic duct, which joins the common hepatic duct to form the common bile duct.

C. PANCREAS

1. The pancreas also serves both an endocrine and an exocrine function.
   a. The **islets of Langerhans** are small clusters of endocrine cells and associated fenestrated capillary glomeruli which secrete **insulin** and **glucagon** directly into the circulatory system.
   b. The remainder of the pancreas is devoted to the production of **digestive enzymes** which are chiefly responsible for **hydrolysis** of ingested foodstuffs.
      (1) Pancreatic acinar cells are remarkably specialized for protein synthesis and secretion.
      (2) The acini are drained by a complex cellular duct system that empties into the duodenum by way of a main and an accessory pancreatic duct.
2. **Pancreatic Ducts.**
   a. The main duct may either join with the common bile duct or enter the duodenum near the entrance point of the common bile duct.
   b. The accessory pancreatic duct enters the duodenum just cranial to the common bile duct.

II. HISTOLOGIC ORGANIZATION OF THE LIVER

A. CAPSULE

1. In most regions, the liver is covered by a thin reflection of the mesothelium that lines the peri-

toneal cavity. Where the liver attaches to the diaphragm, however, this covering is absent.

2. The mesothelium covers a thin but tough connective tissue capsule containing fibroblasts, collagen fibers, and some elastic fibers as well as small blood vessels.

B. PORTA HEPATIS

1. The inferior concave surface of the liver contains a central hilus called the **porta hepatis**, where blood vessels, bile ducts, nerves, lymphatics, and a considerable amount of connective tissue enter the organ.

2. Branches or **trabeculae** of the connective tissue capsule spread throughout most of the liver, carrying within them progressively smaller branches of the blood vessels, nerves, and bile drainage system.

C. PARENCHYMAL CELLS.

C. PARENCHYMAL CELLS. Most of the liver is composed of cords and plates of tightly joined and extremely complex parenchymal cells. These cells have many important components, the histologic aspects of which are discussed below.

1. **Endoplasmic Reticulum and Golgi Apparatus.**
   a. Parenchymal cells produce and secrete a great amount of protein and consequently have an abundant rough endoplasmic reticulum and prominent Golgi apparatus.
   b. They also are equipped with substantial amounts of smooth endoplasmic reticulum. In certain states of acute drug intoxication, the amount of smooth endoplasmic reticulum increases suddenly to remove the offending substance.
   c. The Golgi apparatus contains numerous electron-dense bodies, 25–80 nm in diameter, in the dilated terminal portions of lamellae.
      (1) These bodies are thought to represent the **very low-density lipoproteins** (VLDLs) that play an important role in transport of triglycerides and cholesterol in the blood.
      (2) These VLDLs are synthesized in the liver, presumably in close association with rough endoplasmic reticulum and Golgi.
   d. Many lipid-soluble drugs and most steroids are metabolized in the endoplasmic reticular membranes of the liver, where the **mixed-function oxidase system** is found. Also, glucuronyl transferases in the endoplasmic reticulum are used in the conjugation of bilirubin glucuronides.
   e. The metabolism of compounds in the liver is not always beneficial. For example, certain mild carcinogens contained in cigarette smoke are converted to more potent carcinogens by hepatic microsomal hydroxylation.

2. **Glycogen.**
   a. Liver parenchymal cells are rich in glycogen granules. The liver receives large amounts of glucose from the portal drainage of the gut and rapidly converts much of it into stored glycogen.
   b. This stored glycogen can be mobilized when needed for regulation of blood glucose levels and also for synthesis of the sugar portions of glycoproteins present in cell membranes and in certain serum proteins made in the liver.

3. **Lysosomes.**
   a. Liver parenchymal cells contain many lysosomes, which are involved in the turnover of exhausted components of these active cells.
   b. Lysosomes also are important in several disease processes (e.g., the lysosomal storage diseases).
   c. Lysosomes may be involved in glycogen breakdown, as indicated by the fact that patients with α-glucosidase deficiencies, demonstrating Pompe's disease, have large accumulations of glycogen granules within lysosomes.

4. Hepatic parenchymal cells contain **peroxisomes** (or **microbodies**), which are membrane-delimited structures of unknown function. Peroxisomes are similar to but smaller than lysosomes.

5. Each parenchymal cell has a large nucleus with much euchromatin and a prominent nucleolus.

D. LOBULES. The liver is divided into three histologic units or **lobules**.

1. **Classic Lobules.**
   a. These lobules are roughly hexagonal in outline and are centered about a large blood vessel called the **central vein**.
   b. Arranged around the edges of each classic lobule are **portal canals** containing a small branch of the **bile duct** system, a **hepatic artery**, and a **portal vein**.

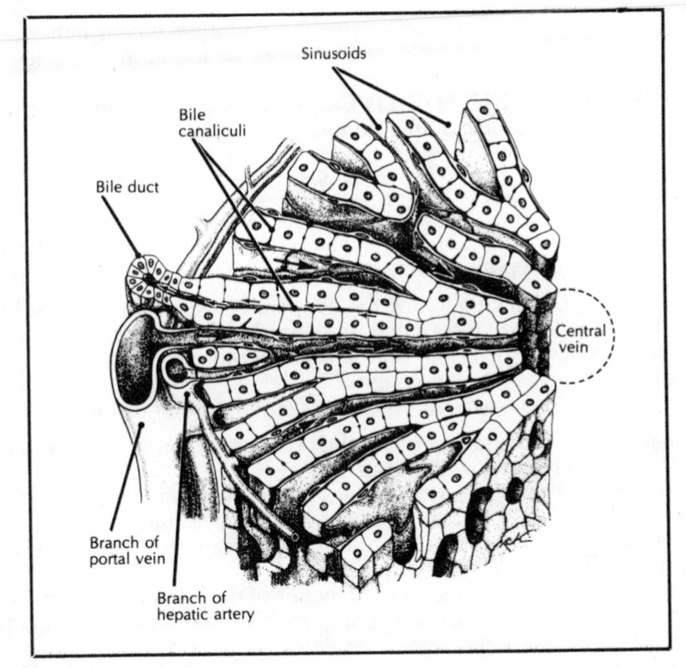

Figure 16-1. Diagram illustrating the relationship between the structures in the portal canal and the central vein. The *arrows* show the directions of flow of blood and bile in separate compartments. (Reprinted with permission from Bloom W, Fawcett D: *A Textbook of Histology.* Philadelphia, WB Saunders, 1975, p 694.)

c. The portal vein carries blood drained from the gastrointestinal tract into the liver parenchyma and communicates directly with the **hepatic sinusoids**—the vascular channels that pass between all liver parenchymal cells.
d. The liver parenchymal cells, in turn, are arranged in anastomosing plates and cords with sinusoids on each side of a plate.
e. Branches of the hepatic artery bring blood from the systemic circulation to nourish the connective tissue of the liver and also to communicate directly with the hepatic sinusoids to provide oxygen for the liver parenchymal cells.

2. **Portal Lobules.**
   a. Portal lobules are roughly triangular structural units.
   b. Each portal lobule is centered on a portal canal and has a central vein at each of its three apices.

3. **Liver Acinus.**
   a. This functional unit of the liver has an irregular shape that sometimes is roughly oval.
   b. The liver acinus has two central veins, one at each end of its horizontal axis, and two portal canals, one at each end of its vertical axis.

E. **SINUSOIDS.** Figure 16-1 shows the relationship between the sinusoids and the blood vessels. The bile space and blood space always are separated.

1. Most of the blood flowing through the liver passes through the sinusoids. These sinusoids are lined by discontinuous capillaries (i.e., there are large intercellular gaps between the cells). These gaps allow serum proteins that are synthesized within parenchymal cells to flow, in bulk, into the circulation.

2. **Kupffer Cells.**
   a. Liver sinusoids are lined by some phagocytic cells called **Kupffer cells**. These sinusoids make up an important part of the mononuclear phagocyte system.
   b. Kupffer cells commonly contain one or more of the following: ingested debris, whole cells, and iron deposits left after the destruction of erythrocytes.

3. Plates of parenchymal cells usually are closely associated with sinusoids. The gap between the sinusoidal endothelial cells and the parenchymal cells is called the **space of Disse**.
   a. Numerous microvilli from the surfaces of the parenchymal cells project into the space of Disse.

b. The basement membrane underlying the endothelial cells either is thin or is absent, and occasional collagen and reticular fibers can be found in the space of Disse.

III. BLOOD FLOW THROUGH THE LIVER.

There is a rich blood supply to the liver. Approximately 25 percent of the cardiac output perfuses the organ; blood pumps through the liver at a rate of about 1 ml/g of tissue every minute.

A. AFFERENT BLOOD SUPPLY (Fig. 16-2)

1. Approximately 75 percent of the hepatic blood is supplied by the **portal vein**; 25 percent is received from the **hepatic artery**. Both vessels enter at the porta hepatis.
2. Branches of the portal vein follow the connective tissue bundles first into the lobes and then into lobules where they form small branches in the portal canals.
   a. These branches of the portal vein send numerous smaller branches to the sinusoids, which in turn drain into the central veins.
   b. Many different sinusoids empty into a central vein of a single classic lobule.
3. Ramifications of the hepatic artery follow the large connective tissue trabeculae and supply these structural elements with blood.
   a. Small branches of the hepatic arteries enter the portal canals and then divide rapidly into capillary beds that supply the cells of the portal canal.
   b. In a few cases, small branches of the hepatic artery drain into sinusoids and eventually empty into central veins.
   c. Most capillary beds derived from branches of the hepatic artery have no direct connection with the sinusoids. Instead, they drain small venules that are tributaries of the portal veins. Branches of these portal veins then enter the sinusoids.

B. EFFERENT BLOOD SUPPLY

1. Sinusoids empty into central veins.
2. Central veins empty into sublobular veins which in turn drain into hepatic veins.
3. Hepatic veins leave through the porta hepatis and join the inferior vena cava.

IV. BILE

BILE is produced by the parenchymal cells perhaps in the Golgi apparatus, which commonly are oriented toward the biliary space. Parenchymal cells secrete bile into bile canaliculi, the small extracellular channels that run between virtually all parenchymal cells.

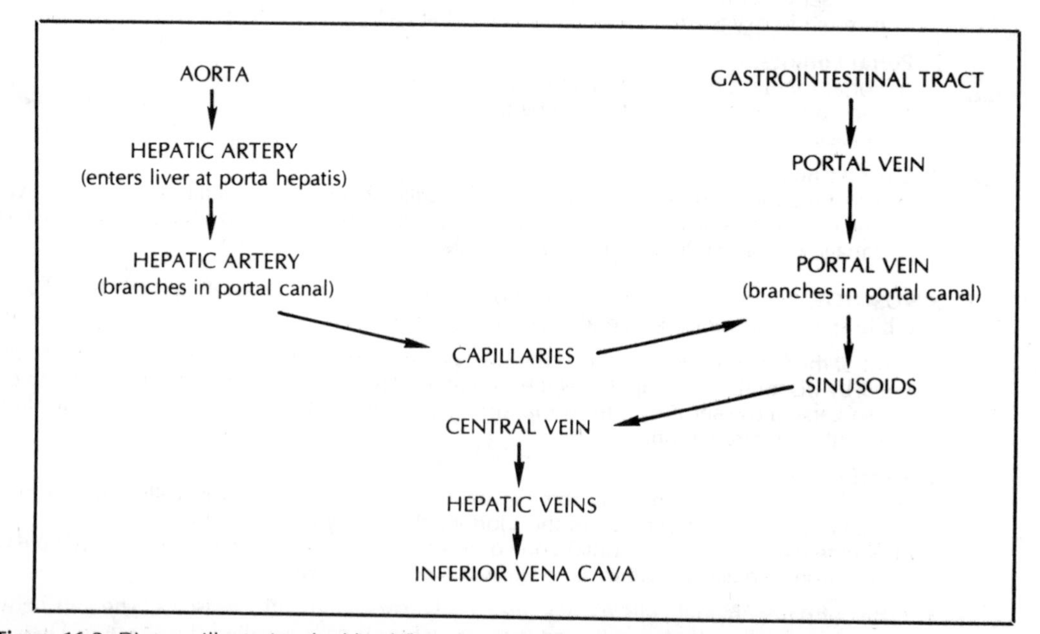

Figure 16-2. Diagram illustrating the blood flow through the liver. (Reprinted with permission from Johnson KE: *Histology: Microscopic Anatomy and Embryology*. New York, John Wiley, 1982, p 237.)

A. COMPOSITION

1. Bile, which aids in digestion of fats, contains bile salts, bile pigments, phospholipids, cholesterol, glucose, and bicarbonate.
   a. Bile salts are metabolites of cholesterol and are synthesized in the liver. They emulsify lipids and aid in the absorption of fatty acids and monoglycerides in digested food.
   b. Many of the bile salts are resorbed in the gastrointestinal tract and returned to the liver, where they again enter a new batch of bile.
2. Bile also contains breakdown products of hemoglobin such as bilirubin, which is excreted from the body.
3. Because bile is a mixture of strong detergents and waste products, it is kept out of the systemic circulation.

B. CANALICULI

1. The walls of the bile canaliculi are formed by the cell membranes of parenchymal cells.
2. At the edge of the canaliculi, individual cells are tightly coupled by junctional complexes that consist of tight junctions and desmosomes. These junctional complexes prevent bile secreted by parenchymal cells from flowing back into the sinusoids that run along the plates of parenchymal cells.
3. Bile canaliculi usually are closely associated with filamentous networks similar to those found in other junctional complexes.
4. Bile canaliculi drain directly into small bile ducts in the portal canal.

C. DUCTS

1. A **small bile duct** is lined by a simple cuboidal epithelium of cells that are joined apically by junctional complexes, rest on a conspicuous basement membrane, and may have a collection of apical microvilli.
2. Small bile ducts empty into large ducts that run in the connective tissue trabeculae scattered throughout the liver. These larger, **intrahepatic** bile ducts may be lined with a columnar epithelium, but the apical junctional complexes and microvilli remain prominent.
3. The intrahepatic bile ducts combine at the porta hepatis to form several **extrahepatic** ducts lined by tall columnar cells. The wall of each extrahepatic duct is arranged much like the rest of the gastrointestinal tract, with a mucosa, submucosa, muscularis, and adventitia.
4. The extrahepatic ducts drain into a **common hepatic duct**, which combines with the **cystic duct** to form the **common bile duct**. All these ducts are histologically similar; each has a tall columnar epithelium, submucosal mucous glands, smooth muscle layers in the muscularis, and connective tissue in the adventitia.

D. GALLBLADDER

1. The gallbladder is a closed sac with a capacity of approximately 40 ml.
2. When filled with bile it has a smooth surface; when empty, the surface shows numerous folds and deep invaginations that are continuous with the surface epithelium.
3. Bile is concentrated in the gallbladder, probably by resorption of water by the microvilli.

V. PANCREAS.

V. PANCREAS. This retroperitoneal gland is surrounded by a thin connective tissue capsule. Branches of the celiac and superior mesenteric arteries enter the pancreas all along its length. Delicate slips of loose areolar connective tissue divide the gland into small lobules composed of numerous **acini**, which serve an exocrine function, and scattered **islets of Langerhans**, which serve an endocrine function. Individual acini are composed of a group of 5–10 pyramidal **acinar cells**.

A. THE EXOCRINE PANCREAS (Fig. 16-3)

1. General Considerations.

a. Pancreatic acinar cells are active in the synthesis of digestive enzymes responsible for most of the hydrolysis of food macromolecules.
   (1) The basal portion of the acinar cell is broad, intensely basophilic, and rests on a basement membrane.
   (2) The apical portion of the cell is constricted and loaded with acidophilic **zymogen granules** that are rich in proteases, lipases, glycosidases, and ribonucleases.

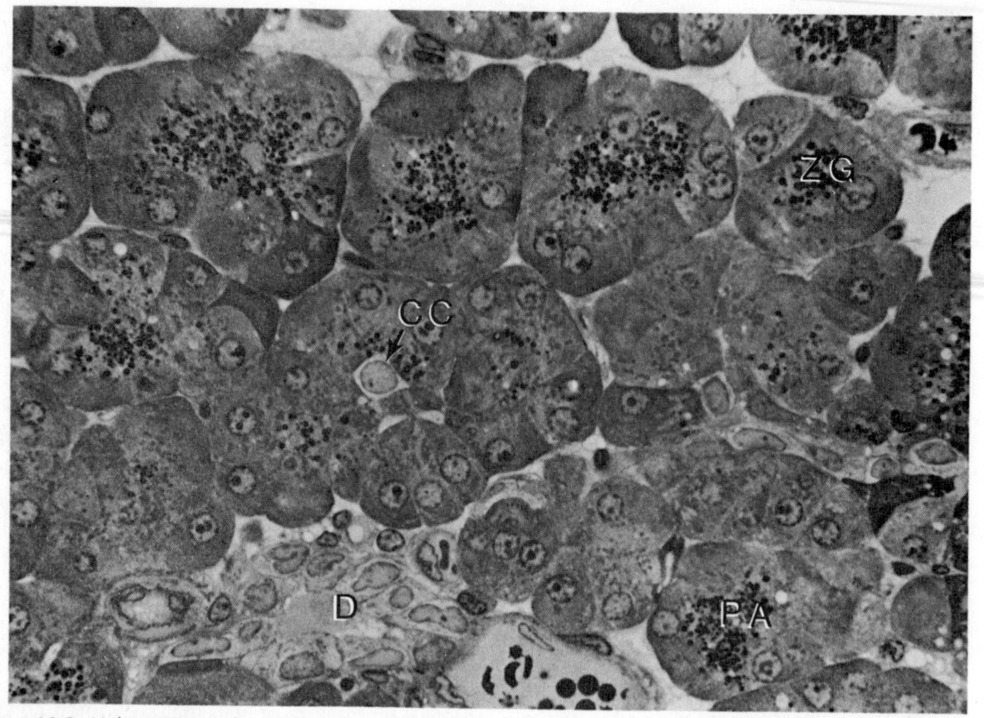

Figure 16-3. Light micrograph revealing the components of the exocrine pancreas. Pancreatic acini (*PA*) have cells that contain zymogen granules (*ZG*). These zymogen granules contain a variety of enzymes, which are discharged into the ducts of the gland. These ducts begin at the centroacinar cells (*CC*) and grow rapidly into larger ducts (*D*). (Courtesy of Dr. Frank J. Slaby, Department of Anatomy, George Washington University.)

b. In addition to secretion of enzymes, the pancreas secretes a substantial amount of a bicarbonate that, along with the secretions of the submucosal Brunner's glands in the duodenum, neutralizes the acidic chyme.

c. Centroacinar cells represent the terminal portion of the duct system that drains the pancreas.

d. These cells form a flange-like structure that abuts directly on the acini and drains into small **intercalated ducts**—thin channels lined by a simple cuboidal epithelium.

e. Intercalated ducts drain into larger, intralobular ducts and eventually into **interlobular ducts**, which are lined by columnar cells and occasional goblet cells.

f. Interlobular ducts follow the connective tissue septa and eventually drain into either the **main** or the **accessory pancreatic duct**.

2. Pancreatic Acinar Cells.

a. Structure (Fig. 16-4).

(1) The basal portion of a pancreatic acinar cell is rich in granular endoplasmic reticulum and free ribosomes. Mitochondria with dense matrices and numerous cristae frequently are located between the layers of endoplasmic reticulum.

(2) The nucleus of the cell is located basally, and there is a prominent nucleolus and an abundance of euchromatin. On the apical side of the nucleus is seen a large Golgi apparatus composed of several dictyosomes of flattened cisternae as well as other, larger, vesicles with electron-dense material in them.

(3) The apical portion of an acinar cell contains numerous membrane-delimited vesicles filled with a high concentration of hydrolytic enzymes contained in zymogen granules. These granules fuse with the apical plasma membrane and with one another during secretion of pancreatic enzymes.

(4) Numerous microvilli coat the apical surface of a pancreatic acinar cell.

b. Function. The polypeptide chains of digestive enzymes are synthesized in the rough endoplasmic reticulum and transported to the Golgi apparatus, where zymogen granules are formed as buds from the membranes of the Golgi cisternae.

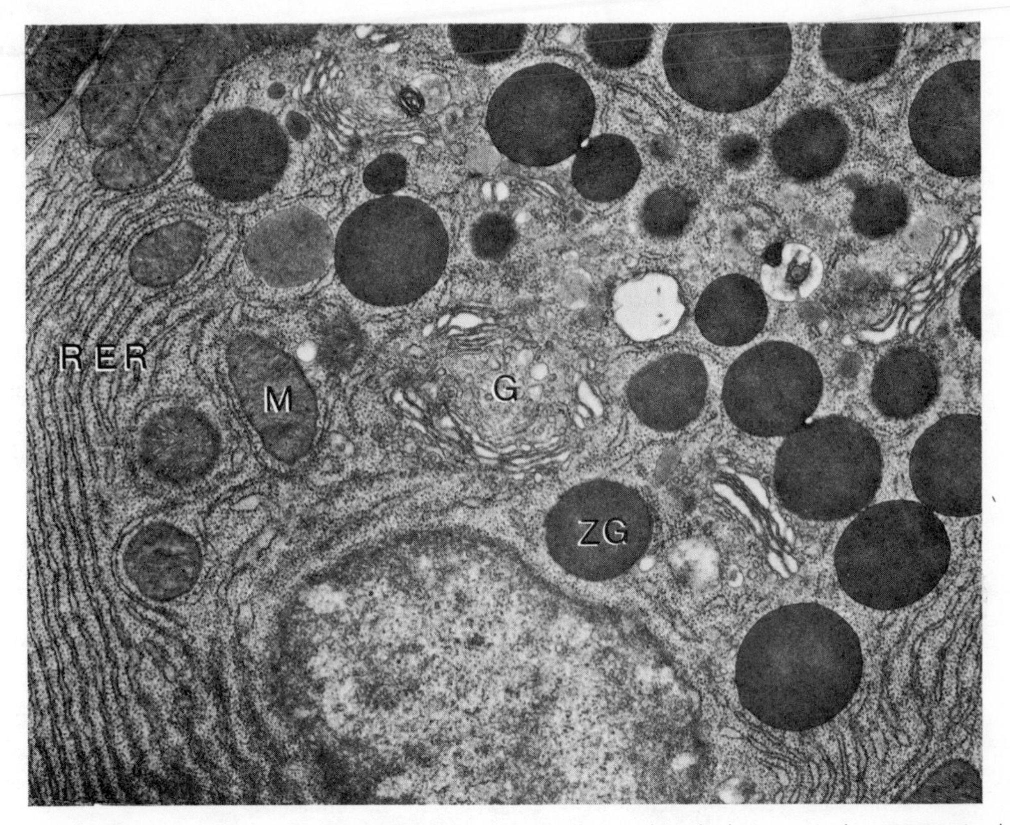

Figure 16-4. Electron micrograph of a pancreatic acinar cell. The rough endoplasmic reticulum (*RER*), mitochondria (*M*), Golgi apparatus (*G*), and zymogen granules (*ZG*) are visible. (Courtesy of Dr. Frank J. Slaby, Department of Anatomy, George Washington University.)

B. THE ENDOCRINE PANCREAS

1. Scattered throughout the lobules of acinar cells are the **islets of Langerhans**. These groups of endocrine cells are surrounded by a delicate connective tissue capsule and contain a rich bed of fenestrated capillaries.
2. Scattered among the capillaries are several types of cells, most notably **alpha** or **A cells** and **beta** or **B cells**.
   a. A cells represent about 20 percent of the nonvascular cells in islet tissue. They are relatively large, have an irregular shape, and when appropriately fixed and stained show numerous cytoplasmic granules. Recent studies have shown that A cells produce and secrete **glucagon**, a polypeptide hormone that inhibits liver glycogenesis and elevates blood sugar.
   b. B cells represent about 75 percent of the nonvascular islet cells. They are relatively small, have a rounded shape, and contain numerous granules of **insulin**, a polypeptide hormone that lowers blood sugar.
   c. The remaining fraction of islet cells is made up of C cells and D cells. Their function is unclear, but there is evidence that they are involved in the production of enteric polypeptide hormones such as **somatostatin**, a hormone that inhibits the release of growth hormone.

VI. DEVELOPMENT OF THE LIVER AND PANCREAS

A. LIVER

1. The liver develops from a small **diverticulum** of the foregut which forms just cranial to the junction between the foregut and the yolk sac.

a. The cranial portion of this diverticulum grows into a mesenchymal mass known as the **septum transversum**.
b. The original diverticulum forms a small bud that becomes the gallbladder, the cystic duct, and the common bile duct.
c. Also, a ventral bud forms from the root of the liver diverticulum; this becomes one part of the pancreas.
d. The bulk of the liver diverticulum grows into the septum transversum and undergoes considerable branching.
e. Eventually, cords of cells push off into the septum transversum and become surrounded by capillaries that arise within this mesenchyme.
f. In the adult, the liver parenchymal cells and the lining of the biliary ducts are endodermal derivatives of the liver diverticulum, whereas the sinusoids and other blood vessels as well as the connective tissue capsules are mesodermal derivatives of the septum transversum.

2. During the period between 6 weeks and 30 weeks gestation, the liver acts as the embryo's chief site of active hematopoiesis.

3. By birth, the hepatic production of blood cells is reduced. In severely anemic patients, however, hepatic hematopoiesis is found occasionally.

B. PANCREAS

1. The pancreas develops from a **ventral pancreatic rudiment** (i.e., a small bud off the liver diverticulum) and a **dorsal pancreatic rudiment**. During rotation of the gut, the two primordia come into close apposition and then fuse together.
   a. The dorsal rudiment grows more substantially than the ventral rudiment, forming the tail, body, and part of the head of the pancreas.
   b. The ventral rudiment gives rise to the rest of the head of the pancreas.
   c. Interestingly, the duct of the smaller, ventral rudiment becomes the main pancreatic duct while the larger, dorsal rudiment forms the smaller accessory pancreatic duct.

2. The pancreatic rudiments first appear during the fourth week gestation. By the third or fourth month, islet tissue and acinar tissue become well-differentiated.

STUDY QUESTIONS

Directions: The group of questions below consists of lettered choices followed by several numbered items. For each numbered item select the **one** lettered choice with which it is **most** closely associated. Each lettered choice may be used once, more than once, or not at all.

Questions 1–5

For each description of components of the portal canal and its associated sinusoids, choose the appropriate lettered structure shown in the diagram below.

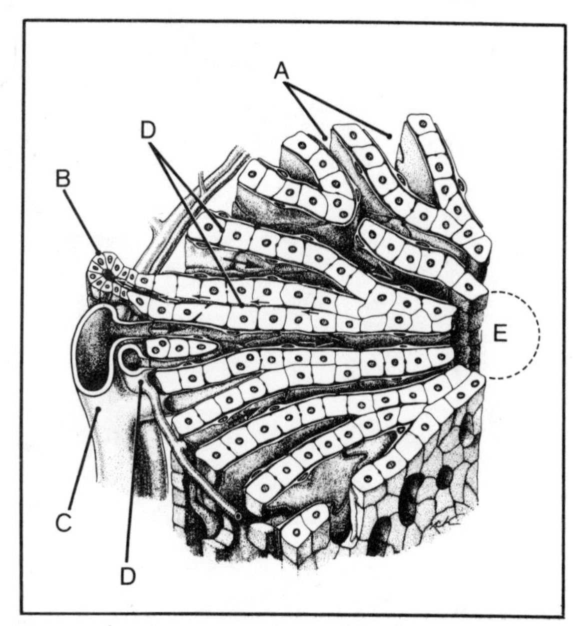

Reprinted with permission from Bloom W, Fawcett D: *A Textbook of Histology*. Philadelphia, WB Saunders, 1975, p 694.

1. Kupffer cells are incorporated in its lining
2. Empties into the hepatic vein
3. Its branches communicate with the hepatic sinusoids
4. Lined by cuboidal epithelial cells
5. Carries blood rich in nutrients from the gastrointestinal tract

ANSWERS AND EXPLANATIONS

1–5. The answers are: 1-A, 2-E, 3-C, 4-B, 5-C. (*II D, E*) This is a diagram of the portal canal, sinusoids, and central vein of a portion of a liver lobule. Sinusoids (A) are lined by phagocytic Kupffer cells. The central vein (E) empties into the hepatic vein. Branches of the hepatic artery (D) nourish hepatic connective tissue and empty into sinusoids. The bile ducts (B) are lined by a cuboidal epithelium. The portal vein (C) drains nutrient-rich blood from the gastrointestinal tract into the liver.

17
Skin

I. INTRODUCTION

A. FUNCTION

1. The entire outer surface of the body, except for the cornea and conjunctiva, is covered by **skin**.
2. Essential to life, this complex covering serves as
   a. A protective barrier against infection
   b. A hydrophobic layer to prevent loss of vital fluids
   c. A mechanism to alter the core body temperature (by regulating blood flow through the capillary beds serving the skin or by sweating)

B. COMPONENTS

1. The skin is composed of an ectodermally derived, stratified squamous keratinized epithelium called the **epidermis** and a mesodermally derived, dense irregular connective tissue called the **dermis**.
2. In several locations the skin produces protective acellular appendages known as **hair** and **nails**. These structures are formed entirely from the epidermis but under the influence of the dermis that is closely associated with hair follicles or nail beds.
3. Cells in the basal layer of the skin proliferate constantly and differentiate into keratinized cells that eventually dry out and desquamate from the surface.
4. Glands in the skin produce **sweat** for cooling and **sebum** for lubricating and conditioning the skin.
5. Skin is colored to a variable extent in different ethnic and racial groups due to the action of neural crest derivatives called **melanocytes**.

II. EPIDERMIS

A. MICROSCOPIC ANATOMY.
The epidermis is divided into four easily recognizable layers.

1. **Stratum Basale** (Stratum Germinativum).
   a. The stratum basale is the most basal layer of cells in the epidermis, and it rests on a thick basement membrane.
   b. The cells of the stratum basale are highly proliferative, undergoing repeated mitosis. This division results in the production of a proliferative stem cell layer that remains on the basement membrane and gives rise to the next generation of cells as well as cells that differentiate into **keratinocytes**.
   c. The layering seen in the epidermis represents the progressive accumulation of **keratin** within the progeny of the proliferative cells in the stratum basale.
2. **Stratum Spinosum** (Fig. 17-1).
   a. The next epidermal layer, the stratum spinosum, varies from two to six cell layers in thickness.
   b. With the light microscope, small spine-like projections appear to join the cells of this layer.
   c. Examination with the electron microscope, however, shows that the cells of the stratum spinosum are interconnected by a large number of **desmosomes**.
   d. In addition, the cells of the stratum spinosum contain a peculiar organelle known as the **lamellated granule**. This small membrane-delimited ellipsoid granule is 0.3 μm in diameter and has lamellae running perpendicular to its long axis. Lamellated granules often are closely associated with the Golgi apparatus.

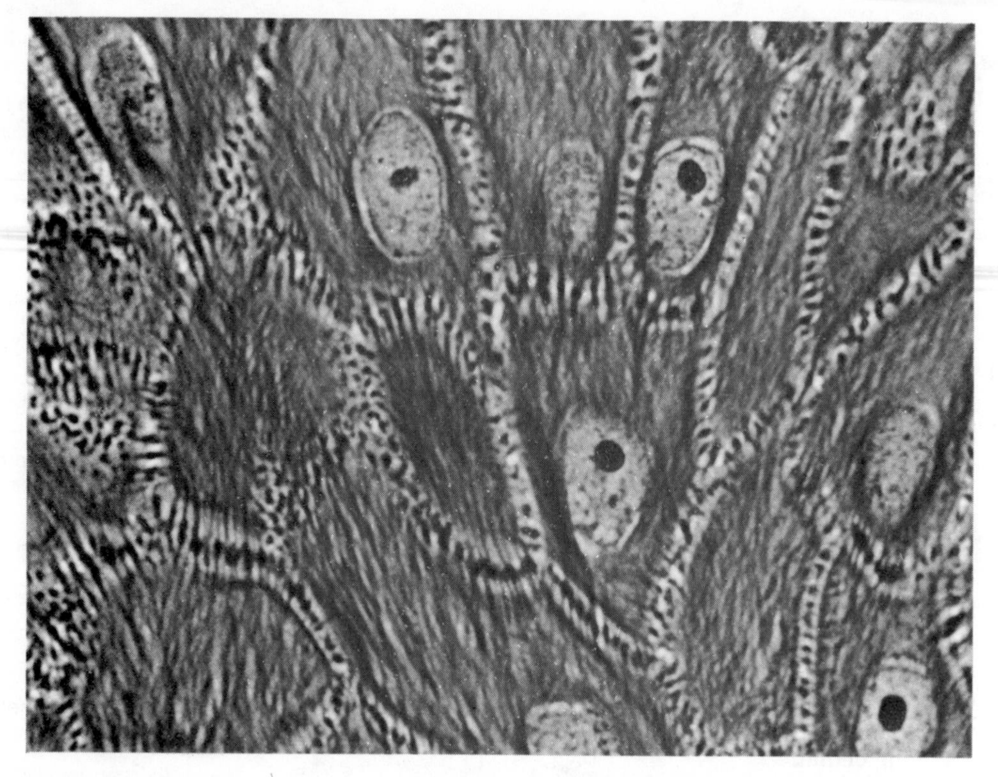

Figure 17-1. High-power light micrograph of the epidermal stratum spinosum.

3. Stratum Malpighii.

a. Some authors describe the stratum basale and stratum spinosum together as one layer known as the stratum malpighii.

b. Electron-microscopic examination of cells in the "stratum malpighii" reveals numerous mitochondria, melanin granules, abundant rough endoplasmic reticulum, and numerous bundles of 6–8 nm filaments.

4. Stratum Granulosum.

a. The next layer up from the basement membrane, the stratum granulosum, consists of three to five layers of flattened, irregularly shaped cells loaded with **keratohyalin granules**.

(1) With the light microscope, these granules reveal the ability to stain intensely with various basic dyes including hematoxylin.

(2) Classical histologists consider these to be an intermediate stage in the keratinization process.

b. Electron-microscopic investigation reveals that cells in the stratum granulosum are filled with small granules that have an irregular amorphous core and filaments running up to or through them.

c. In addition, cells of this layer have lamellated granules that migrate from the Golgi apparatus toward the cell periphery. In some cases, lamellated granules discharge their contents into the intercellular space.

5. Stratum Corneum.

a. The stratum corneum is the most apical layer of the epidermis, ranging in thickness from as few as five to ten cell layers to as many as several hundred cell layers.

(1) For example, the stratum corneum on the back of the hand or on the eyelid is extremely delicate. In contrast, the stratum corneum on the sole of the foot is extremely thick.

(2) Furthermore, this layer can become even thicker if circumstances warrant (e.g., in the case of callus formation).

b. With the light microscope, the stratum corneum appears as a poorly stained layer of dead and desiccated cells that can flake easily from the epidermis. Where the stratum corneum is particularly thick, there is a thin transitional zone between the stratum granulosum and the stratum corneum known as the **stratum lucidum**.

c. With the electron microscope, the cells of the stratum corneum appear to be dead or inactive.

(1) The usual nucleus, mitochondria, and rough endoplasmic reticulum are completely replaced by masses of 6–8 nm filaments that are intermingled with masses of amorphous material.

(2) The inner aspect of the plasma membrane of these cells is considerably thickened, and the intercellular clefts are filled by the discharged contents of the lamellated granules.

(3) Desmosomes still are evident between cells of the stratum corneum. Typically, the outermost layers of this epidermal component are loosely attached to and appear ready to fall from the underlying cells. (A cross-sectional view of the layers of the epidermis is shown in Figure 17-2.)

B. KERATINIZATION

1. Stratified squamous epithelium is distributed widely throughout the body and found in most locations where friction commonly is encountered (e.g., on the conjunctiva and anterior surface of the cornea, oral cavity, esophagus, vagina, and anal canal). The surface of the body is covered by such an epithelium, but it is different from all others of the body in that it is **keratinized**.

2. Keratinization is the process whereby the living cells undergoing mitosis in the stratum basale and one of the daughter cells become converted to the dead cells of the stratum corneum. These dried cells filled with hydrophobic proteins limit water loss from the body and limit the entry of external noxious substances and microorganisms into the body.

3. Keratinization involves three processes.
   a. First, the cells that will become committed to this "fatal" differentiation must be produced. This is accomplished by repeated mitosis of proliferative cells in the stratum basale.
   b. Second, cells in the stratum basale and stratum spinosum become committed to produce large amounts of different types of proteins. This is done by selecting certain genes to become activated for the synthesis of messenger RNAs (mRNAs) and eventually for the synthesis of proteins in the filaments and keratohyalin granules present at a later stage in the differentiation.
   c. Third, these proteins accumulate within the cells to such an extent that they kill the cells and exclude all other organisms.

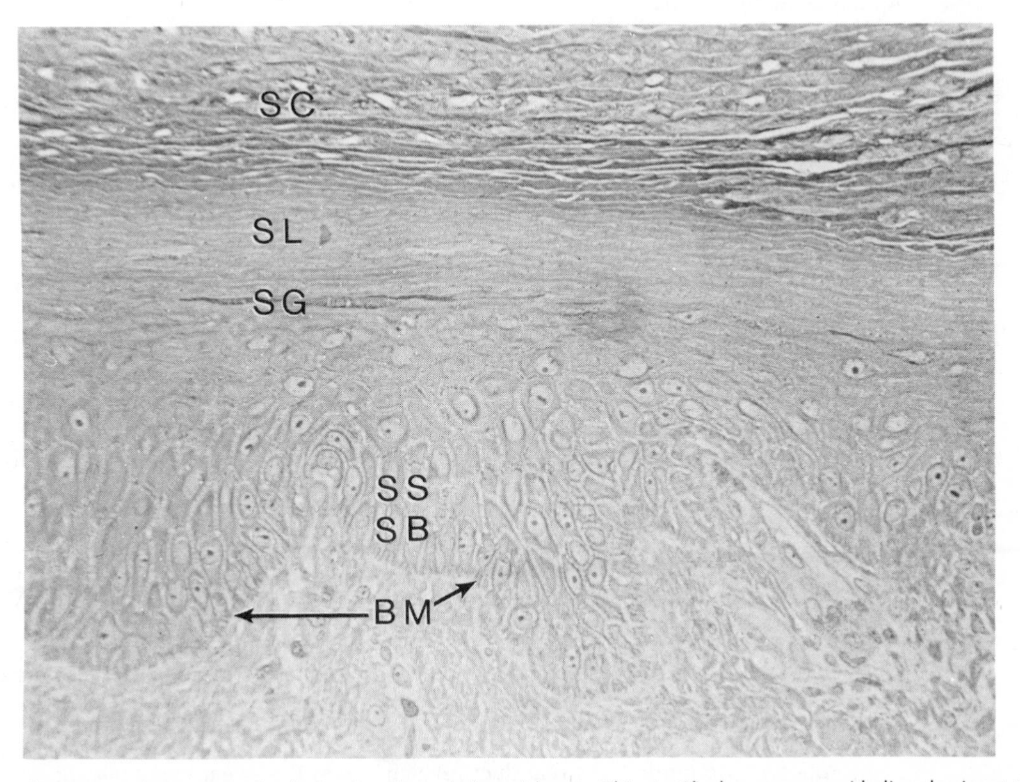

Figure 17-2. Light micrograph of a section through the epidermis. This stratified squamous epithelium begins at the basement membrane (*BM*) and extends through the top of the picture. Cells of the stratum basale (*SB*), stratum spinosum (*SS*), stratum granulosum (*SG*), stratum lucidum (*SL*), and stratum corneum (*SC*) all are visible.

4. The death of a cell in the epidermis is more rapid but no less inevitable than the death of a highly differentiated red blood cell or pyramidal neuron in the cerebrum. All three types become highly differentiated and lose the ability to proliferate, thus ensuring their shortened life span.

5. In contrast, some cells do not differentiate but remain proliferative throughout the life of the organism. The cells in the stratum basale of the epidermis as well as spermatogonia, hematogenous cells, and cells lining the crypts in the intestine all are proliferative throughout life.

6. The stratifications in the epidermis simply represent different stages in the process of keratinization, with completely proliferative unkeratinized cells near the basement membrane and nonproliferative, completely keratinized cells comprising the outermost layers of the epidermis. In between these two extremes, intermediate stages in the accumulation of keratin are evident.

C. BIOCHEMISTRY OF KERATIN

1. Keratin is a peculiar protein prevalent in the cells of the stratum corneum. It contains some sulphur with cross-linking by disulfide bridges. These bonds result in keratin being a very stable protein that can only be dissolved by strong acid.

2. The proteins derived from keratohyalin granules also are extensively cross-linked, via disulfide bonds, to each other as well as to the proteins of the keratin.

3. Such extensive cross-linking makes proteins of the stratum corneum very resistant to dissolution.

III. GLANDS

A. SWEAT GLANDS. These epithelial modifications are widely distributed throughout the body. They are specialized for the production of sweat, which cools the body by evaporation, and other complex secretions.

1. Apocrine Sweat Glands.

a. These produce a milky white secretion from the cuboidal and columnar eosinophilic secretory cells in the deeper portions of the glands.

b. The secretion is expelled from the glands following adrenergic stimulation of myoepithelial cells. There is at least a one-day lapse between secretions, during which apocrine glands must resynthesize and store new secretion.

c. The ducts of apocrine sweat glands empty into hair follicles. In lower animals, these glands are distributed uniformly throughout the surface of the body, but in humans they are restricted to the axilla, the area around the genitalia, and the nipples on the mammary glands.

d. Apocrine glands start secreting after puberty, under the influence of sex steroids.

e. Initially, the milky secretion is odorless; later, it develops an odor due to the action of cutaneous bacteria.

f. In lower animals, the apocrine glands play an important role in sexual behavior and in marking territorial zones. In humans, the role of these glands is not well-defined but may be involved in temperature regulation.

g. The ceruminous glands in the external auditory meatus and glands of Moll in the eyelid are modified apocrine sweat glands.

h. Recent research shows that apocrine sweat glands release their secretions by a **merocrine mechanism**.

2. Eccrine Sweat Glands.

a. Eccrine sweat glands are distributed widely throughout the entire surface of the body. They are found everywhere but on the lips, penis, clitoris, and labia minora.

b. These elaborately coiled glandular structures connect to the surface of the body by a coiled duct.

c. In the glandular portion of the eccrine sweat gland are **myoepithelial cells**; also present are large eosinophilic **clear cells**, thought to secrete sodium, and smaller basophilic **dark cells**, thought to produce a glycoconjugate-rich material.

d. Sweat contains sodium, water, chloride, and other substances. It is formed by active pumping of sodium out of clear cells, passive diffusion of water out of clear cells, and a resorption of sodium in the sweat ducts, which makes the sweat hypotonic.

e. Eccrine glands respond to cholinergic stimulation and, under certain circumstances, to adrenergic stimulation; the sweat produced varies in composition according to stimuli.

f. Whenever heat radiation from the body is inadequate and the core body temperature rises, sweating begins. As the water in sweat evaporates, the body is cooled.

B. SEBACEOUS GLANDS (Fig. 17-3)

1. Sebaceous glands are associated with hair follicles and are found on all body surfaces except the palms of the hands and the soles of the feet.
2. A short duct connects the gland to the shaft of a hair follicle.
3. The gland is composed of two types of cells: an outer layer of stem cells called **basal cells** and a central group of cells that accumulate lipid droplets in their cytoplasm.
   a. The largest cells in the central portion of a sebaceous gland acinus are full of lipid droplets.
   b. The entire lipid-laden cell bursts and becomes part of the secretion of the gland. This type of secretion is known as **holocrine secretion**.
4. **Sebum** contains large amounts of triglycerides and free fatty acids, and its secretion is partially controlled by sex hormones. The high lipid content of sebum suggests that it may serve to condition the skin or that it may represent some part of the hydrophobic barrier that exists in the epidermis.

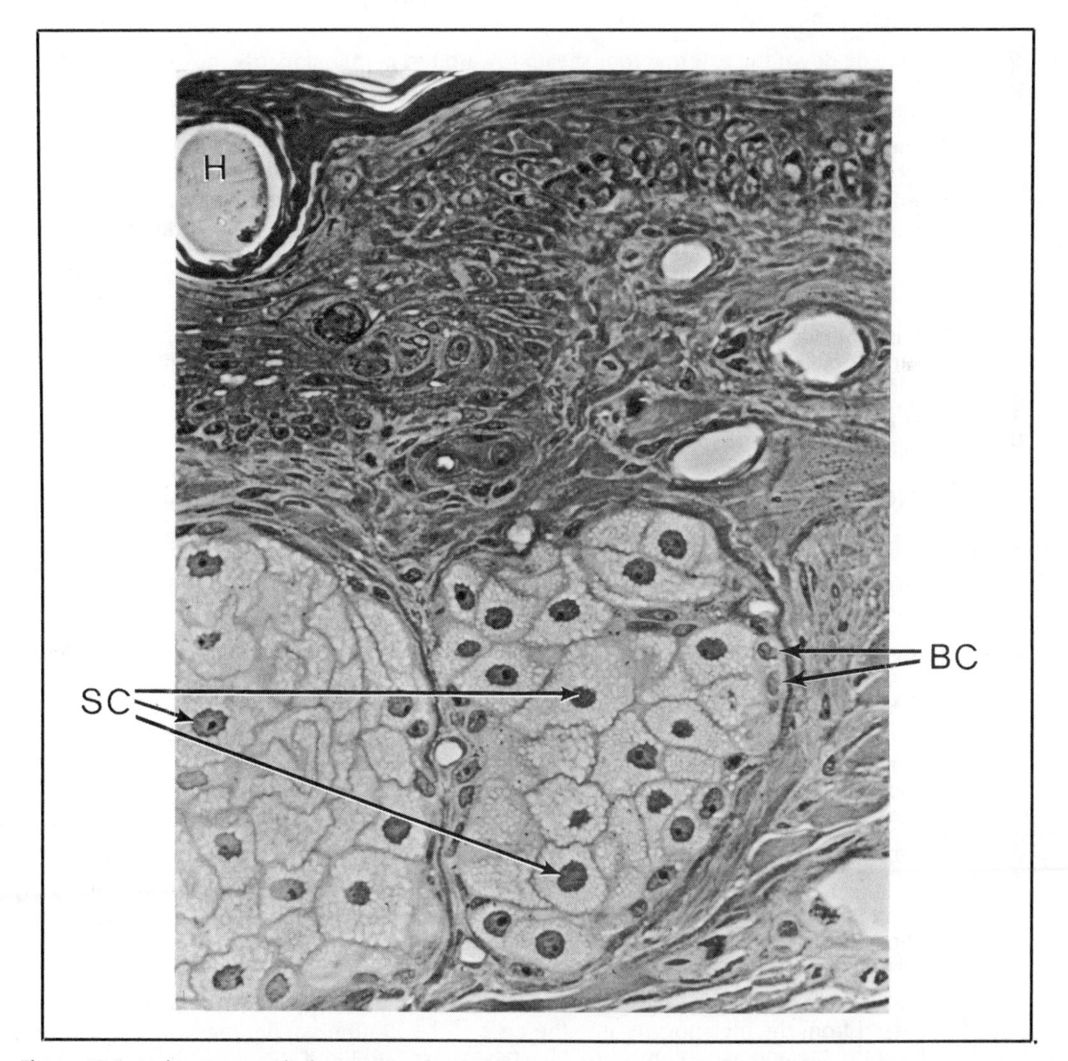

Figure 17-3. Light micrograph showing two large sebaceous glands. Most cells are lipid-laden secretory cells (*SC*), but some basal cells (*BC*) also are visible. A section of a shaft of a hair (*H*) also is visible, but the remainder of the pilosebaceous unit is out of this plane of section. (Reprinted with permission from Johnson KE: *Histology: Microscopic Anatomy and Embryology*. New York, John Wiley, 1982, p 253.)

IV. HAIR

A. GENERAL CONSIDERATIONS

1. Hairs are appendages of the skin; they are produced by deep epithelial ingrowths that form hair follicles.
2. Hair and nails both are composed largely of hard keratin, a substance that is chemically similar to soft keratin of the skin but is more extensively cross-linked by disulfide bridges.

B. HAIR FOLLICLES

1. In hair follicles, a deep ingrowth of the epithelium of the skin terminates as a bulbous core indented by a connective tissue papilla. The papilla carries blood vessels and, thus, nutrients to the forming hair. The bulbous portion contains a group of cells known as the **germinal matrix**, which produces the hair much as the stratum germinativum produces the stratum corneum.
2. The portion of the epithelial ingrowth that connects the germinal matrix to the skin surface is called the **external root sheath**. Near the surface of the skin, the external root sheath resembles an invagination of the stratified squamous epithelium of the skin and, furthermore, exhibits all of the same strata.
3. Moving down the external root sheath toward the germinal matrix, however, the layering of the skin is less apparent. Eventually only a stratum germinativum-like layer exists, mingled in the jumbled mass of cells in the germinal matrix.
4. The proliferative cells in the germinal matrix produce two structures: **hair** and a **cellular collar** that surrounds the hair shaft between the shaft and the external root sheath.
   a. This tubular collar, called the **internal root sheath**, is made of soft keratin.
   b. The soft keratin in the internal root sheath arises from **trichohyalin granules**, structures similar to keratohyalin granules.
   c. The internal root sheath extends only partly up the hair shaft; above this collar a naked shaft is exposed.
5. **Melanocytes** in the germinal matrix add melanin granules to the growing hair. Hair color varies from light blond to dark black, depending on the amount of melanin deposited in the hard keratin of the hair shaft.
6. The ducts of sebaceous glands and apocrine sweat glands empty their secretions into the hair follicle.

V. MELANOCYTES AND MELANIN

A. GENERAL CONSIDERATIONS

1. Melanocytes synthesize **melanin**, the pigment that is responsible for skin, hair, and iris coloration.
2. Melanocytes are derived from the neural crest and, therefore, originate first in the ectoderm.

B. MELANOCYTE STRUCTURE AND FUNCTION

1. Melanocytes rest on the apical side of the basement membrane of the epidermis, usually between cells of the stratum basale, and they produce long processes that are pushed around and between the basal cells of the epidermis.
2. Melanocytes have an enzyme called **tyrosinase**, which catalyzes one step in the reaction that converts tyrosine into the polymeric oxidation product known as melanin.
3. Melanocytes can be distinguished from other surrounding cells of the stratum basale by the **dopa reaction**. (In essence, these cells produce extraordinary amounts of melanin when supplied with 3,4-dihydroxyphenylalanine or **dopa** by the action of their tyrosinase; they then "mark" themselves by accumulating a dark reaction product.)
4. Melanocytes synthesize melanin in membrane-delimited **melanin granules** which then are passed from the melanocytes into the cells of the stratum basale, presumably by exocytosis from the melanocyte followed by phagocytosis by keratinocytes.

C. MELANIN FUNCTION

1. The nucleus of a cell in the stratum basale is shielded from ionizing (i.e., mutagenic)

ultraviolet solar radiation by an apical cap of melanin granules. The melanin content of this cell increases rapidly in response to exposure to the sun—the "tanning reaction."

2. Racial differences in skin color are due to several factors that involve not just the numbers of melanocytes but also the types and sizes of melanin granules.

VI. DERMIS

A. MICROSCOPIC ANATOMY

1. The dermis is a dense, irregular connective tissue underlying epidermis everywhere in the body. The epidermal-dermal boundary is at the basement membrane of the epidermis.

2. The dermis contains a series of peg-like folds called **dermal papillae**. These increase the contact area and thus strengthen the adhesion between the epidermis and dermis. Skin is subjected to the most profound and constant abrasion in the palms and soles, and it is here that the epidermal-dermal contact is most complex.

3. The dermis can be divided into two parts, based largely on differences in the texture and arrangement of the collagen fibrils present.
   a. The layer associated with the papillae is called the **papillary layer**. Here collagen fibers are arranged in an irregular meshwork.
   b. In the **reticular layer**, which is deep to the papillary layer, collagen bundles are arranged in coarser bundles that crisscross one another.
   c. The boundary between these two layers is indistinct, but in histologic section it is relatively easy to differentiate the papillary and the reticular areas.

4. Beneath the dermis is a thick layer of subcutaneous connective tissue that contains considerable adipose tissue.

B. EXTRACELLULAR MATRIX

1. The extracellular matrix of the dermis is rich in type I collagen but also contains a smaller amount of type III collagen.
   a. Type I collagen is composed of two $\alpha1$ (I) chains and one $\alpha2$ chain, whereas type III is composed of three $\alpha1$ (III) chains.
   b. There also are other minor biochemical differences between these two types of collagen.

2. The extracellular matrix also is rich in **elastic fibers** and the glycosaminoglycans **hyaluronic acid**, **chondroitin sulfate**, and **dermatan sulfate**.

C. CELLULAR ELEMENTS

1. The important cellular element of the dermis is the **fibroblast**, which is responsible for synthesizing and secreting the complex extracellular matrix of the dermis.

2. In addition to fibroblasts, mast cells, macrophages, and other formed elements of the blood are found in the dermis.

3. Small lesions of the skin may result in localized infections, with a concomitant increase in the number of phagocytic polymorphonuclear leukocytes in the dermis.

D. BLOOD SUPPLY

1. The blood supply of the skin comes from the dermis and is somewhat complex as a result of its role in thermoregulation.

2. There is a deep network of branches of the main arteries that supply the skin. This network, the **rete cutaneum**, is located just deep to the dermis in the subcutaneous adipose tissue.

3. Some branches from the rete cutaneum project toward the epidermis and form a second **rete subpapillare** at the junction between the papillary and reticular layers of the dermis.

4. Small arterioles in the rete subpapillare then send branches into dermal papillae, where extensive capillary beds in the papillae are bypassed in favor of direct **arteriovenous shunts**.
   a. When the body becomes overheated, the blood supply to the papillary capillaries increases and sweating commences. The evaporation of the water in sweat cools the blood slightly, reducing body temperature.
   b. In contrast, when the body becomes chilled, the blood is directed away from the capillary beds by way of the arteriovenous shunts, conserving body heat.

E. VARIATION WITH SKIN THICKNESS

1. **Thick skin**, such as that found on the sole of the foot, has a thick epidermis with a thick stratum corneum but a relatively **thin** and compact **dermis** with numerous dermal papillae.
2. **Thin skin**, such as that found on the forearm, has a thin epidermis with a thin stratum corneum but a relatively **thick** and more diffuse **dermis** without dermal papillae.

VII. EMBRYOLOGY OF THE SKIN

A. PRIMARY GERM LAYERS

1. The epidermis of the skin is derived from the ectoderm.
2. The dermis is derived from the mesoderm.
3. The glands and appendages of the skin, including hair and nails, all are derived from ectoderm, but their formation is controlled in part by the inductive influence of the underlying mesoderm.

B. SKIN DEVELOPMENT

1. At five weeks gestation, the skin is a simple cuboidal epithelium overlying a spongy mesenchyme.
2. By seven weeks, there are two epidermal layers: a basal layer of cuboidal cells and an outer squamous layer known as the **periderm** or **epitrychium**.
3. During the first three months gestation, the epidermal component of the skin also is invaded by melanoblasts derived from the neural crest. These begin to synthesize melanocytes and pass them on to neighboring cells.
4. An intermediate layer, consisting of three to five layers of polygonal cells, by the fourth month gestation becomes interposed between the cuboidal basal cells and the periderm.
5. After four months, the skin continues to increase in thickness until, at birth, it has the essential histologic organization of adult skin.

C. DEVELOPMENT OF GLANDS AND APPENDAGES

1. **General Considerations.**
   a. Hair follicles, sebaceous glands, and eccrine sweat glands all develop in the skin as ingrowths from the surface epithelium.
   b. The ability to form appendages from the skin is an innate property of the epidermis, but the specific type of appendage formed is dictated by the underlying dermal cells.
2. **Sebaceous Glands and Fetal Development.**
   a. During gestation, the secretions of sebaceous glands mix with desquamated epithelial cells to form a whitish protective material known as the **vernix caseosa**.
   b. This coats the fetus and probably protects fetal skin from potential harm within the fluid-filled amniotic cavity.
3. **Hair Development.**
   a. Hair develops in the skin early in fetal life and becomes visible on the surface of the body after the fifth month gestation.
   b. The first fine hair to form is known as **lanugo hair**.
   c. At about the time of birth, lanugo hair is replaced by larger **vellus hair**.
   d. Under hormonal control at puberty, coarse hair grows in the axilla, in the pubic region, and in other parts of the body such as the legs, chest, and stomach. This hair varies in amount and distribution depending on sex.

STUDY QUESTIONS

Directions: Each question below contains four suggested answers of which **one or more** is correct. Choose the answer

A if **1, 2, and 3** are correct
B if **1 and 3** are correct
C if **2 and 4** are correct
D if **4** is correct
E if **1, 2, 3, and 4** are correct

1. True statements concerning hair follicles include which of the following?

(1) They usually are associated with sebaceous glands and smooth muscle cells
(2) They contain melanosomes in the matrix near the papilla
(3) Apocrine sweat glands empty into them at some places in the body
(4) The germinal matrix produces the hair of a follicle

2. True statements concerning sebaceous glands include which of the following?

(1) They secrete directly into hair follicles
(2) They contain myoepithelial cells and lipid-laden secreting cells but no proliferative cells
(3) Their secretion product includes whole degenerating cells
(4) They release their product by apocrine secretion

Continued on next page

Directions: The group of questions below consists of lettered choices followed by several numbered items. For each numbered item select the **one** lettered choice with which it is **most** closely associated. Each lettered choice may be used once, more than once, or not at all.

Questions 3–8

For each description of a component of the skin, choose the appropriate lettered structure shown in the skin micrograph below.

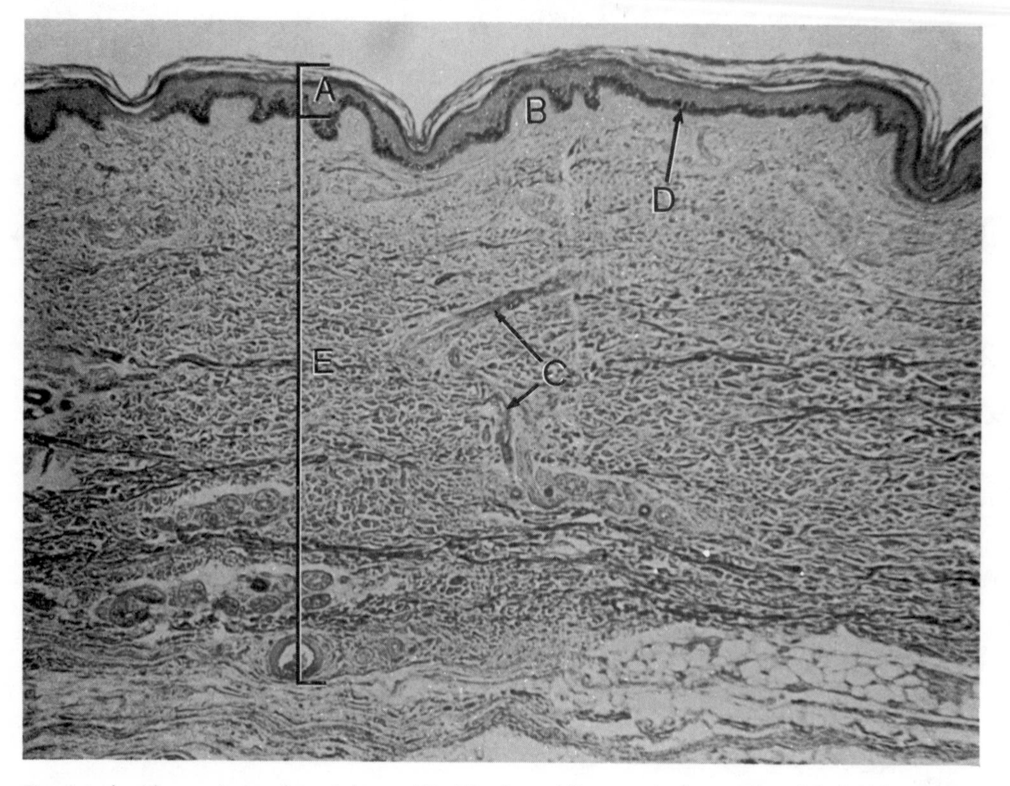

Reprinted with permission from Johnson KE: *Histology: Microscopic Anatomy and Embryology*. New York, John Wiley, 1982, p 256.

3. Deep to this location lie blood vessels and fibroblasts; superficial to it is the epithelium

4. This dense, irregular connective tissue is rich in fibroblasts and collagen

5. Important in temperature homeostasis

6. This avascular layer of proliferative cells is nourished by diffusion from nearby connective tissue

7. The basement membrane of the epidermis

8. Increases the contact area between the epidermis and the dermis and is more prominent in thick skin

ANSWERS AND EXPLANATIONS

1. The answer is E (all). (*IV B*) Apocrine sweat glands are restricted primarily to the axilla and the anogenital region; where present, they empty into hair follicles. Hair follicles usually have sebaceous glands and arrectores pilorum (i.e., smooth muscles) associated with them. The melanosomes in hair follicles are responsible for hair color and are derived from melanocytes.

2. The answer is B (1, 3). (*III B*) The basal cells of sebaceous glands are proliferative. Sebaceous gland cells secrete via a holocrine mechanism, not an apocrine mechanism. They release sebum directly into the hair follicle.

3–8. The answers are: 3-D, 4-E, 5-C, 6-A, 7-D, 8-B. (*II A; VI A*) This is a light micrograph of skin. Skin is composed of an epithelial epidermis (A) which rests on a basement membrane (D). The dermis (E) is a dense irregular connective tissue, sometimes thrown into dermal papillae (B), which underlies the epidermis. Eccrine sweat glands and their ducts (C) are a common feature of skin.

18 Thyroid and Parathyroid

I. INTRODUCTION

A. ENDOCRINE GLANDS. The thyroid and parathyroid glands have exclusively endocrine functions. Like all endocrine organs, they share certain common features.

1. They elaborate **hormones** that have profound effects on distant and varied parts of the body.
2. These hormones are secreted into the systemic circulation where they dissolve and are carried to target organs.
   a. For this reason, endocrine organs are supplied with a rich network of capillaries that are closely associated with the cells that elaborate the hormones.
   b. The capillaries in endocrine organs usually are fenestrated to facilitate passage of hormones into the systemic circulation.
3. The function of most endocrine organs is under the strict control of the hypothalamus and hypophysis.
   a. These glands form releasing hormones and trophic hormones that participate in a **feedback-loop system**. This system regulates hormone blood levels.
   b. The level of one hormone secreted by the thyroid is controlled by such a system.

B. THYROID AND PARATHYROID

1. The thyroid elaborates **thyroxine** and **calcitonin**. Thyroxine has a profound effect on the basal metabolic rate and growth rate.
2. The antagonist of calcitonin, **parathormone**, is secreted by the parathyroid glands. Calcitonin and parathormone are crucial in the regulation of the body's serum calcium level and, thus, are essential to life.

II. THYROID

A. EMBRYOLOGY

1. The thyroid epithelium originates from an endodermal invagination near the root of the tongue.
2. The thyroid descends into the neck region by a downward growth of a **thyroglossal duct**. Sometimes, remnants of thyroid tissue may persist in the root of the tongue or anywhere along the pathway of the thyroglossal duct.
3. The parafollicular cells are thought to be derived from the neural crest via the ultimobranchial body of the fourth pharyngeal pouch.

B. HISTOLOGY (Fig. 18-1)

1. The entire thyroid gland normally weighs 15–30 g in an adult. It lies over the larynx and has two lateral lobes of variable shape connected by an **isthmus**.
2. The thyroid has a simple **follicular epithelial layer** that surrounds a cavity filled with colloidal thyroglobulin and a connective tissue capsule that surrounds the entire organ. Projecting from the capsule are trabeculae that divide the thyroid into lobules.
3. Thyroid follicles are lined by a simple epithelium that varies in height from squamous to columnar, depending on the functional activity of the follicle.
   a. This epithelium is bounded by a well-formed basement membrane and is surrounded by a complex network of capillaries.
   b. The follicular epithelium is composed primarily of two cell types.

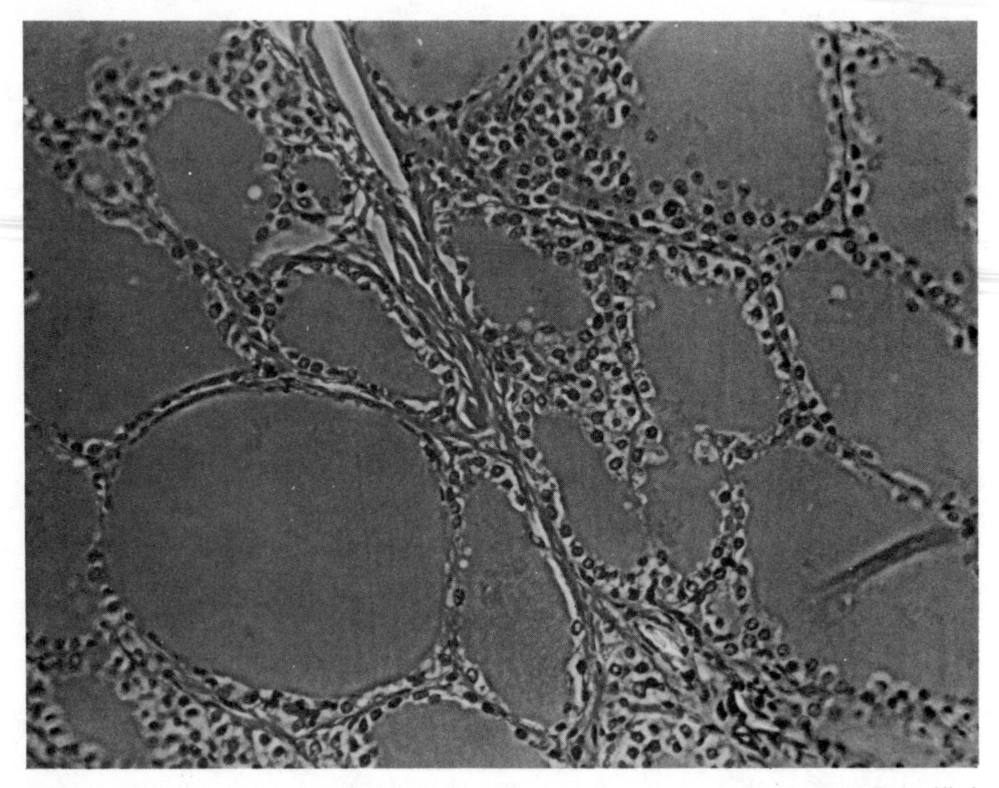

Figure 18-1. Light micrograph of a human thyroid gland. The large round structures are thyroid follicles filled with thyroglobulin. Each of the follicles is lined by a layer of follicular epithelial cells. (Reprinted with permission from Johnson KE: *Histology: Microscopic Anatomy and Embryology*. New York, John Wiley, 1982, p 266.)

(1) **Follicular epithelial cells** secrete thyroglobulin into the colloid-filled follicles and thyroxine into the capillaries near the basement membrane.
(2) **Parafollicular** or **C cells** secrete calcitonin.

4. Thin slips of connective tissue are found among the follicles and serve to bind them into a cohesive mass of thyroid tissue.

C. ULTRASTRUCTURE

1. Follicular Epithelial Cells.

a. These cells are strikingly polarized. Their apical surfaces (i.e., those facing the lumen of the follicle) are covered by numerous microvilli that presumably are involved in engulfment of stored thyroglobulin. Stimulation by thyroid-stimulating hormone (TSH) results in increased apical surface activity in these cells.
b. The epithelial cells are joined by well-developed apical junctional complexes, preventing thyroglobulin leakage from the follicle into the surrounding stroma in the gland.
c. The cytoplasm has a well-developed machinery for protein synthesis, including a prominent rough endoplasmic reticulum (RER) and a well-developed Golgi apparatus. In addition, numerous lysosomes and phagolysosomes exist in the cytoplasm.
d. In normal function, the follicular epithelium either secretes thyroglobulin to be stored as colloid in the lumen of the follicle or engulfs thyroglobulin from the storage site and breaks it down into active thyroid hormones via lysosomes.
e. The basal portions of the follicular epithelial cells are deeply infolded under certain circumstances, and they lie on a delicate basement membrane closely associated with fenestrated capillary endothelial cells.

2. Parafollicular Cells.

a. Parafollicular cells may occur in isolated nests between follicles or may be included within the basement membrane of the follicle.
b. There usually is a follicular epithelial cell located between the lumen of the follicle and parafollicular cells, and this cell rests on the follicular basement membrane.

c. Parafollicular cells have an abundant RER, a well-developed Golgi apparatus, and numerous large vesicles. These vesicles range in diameter from 100 to 200 nm and are filled with a flocculent material that presumably represents stored calcitonin or a precursor of it.

D. **HISTOPHYSIOLOGY.** The control of thyroid hormone secretion is not well understood. In essence, it involves two steps: an **exocrine** one, in which thyroglobulin is secreted into the colloid-filled follicles; and an **endocrine** one, in which this resorbed thyroglobulin is degraded into active **thyroxine** and then released into the blood.

1. **Exocrine Phase.**
   - a. Thyroglobulin is synthesized in the follicular epithelial cells from amino acids and iodine that accumulate within these cells by active transport.
     - (1) The amino acids are assembled into polypeptide chains in the RER and have sugar residues added to them.
     - (2) Mannose is added while they are in the RER, and fucose and galactose are added when they are in the Golgi.
   - b. Once synthesized, the thyroglobulin is expelled into the lumen of the follicle.
   - c. Interestingly, the thyroglobulin is iodinated after it is secreted, presumably in the lumen of the follicle by a **peroxidase** enzyme elaborated and secreted by the same follicular epithelial cells. The control of the secretion-iodination process is poorly understood.

2. **Endocrine Phase.**
   - a. When there is a need to increase the levels of thyroxine in the blood, the follicular epithelial cells are stimulated to engulf stored thyroglobulin.
     - (1) Phagosomes containing thyroglobulin fuse with lysosomes to produce phagolysosomes.
     - (2) In these phagolysosomes, thyroglobulin is degraded into its constituent amino acids.
   - b. The iodinated amino acids, **L-thyroxine (T_4)** and **3,5,3′-triiodo-L-thyronine (T_3)**, are released from the epithelial cells into the capillaries surrounding them, and the other amino acids of thyroglobulin are retained by the follicular epithelial cells for reuse in new thyroglobulin synthesis.

3. **Feedback Regulation of Thyroxine Levels.**
   - a. Both the hypothalamus and the hypophysis are thought to be involved in the regulation of thyroxine in the blood.
     - (1) Low blood thyroxine levels result in an indirect stimulation of the hypophysis by **thyrotropin releasing hormone (TRH)**, which is secreted by the hypothalamus.
     - (2) This TRH secretion stimulates direct secretion of **thyroid-stimulating hormone (TSH)** by the adenohypophysis.
   - b. TSH stimulates thyroid follicular epithelial cells by binding to their cell-surface TSH receptors and increasing the cyclic adenosine monophosphate levels inside the epithelial cells.
     - (1) TSH-stimulated cells show increased apical phagocytic activity and apical migration of lysosomes, eventually resulting in increased thyroxine levels in the blood.
     - (2) TSH stimulation also causes increased thyroglobulin production and even thyroid hypertrophy, if it continues indefinitely.
   - c. When thyroxine levels rise, TRH and TSH production falls, again resulting in a decrease in thyroxine levels.

4. **C-Cell Function.**
   - a. Parafollicular cells secrete **calcitonin**, which inhibits the resorption of bone calcium by osteocytes and osteoclasts. Calcitonin is antagonistic to **parathormone**, the secretion of the parathyroid glands.
   - b. When blood calcium levels are high, calcitonin is secreted and bone resorption decreases, eventually resulting in a fall in blood calcium.
   - c. If animals are artificially administered calcium by injection, their parafollicular cells rapidly degranulate, presumably due to their loss of calcitonin.

III. PARATHYROIDS

A. DISTRIBUTION

1. Usually there are four parathyroid glands embedded on the dorsal (or posterior) surface of the thyroid gland.

2. In many individuals, there is a set of one superior parathyroid and one inferior parathyroid on both the left and right sides of the body.

3. The distribution and even the number of parathyroids, however, are highly variable.

B. EMBRYOLOGY

1. The superior parathyroids are thought to arise from the left and right **fourth pharyngeal pouches**.
2. The inferior parathyroids, which lie immediately adjacent to the thymus, are thought to arise from the left and right **third pharyngeal pouches**, along with the thymic rudiment.

C. HISTOLOGY

1. Each parathyroid has a delicate connective tissue capsule and thin trabeculae that divide it into a series of poorly defined lobules.
2. Beginning at puberty, the central portion of a parathyroid may become infiltrated with **fat cells**.
3. **Chief cells** and **oxyphil cells** comprise the rest of the parathyroid parenchyma, along with **transitional cells** that are intermediate between chief cells and oxyphil cells.

D. PARATHYROID CELLS (Fig. 18-2)

1. **Chief Cells.**
   a. Chief cells outnumber oxyphil cells.
   b. A chief cell contains a large nucleus with several nucleoli, abundant mitochondria, RER, and glycogen granules.
   c. This cell is believed to have other cytoplasmic granules that are thought to represent the **parathormone** secretion of the parathyroid.
2. **Oxyphil Cells.**
   a. Oxyphil cells first appear during puberty, and they increase in number with an individual's increasing age.

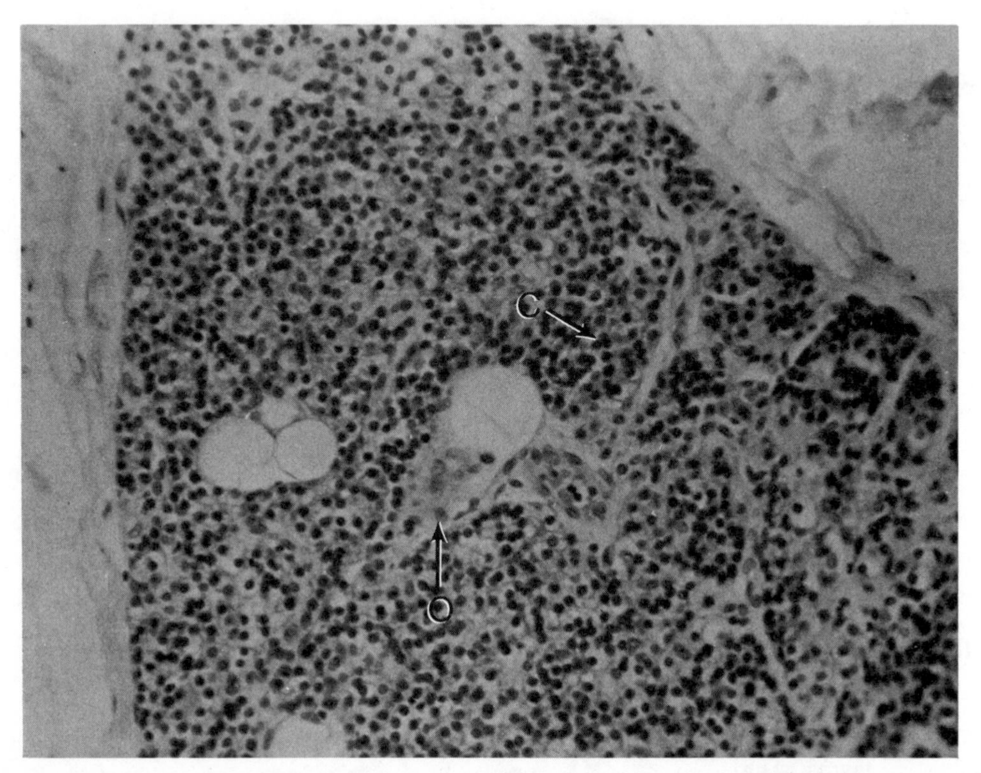

Figure 18-2. Light micrograph of a human parathyroid gland. Chief cells (C) and oxyphil glands (O) are visible. (Reprinted with permission from Johnson KE: *Histology: Microscopic Anatomy and Embryology*. New York, John Wiley, 1982, p 269.)

b. They are larger than chief cells and have an abundance of large mitochondria, giving them an eosinophilic cytoplasm.

c. Oxyphils exist in three forms: as solitary cells, as small bunches, and as small distinct nodules surrounded by chief cells.

3. Transitional Cells.

a. The wide variety of cell types intermediate between chief cells and oxyphil cells are described as transitional cells.

b. It is currently believed, however, that the different cell types within the parathyroid simply represent different functional states of chief cells.

E. PARATHORMONE

1. Parathyroid hormone, parathormone (PTH), is a small polypeptide with a molecular weight just under 10,000 daltons.

2. PTH has a crucial role in maintaining the serum calcium level.

a. PTH causes increased resorption of bone by osteocytes and osteoclasts and decreases renal tubule calcium excretion.

b. The net effect of this hormone is antagonistic to calcitonin (i.e., it raises serum calcium levels).

3. The production of PTH is **directly responsive** to serum calcium levels (i.e., PTH secretion increases when serum calcium falls and decreases when serum calcium rises).

4. PTH also inhibits renal tubule resorption of phosphate ions. In this manner it serves to regulate serum phosphate levels.

STUDY QUESTIONS

Directions: Each question below contains four suggested answers of which **one or more** is correct. Choose the answer

A if **1, 2, and 3** are correct
B if **1 and 3** are correct
C if **2 and 4** are correct
D if **4** is correct
E if **1, 2, 3, and 4** are correct

1. Components of thyroid follicular epithelial cells include

(1) abundant lysosomes
(2) abundant rough endoplasmic reticulum
(3) apical microvilli
(4) deep basal infoldings

2. True statements concerning parathyroid glands include which of the following?

(1) The inferior glands are derived from the third pharyngeal pouch
(2) The superior glands are derived from the fourth pharyngeal pouch
(3) The inferior glands are derived from the same pharyngeal pouch as the thymus
(4) Oxyphils decrease in number as an individual ages

Continued on next page

Directions: The group of questions below consists of lettered choices followed by several numbered items. For each numbered item select the **one** lettered choice with which it is **most** closely associated. Each lettered choice may be used once, more than once, or not at all.

Questions 3–6

For each description of a site or structure of a thyroid cell, choose the lettered component shown in the diagram below.

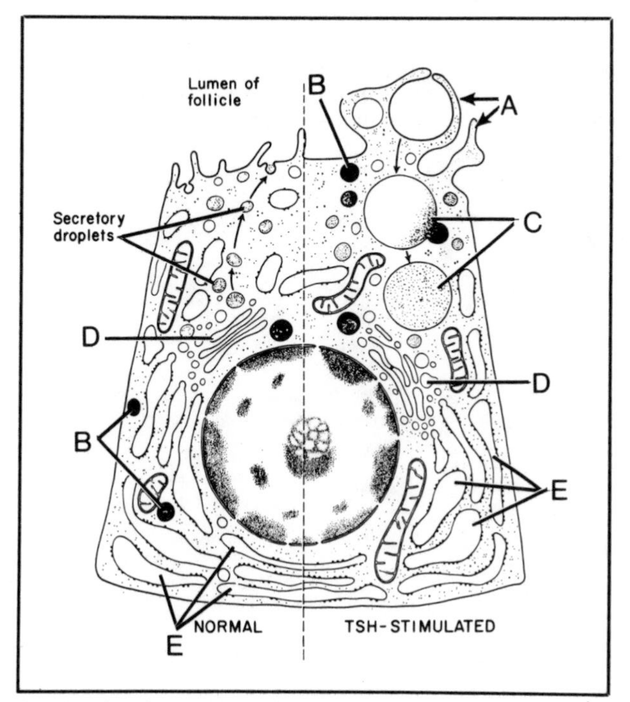

Reprinted with permission from Long DW, Jones AL: *Rec Progr Hormone Res* 25:315–380, 1969. ©Academic Press.

3. Results from fusion of a phagosome and a lysosome

4. Surface elaboration on the follicular apex which increases in size in response to thyroid-stimulating hormone (TSH)

5. Lysosomes are synthesized here, and thyroglobulin synthesis is completed here

6. Membrane-delimited packet of hydrolytic enzymes

ANSWERS AND EXPLANATIONS

1. The answer is E (all). (*II C 1*) Thyroid follicular epithelial cells secrete thyroglobulin into the follicles and thyroxine into capillaries surrounding the follicles. These epithelial cells have abundant rough endoplasmic reticulum to aid in thyroglobulin synthesis. They have apical microvilli and lysosomes for converting stored thyroglobulin into thyroxine. Their deep basal infoldings probably are used for active transport of substances into and out of these cells.

2. The answer is A (1, 2, 3). (*IV B*) The superior parathyroids are derived from the fourth pharyngeal pouch. The thymus and inferior parathyroids are derived from the third pharyngeal pouch. Oxyphils increase in number as an individual ages.

3–6. The answers are: 3-C, 4-A, 5-D, 6-B. (*II C 1; II D 2*) Thyroid-stimulating hormone (TSH) stimulation causes an elaboration of apical microvilli (A) and production of phagosomes. When phagosomes fuse with lysosomes (B), phagolysosomes (C) result. The Golgi apparatus (D) is involved in lysosome synthesis. Thyroglobulin synthesis is begun in the rough endoplasmic reticulum (E) and completed in the Golgi apparatus.

19
Adrenal Glands

I. INTRODUCTION

A. GROSS ANATOMY

1. Adrenal glands are paired organs that rest on top of the kidneys. In humans, they also are referred to as the **suprarenal glands**.
2. They are roughly triangular and lie embedded in the retroperitoneal fat pads with the kidneys.
3. There are approximately 10 g of adrenal tissue in a normal adult, although size and weight of an adrenal gland can vary enormously depending on its functional state.
4. Upon gross inspection of a cut specimen, an adrenal gland shows an outer **adrenal cortex** and an inner **adrenal medulla**.

B. EMBRYOLOGY

1. Adrenal cortical tissue arises from **mesothelial cells** lining the coelomic cavity medial to the urogenital ridges.
2. The adrenal medulla is derived from the **neural crest**.

C. BLOOD SUPPLY

1. Blood comes to these glands from three different sources.
   a. The superior suprarenal arteries (i.e., branches of the inferior phrenic arteries) are the primary blood source.
   b. The middle suprarenal arteries (i.e., branches of the aorta) and inferior suprarenal arteries (i.e., branches of the renal arteries) also contribute to the blood supply.
2. Suprarenal veins empty from the right adrenal gland into the inferior vena cava and from the left adrenal gland into the renal vein.

D. FUNCTION

1. The cortex actively secretes a variety of **steroid hormones**.
2. The adrenal medulla actively secretes **catecholamines** into the systemic circulation.
3. The adrenal cortex and adrenal medulla interact with the kidneys, together functioning to maintain blood chemistry.

II. HISTOLOGY AND EMBRYOLOGY OF ADRENAL CORTEX

A. HISTOLOGY

1. The adrenal cortex is divided into three conspicuous layers (Fig. 19-1).
   a. **Zona Glomerulosa.**
      (1) The outer zona glomerulosa represents about 15 percent of the thickness of the cortex and is composed of round clusters of cells that are 12–15 μm in diameter.
      (2) Each of these cells has a small round nucleus and a small amount of cytoplasm with a few scattered lipid droplets.
   b. **Zona Fasciculata.**
      (1) The zona fasciculata comprises about 75 percent of the cortical thickness and is composed of relatively straight rows of cells that are about 20 μm in diameter.
      (2) These cells have large numbers of lipid vacuoles and often present a washed-out, spongy appearance in routine histologic preparations due to the extraction of lipids by organic solvents.

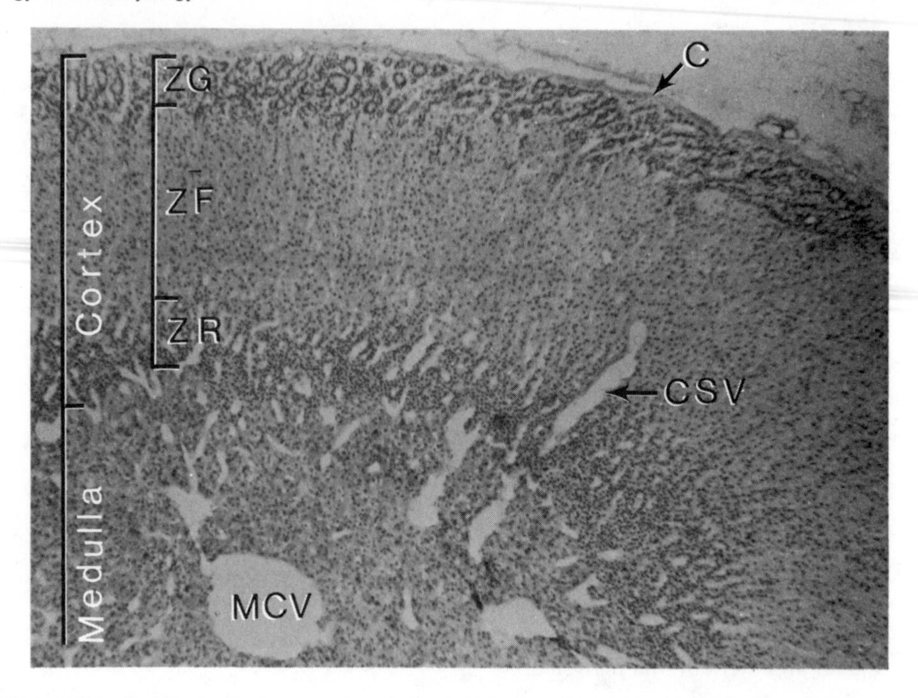

Figure 19-1. Light micrograph of a human adrenal gland. This gland has a thin connective tissue capsule (*C*) and a cortex and medulla. The cortex is divided into a zona glomerulosa (*ZG*), zona fasciculata (*ZF*), and zona reticularis (*ZR*). Examples of a cortical shunt vessel (*CSV*) and a medullary central vein (*MCV*) are present. (Reprinted with permission from Johnson KE: *Histology: Microscopic Anatomy and Embryology.* New York, John Wiley, 1982, p 276.)

c. Zona Reticularis.

(1) In the zona reticularis, cells are arranged in anastomosing cords rather than in straight columns.

(2) These cells are smaller and contain fewer lipid droplets than the cells of the zona fasciculata.

2. Thin capillaries abound in the spaces between and around cells all through the adrenal cortex. Branches of the suprarenal arteries ramify into capillaries that filter down through cortical sinuses and drain into medullary sinuses, which in turn drain into the systemic circulation.

3. A few cortical arteries take a direct course to the medullary sinuses by a kind of cortical shunt.

B. EMBRYOLOGY

1. At about five weeks gestation, **mesothelial cells** lining the coelomic cavity invade the mesenchyme of the body wall and form the **fetal cortex**.

2. Mesothelial cells continue to leave the lining of the coelom and encapsulate the fetal cortex. This second wave of invading cells eventually give rise to the **definitive** or **adult cortex.**

3. The fetal cortex undergoes a gradual regression; at birth, there is a presence of both fetal cortex and definitive cortex.

4. By one year of age, the fetal cortex has been almost entirely replaced by the definitive cortex.

5. It is not until about four years of age, however, that the fetal cortex is completely replaced by a definitive cortex that shows the characteristic zona glomerulosa, fasciculata, and reticularis.

III. ULTRASTRUCTURE OF ADRENAL CORTEX

A. COMMON FEATURES

1. The cells of the adrenal cortex are involved in the secretion of steroids and, thus, are similar in ultrastructure to the cells of the corpus luteum and the Leydig cells in the testis. For example, these large cells of the adrenal cortex often have numerous lipid droplets, which probably represent stored cholesterol esters and other steroid precursors.

2. They also have an abundance of smooth endoplasmic reticulum (SER) and numerous large mitochondria with peculiar tubular cristae. The significance of this mitochondrial ultrastructure is obscure, but it is widely distributed in cells engaged in steroid synthesis and secretion.

B. ZONAL VARIATION

1. The cells in the **zona glomerulosa** have few lipid droplets, long thin mitochondria, an abundance of SER, and little rough endoplasmic reticulum (RER).
2. In contrast, the cells of the **zona fasciculata** have huge numbers of lipid droplets, round fat mitochondria, and a considerable amount of both SER and RER (Fig. 19-2).
3. The cells in the **zona reticularis** are much like those in the zona fasciculata except that they have fewer lipid droplets, smaller mitochondria, and substantial accumulations of **lipofuscin** pigments that give this layer a brown tint.
   a. Lipofuscin granules may represent accumulated by-products of lipid metabolism.
   b. They increase in size and number with age and often have been referred to as accumulated break-down products of cells related to cellular wear and tear over time.

C. VARIATION WITH RESPECT TO FUNCTION

1. When a cortical cell is stimulated by injections of adrenocorticotropic hormone (ACTH), its number of lipid droplets decreases and quantity of SER increases, presumably as a reflection of increased steroid synthesis.
   a. The exact intracellular routes of steroid synthesis and secretion currently are under investigation.
   b. Presently, most researchers believe that steroid secretion by cortical cells is continuous and does not involve the formation and release of specific granules of hormone, as is the case in the adenohypophysis and in islet tissue of the pancreas.
2. The endothelial cells of capillaries in the adrenal cortex have fenestrae that are closed by electron-dense diaphragms. These diaphragms have a fine structure different from that of a unit membrane.
   a. The capillaries are intimately associated with a continuous thin basement membrane, and they have numerous fine reticular fibers near them.
   b. The reticular fibers are components of the connective tissue that segregates individual groups of cortical cells.

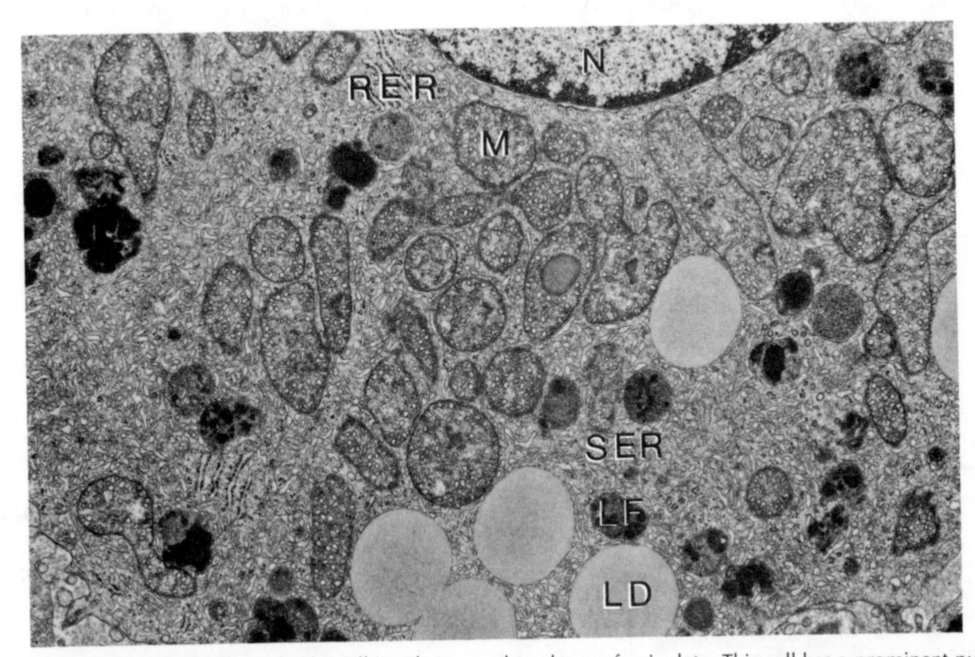

Figure 19-2. Electron micrograph of a cell in a human adrenal zona fasciculata. This cell has a prominent nucleus (*N*), a considerable amount of rough endoplasmic reticulum (*RER*) and smooth endoplasmic reticulum (*SER*), mitochondria (*M*), lipid droplets (*LD*), and lipofuscin droplets (*LF*). (Courtesy of Dr. John A. Long, Department of Anatomy, University of California, San Francisco.)

3. The steroid-secreting cells of the cortex also are surrounded by a continuous basement membrane. This is not so peculiar, since they are epithelial derivatives and so retain some features of epithelial organization.

IV. HISTOPHYSIOLOGY OF ADRENAL CORTEX

A. ADRENOCORTICAL HORMONES

1. The cells of the adrenal cortex secrete a variety of steroid hormones usually classified as **glucocorticoids** (e.g., cortisol and corticosterone), **mineralocorticoids** (e.g., aldosterone), and weak **androgens** (e.g., dehydroepiandrosterone).
   a. Glucocorticoids have profound effects on carbohydrate metabolism and, thus, affect blood-sugar level.
   b. Mineralocorticoids have effects on renal tubular function and, thus, are involved in regulating blood ion composition.
   c. Androgens are sex steroids and have effects on secondary sex characteristics and on the gonads.

2. In addition to an anatomic zonation of the cortex, there also is a functional zonation.
   a. The zona glomerulosa is primarily responsible for the elaboration of **aldosterone**, one of the mineralocorticoids.
   b. In contrast, the zona fasciculata and zona reticularis are primarily responsible for the elaboration of **cortisol** and **dehydroepiandrosterone**. There also is evidence that the zona reticularis is more active in androgen secretion than is the zona fasciculata.

3. Steroids are a closely related family of molecules, which are synthesized from cholesterol precursors.

4. The **enzymes** required for steroid biosynthesis are located either in mitochondria or in SER membranes, and they may have specific zonal distributions. The following are examples.
   a. The enzymes used in the initial steps of steroid biosynthesis (e.g., **pregnenolone synthetase**, which converts cholesterol to pregnenolone) are found in all cortical zones.
   b. The enzymes used for the final steps of aldosterone synthesis are restricted to the zona glomerulosa, as is aldosterone itself.
   c. A key enzyme in the cortisol pathway, 17-α-hydroxylase, is restricted to the zona fasciculata and zona reticularis.

B. CONTROL OF ADRENOCORTICAL FUNCTION comes from two distinct sources.

1. The zona fasciculata and the zona reticularis are under the control of the hypothalamus and the adenohypophysis via corticotropin releasing factor (CRF) and ACTH.

2. The zona glomerulosa, which produces aldosterone, is regulated by the renin-angiotensin system and the kidneys.
   a. The juxtaglomerular apparatus releases **renin**, a hydrolytic enzyme that converts serum **angiotensinogen** to **angiotensin I**.
   b. In the lungs, angiotensin I is converted to **angiotensin II.** This hormone stimulates the zona glomerulosa to secrete aldosterone.
   c. Aldosterone promotes sodium ion and water retention in the kidneys and, thus, increases blood pressure.
   d. An increase in blood pressure then prevents further renin production.

C. ACTH

1. The remainder of the adrenal cortex is under control of ACTH. In hypophysectomized animals, the zona glomerulosa remains unchanged but the zona fasciculata and zona reticularis undergo involution.
   a. If ACTH is administered, the involution is reversed.
   b. If ACTH is administered in excess, there is a striking loss of lipid droplets and a striking increase in steroid levels in the serum.

2. The levels of glucocorticoids in the circulation have a direct effect on both hypothalamic secretion of CRF and hypophyseal secretion of ACTH.
   a. When glucocorticoid levels are high, CRF and ACTH secretion are suppressed. As a result, glucocorticoid levels drop.
   b. When glucocorticoid levels fall, the feedback inhibition is removed and CRF is secreted, stimulating ACTH secretion.
   c. The ACTH then is released and carried to the adrenal cortex where it causes glucocorticoid secretion.

d. Glucocorticoids have complex systemic effects, the net result of which is a raised glucose level in the blood.

V. ADRENAL MEDULLA

A. HISTOLOGY

1. The adrenal medulla is composed of masses of catecholamine-containing cells, which are highly modified postganglionic cells surrounded by complex networks of blood vessels. Cell bodies of preganglionic sympathetic neurons also exist in this region. These neurons innervate the catecholamine-containing cells and control catecholamine secretion.

2. The catecholamine-containing cells sometimes are called **chromaffin cells** because they stain somewhat brown when exposed to solutions of dichromate ions.
a. Chromaffin granules are probably caused by the reaction of the dichromate with catecholamine.
b. The granules also have a specific **autofluorescence** that makes them detectable in an appropriately equipped microscope after aldehyde fixation.

B. ULTRASTRUCTURE

1. With the electron microscope, cells of the adrenal medulla reveal prominent membrane-delimited granules that are approximately 200 nm in diameter.
a. These contain a moderately electron-dense material that is thought to represent a storage form of catecholamine.
b. In addition to catecholamines, these granules are thought to contain enzymes responsible for the synthesis of catecholamines from dopamine.
c. These granules exist in two distinct forms.
(1) One has a high electron density and is thought to store **norepinephrine**.
(2) The other has a lower electron density and is thought to store **epinephrine**.
d. These granules are formed in a prominent perinuclear Golgi apparatus.

2. The parenchymal cells of the adrenal medulla also have modest amounts of RER and unremarkable mitochondria.

3. Recent electron-microscopic studies have revealed synaptic contacts between preganglionic cholinergic sympathetic fibers and the catecholamine-containing medullary cells.

D. HISTOPHYSIOLOGY

1. Catecholamine secretion by the cells of the adrenal medulla has dramatic systemic effects.
a. Epinephrine secretion leads to dramatic increases in heart rate and cardiac output.
(1) It increases blood flow in the skeletal muscles of the limbs and decreases splanchnic perfusion.
(2) It increases basal metabolic rate and stimulates glycogenolysis in the liver.
(3) Epinephrine secretion also can stimulate ACTH release from the adenohypophysis, which promotes glucocorticoid secretion from the adrenal cortex and thereby indirectly stimulates gluconeogenesis.
b. Norepinephrine increases blood pressure by its vasoconstrictive action.

2. Epinephrine and norepinephrine release is under the control of the central nervous system by way of the preganglionic sympathetic fibers. During periods of sudden fright or stress, epinephrine secretion occurs suddenly, preparing the person to deal with the stress situation by increasing cardiac output, respiratory rate, and skeletal muscle perfusion in anticipation of physical stress.

3. Blood vessels that percolate down through the cortex (and thus contain large amounts of corticosteroids) are closely associated with the epinephrine-secreting medullary cells.

4. In contrast, vessels that shunt the cortex and drain directly into the medulla are closely associated with norepinephrine-secreting medullary cells.

STUDY QUESTIONS

Directions: Each question below contains five suggested answers. Choose the **one best** response to each question.

1. All of the following statements concerning the zona fasciculata are true EXCEPT

(A) cells have little smooth endoplasmic reticulum
(B) mitochondria are ovoid
(C) cells are sensitive to adrenocorticotropic hormone (ACTH) stimulation
(D) cells have numerous large lipid droplets
(E) it is the thickest cortical zone

2. All of the following statements concerning the embryonic development of the adrenal gland are true EXCEPT

(A) the fetal cortex is formed from proliferating coelomic mesothelial cells
(B) the adrenal medulla forms from the neural crest
(C) definitive adult-type zonation is present at birth
(D) parts of the fetal cortex persist after birth
(E) the definitive cortex is a mesodermal derivative

Directions: The question below contains four suggested answers of which **one or more** is correct. Choose the answer

A if **1, 2, and 3** are correct
B if **1 and 3** are correct
C if **2 and 4** are correct
D if **4** is correct
E if **1, 2, 3, and 4** are correct

3. True statements concerning adrenocortical function include which of the following?

(1) Renin secretion promotes the formation of angiotensin I
(2) Aldosterone secretion causes sodium and water retention
(3) A decrease in blood pressure results in renin secretion
(4) Aldosterone secretion increases blood pressure

Directions: The group of questions below consists of lettered choices followed by several numbered items. For each numbered item select the **one** lettered choice with which it is **most** closely associated. Each lettered choice may be used once, more than once, or not at all.

Questions 4–8

For each description of a site or structure of a cell from the zona fasciculata of the adrenal cortex, choose the appropriate lettered component shown in the drawing below.

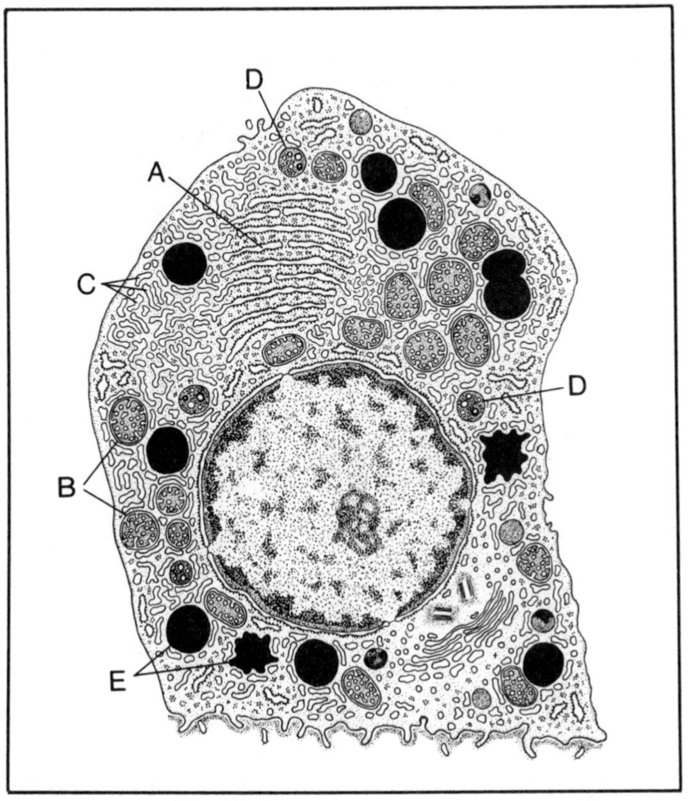

Reprinted with permission from Lentz TL: *Cell Fine Structure*. Philadelphia, WB Saunders, 1971.

4. Storage form of cholesterol and cholesterol esters

5. Has tubular cristae and is prominent in the cells of steroid-secreting tissue

6. Lipofuscin

7. Rough endoplasmic reticulum

8. Many enzymes for steroid biosynthesis are located here

ANSWERS AND EXPLANATIONS

1. The answer is A. (*III A, B*) All cells involved in steroid biosynthesis have an abundance of smooth endoplasmic reticulum (SER). The SER membranes contain enzymes for steroid synthesis.

2. The answer is C. (*II B*) The fetal cortex and definitive cortex both are formed from coelomic mesothelial cells (mesoderm). The adrenal medulla comes from the neural crest. Definitive, adult-type zonation does not develop until well after birth.

3. The answer is E (all). (*IV B 2*) Decrease in blood pressure, sensed in the juxtaglomerular apparatus of the kidney, causes renin release. The renin converts angiotensinogen to angiotensin I. Angiotensin I is converted to angiotensin II, and this stimulates aldosterone secretion. Aldosterone causes increased salt retention and increases blood pressure.

4–8. The answers are: 4-E, 5-B, 6-D, 7-A, 8-C. (*III A, B*) This steroid-secreting cell contains both rough (A) and smooth (C) endoplasmic reticulum. The latter is the site of many enzymes involved in steroid synthesis. This cell also has many lipid droplets (E), which represent cholesterol and cholesterol esters used in steroid synthesis, and mitochondria with tubular cristae (B). Lipofuscin (D) is common in these cells.

20 Pituitary Gland

I. INTRODUCTION

A. MAJOR DIVISIONS

1. The **hypophysis** or **pituitary gland** is a small complex organ located at the base of the brain. It is divided into two major components or **lobes**.
   a. The **anterior lobe** or **adenohypophysis** is derived from the roof of the oral cavity and loses all connection to its primordial epithelium.
   b. The **posterior lobe** or **neurohypophysis** is derived from the infundibulum, a process projecting down from the floor of the diencephalon. In the adult, the neural character of the neurohypophysis is maintained.

2. These lobes also are divided into structural components.

B. ENDOCRINE REGULATION

1. **General Considerations.**
   a. The pituitary gland is essential to the regulation of body functions. It serves as the master gland of the endocrine system, producing hormones that regulate hormone production in many other endocrine organs.
   b. In most cases, hormone production by a target organ alters the function of the pituitary gland, thus regulating the production of the pituitary hormone by a feedback mechanism.

2. **The Feedback System.**
   a. Releasing hormones are carried, by a portal circulation, from the neurohypophysis to the adenohypophysis, where they effect the release of hormones from the cells of the adenohypophysis.
   b. The hormones of the adenohypophysis then are secreted from epithelial cells through a dense network of fenestrated capillaries, and from there they are transported by way of the systemic circulation to such distant target organs as the thyroid and the testis.
   c. The hormones released from target organs of the pituitary regulate everything from basal metabolic rate to growth rate and also feed back on the neurohypophysis, creating a complete, self-regulatory cybernetic loop.

II. GROSS ANATOMY AND SUBDIVISIONS

A. GROSS ANATOMY

1. The pituitary gland is a roughly mushroom-shaped gland and is approximately 1.5 cm in length. It weighs approximately 1 g but may be considerably heavier in women who have had several children.

2. The pituitary gland lies at the base of the third ventricle of the brain in a small depression deep within the sphenoid bone. This depression, shaped like a Turkish saddle, is appropriately named the **sella turcica.**

3. The dura mater is discontinuous at the hypophysis but surrounds the stalk of the pituitary gland and closes the sella with a membranous covering called the **diaphragma sellae.**

4. The anterior lobe of the pituitary gland appears pinkish, is vascularized, and is relatively easy to macerate.

5. The posterior lobe is whitish, contains nerve fibers, and is considerably tougher than the anterior lobe.

B. SUBDIVISIONS (Fig. 20-1)

1. **Adenohypophysis.**
   a. The bulk of the adenohypophysis is occupied by the **pars distalis**.

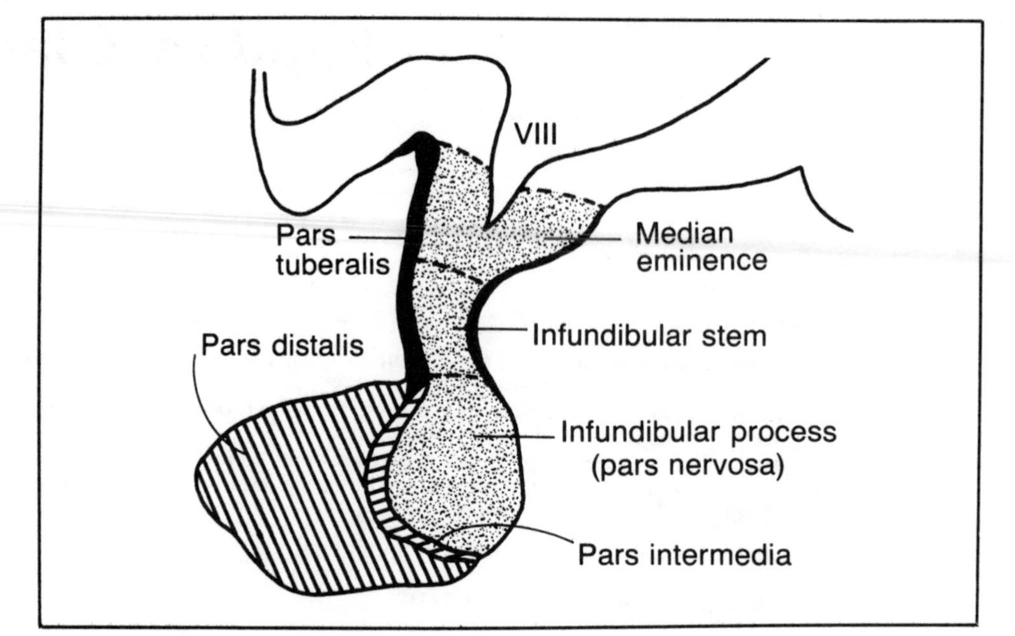

Figure 20-1. Diagram illustrating the major subdivisions of the neurohypophysis (*stippled area*) and adenohypophysis (*black and striped areas*). The median eminence is near the third ventricle (*VIII*) of the brain. (Reprinted in modified form with permission from Bloom W, Fawcett DW: *A Textbook of Histology*. Philadelphia, WB Saunders, 1975.)

b. The **pars intermedia** lies between the pars distalis and parts of the neurohypophysis.

c. The **pars tuberalis** is a collar of tissue surrounding the stalk that connects the neurohypophysis to the rest of the brain.

2. **Neurohypophysis.**

a. The **pars nervosa** (also called the **infundibular process**) is the portion adjacent to the adenohypophysis.

b. The **infundibular stem** is an intermediate connecting piece joining the pars nervosa to the floor of the brain.

c. The **median eminence** is another neurohypophyseal component.

(1) It is part of the **tuber cinereum**. Neurons with their cell bodies in brain nuclei located in or near the tuber cinereum (e.g., supraoptic and paraventricular nuclei) have axons terminating in the median eminence and other parts of the neurohypophysis.

(2) The median eminence comprises the floor of the third ventricle and is poorly developed in humans.

III. EMBRYOLOGY

A. ORIGIN. The pituitary gland develops from two separate rudiments, both of which are derived from ectoderm.

1. The adenohypophysis is derived from Rathke's pouch.

2. The neurohypophysis is derived from the infundibulum, which is part of the diencephalon.

B. DEVELOPMENT OF INFUNDIBULUM

1. The infundibulum is a small evagination of the floor of the diencephalon.

2. As the brain grows, the infundibulum remains connected to it. Later, this process elongates and divides into the infundibular stem and median eminence.

C. DEVELOPMENT OF RATHKE'S POUCH

1. Rathke's pouch is an evagination of the roof of the mouth. After it forms, Rathke's pouch comes in close association with the infundibulum. These two separate rudiments join long before the formation of the sphenoid bone and the secondary palate.

2. Next, the connection between the roof of the mouth and Rathke's pouch is rapidly

obliterated. (In certain pathologic cases, however, small functioning strands of pituitary-like cells may be seen in the pharynx and possibly as nests deep within recesses of the sphenoid bone.)

3. Rathke's pouch becomes a closed vesicle lined by epithelium and closely associated with the infundibulum. Three changes then occur.
   a. The anterior portion of this vesicle grows enormously, eventually giving rise to all of the anterior lobe. The pars distalis of the adenohypophysis develops from the anterior wall of this vesicle.
   b. The posterior wall of this vesicle fuses with the infundibulum and becomes incorporated into it. This explains the later intimate association between the pars intermedia and the pars nervosa.
   c. The walls of the vesicle grow dorsolaterally to surround the infundibulum nearest the median eminence, forming the pars tuberalis.

IV. CIRCULATION

A. AFFERENT BLOOD SUPPLY

1. The pituitary receives blood from the internal carotid arteries. **Inferior hypophyseal arteries** branch from the internal carotids and carry blood to the bulk of the pars nervosa.

2. **Superior hypophyseal arteries**, branching from both the internal carotids and from the posterior communicating artery of the circle of Willis, supply the pars tuberalis, median eminence, and pars distalis. Superior hypophyseal arteries also supply capillaries in the pars tuberalis. These capillaries penetrate into the median eminence and infundibular stem of the neurohypophysis.

B. EFFERENT BLOOD SUPPLY

1. The capillaries of the median eminence are drained by the **hypophyseoportal system** which, in turn, drains into a capillary plexus serving most of the adenohypophysis, particularly the pars distalis. (A portal system connects two capillary beds directly.)

2. The capillary beds in the neurohypophysis and the adenohypophysis drain by separate hypophyseal veins into the cavernous sinus and eventually into the systemic circulation.

3. This vascular link between the median eminence and the pars distalis, via the hypophyseoportal system, is the anatomic basis for the functional integration of these two key endocrine tissues.

V. ADENOHYPOPHYSIS

A. **Hormones.** In spite of its size, the adenohypophysis produces seven different hormones. These hormones are

1. Adrenocorticotropic hormone (ACTH)
2. Luteinizing hormone (LH)
3. Follicle-stimulating hormone (FSH)
4. Growth hormone (GH), which is also known as somatotropin (STH)
5. Prolactin
6. Thyrotropin (TSH)
7. Lipotropin, which is a precursor of melanocyte-stimulating hormone (MSH)

B. HISTOLOGY

1. **Pars Tuberalis and Pars Intermedia.**
   a. The pars tuberalis is rather thin and unimportant in humans.
   b. The pars intermedia is poorly developed.
      (1) It has a few hormone-producing cells, specifically, chromophobic and basophilic cells.
      (2) Usually, the cells of the pars intermedia merge into the tissue of the pars nervosa.
      (3) When a remnant of the cleft of Rathke's pouch persists in the adult, the cells of the pars intermedia lie posterior to the cleft.

2. **Pars Distalis.**
   a. The pars distalis is the most complicated component of the pituitary gland. It is a large col-

lection of epithelial cells that are arranged in cords or clusters or that surround closed, colloid-filled follicles.

b. The pars distalis has little or no connective tissue, which explains why the tissue is so fragile. It does, however, have a rich network of large fenestrated capillaries surrounding the clusters and cords of epithelial cells. Reticular fibers and collagenous fibers are interspersed between cells and blood vessels.

c. Early on, scientists recognized that there were several cell types in the pars distalis, and they suspected that each type produced one of the hormone classes elaborated in this lobe.

d. Labeled antibodies to these hormones have been used as histochemical reagents and, coupled with the electron microscope, have allowed a functional identification of the different cell types in the pars distalis. Furthermore, these methods have shown that different cell types predominate in different regions of the lobe.

e. Classical histologists have distinguished the following three cell types and their functional subclasses, based on dye-binding affinities (Fig. 20-2).

(1) **Acidophils** and their two subclasses, **lactotrops** and **somatotrops**

(2) **Basophils** and their three subclasses, **corticotrops**, **thyrotrops**, and **gonadotrops**

(3) **Chromophobes**, which are functionally degranulated or inactive cells

3. Hormones. Each of the hormones of the pars distalis is produced by a specific cell type.

a. Corticotrops and ACTH.

(1) ACTH is a polypeptide hormone in the corticotrops in the pars distalis. It stimulates the glucocorticoid secretion from the adrenal cortex.

(2) ACTH is produced by one type of basophil that also produces a precursor to MSH known as lipotropin (β-LPH). Both β-LPH and ACTH share a large number of amino acids and have the same sequence.

(3) With the electron microscope, the basophils that produce ACTH and β-LPH are seen as large ovoid cells with prominent granules that stain immunohistochemically for their hormones (Fig. 20-3). Also, the cytoplasm of these basophils contains distinctive filaments that are 6–8 nm in diameter.

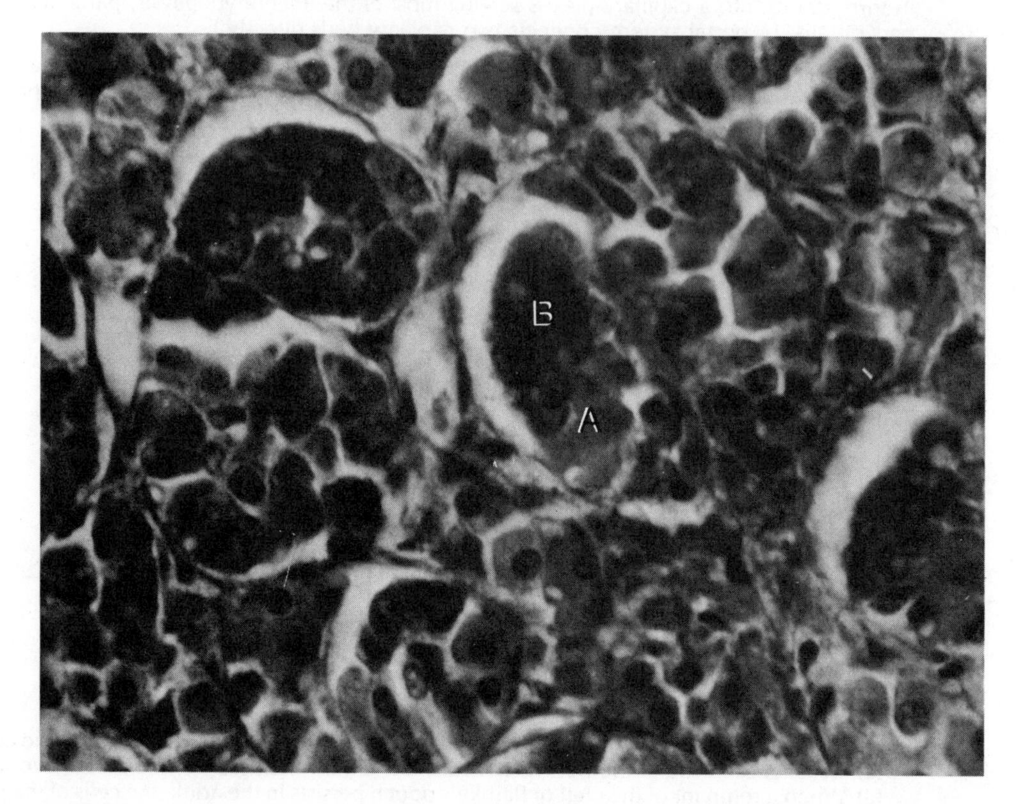

Figure 20-2. Light micrograph of clusters of acidophils (*A*) and basophils (*B*) in the human adenohypophysis. Note that thin slips of connective tissue surround these clusters. (Reprinted with permission from Johnson KE: *Histology: Microscopic Anatomy and Embryology*. New York, John Wiley, 1982, p 290.)

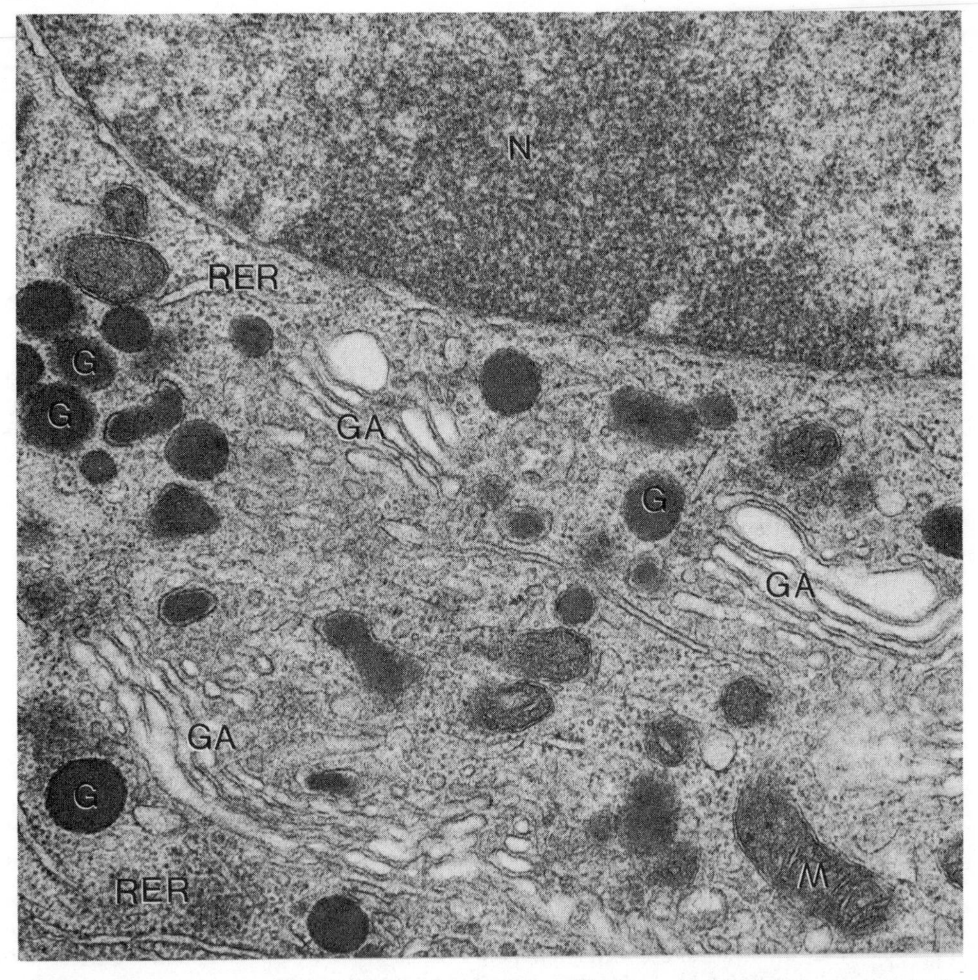

Figure 20-3. Electron micrograph of a large granule-containing cell. The nucleus (*N*) is in the upper right. Granules (*G*) of hormone, the Golgi apparatus (*GA*), mitochondria (*M*), and some rough endoplasmic reticulum (*RER*) are visible. Without specific immunochemical techniques, it is impossible to identify precisely this type of cell. (Courtesy of Dr. Raymond J. Walsh, Department of Anatomy, George Washington University.)

(4) These cells are the most common of the basophils and are most frequently encountered in the anterocentral part of the pars distalis.
(5) Tumors of these cells result in hypersecretion of ACTH and glucocorticoids and are commonly associated with Cushing's disease.
(6) Bilaterally adrenalectomized patients show hypertrophy of basophils. With the electron microscope these basophils reveal hypertrophic masses of filaments as well as small numbers of granules that still stain for ACTH and β-LPH.
(7) These changes are thought to be related to the loss of adrenal feedback upon the corticotrops that in turn secrete ACTH and β-LPH.

b. Gonadotrops, FSH, and LH.

(1) Gonadotropins (i.e., FSH and LH) are secreted by another type of basophil. FSH and LH are glycoprotein hormones that stimulate the gonads.
   (a) FSH stimulates the growth of follicles in the female and stimulates the spermatogenic epithelium in the male.
   (b) LH is required for ovulation and stimulates androgen secretion by testicular Leydig cells. For this reason, LH also is called interstitial cell-stimulating hormone.
(2) With the electron microscope, gonadotrops reveal moderately large granules. (They are intermediate in size between the large granules of the GH, ACTH, and β-LPH cells and the small granules of the TSH cells.)
(3) There is no evidence suggesting that gonadotrops are subdivided into cells that make either FSH or LH. Both appear to be made by a single morphologic entity.

(4) These basophils are small and sparse during childhood when there is little gonadal function.

(5) Gonadal steroid levels exert a negative feedback upon gonadotrops. Thus, during pregnancy when placental estrogen secretion is high, the production of FSH and LH is low and the gonadotrops themselves show functional regression.

c. **Thyrotrops and TSH.**

(1) TSH is secreted by a third class of basophils.

(2) TSH or **thyrotropin** is a glycoprotein contained in small granules within the thyrotrops. The granules of thyrotrops are smaller than the large granules of somatotrops and the intermediate granules of gonadotrops.

(3) Thyrotrops are enlarged and filled with drops of colloid in patients with primary thyroid failure.

(a) There is a chronic stimulation of thyrotrops by hypothalamic releasing hormones because the lack of thyroid function removes the feedback inhibition of thyrotropin releasing hormone (TRH) production. Thus, releasing hormones constantly are produced, and thyrotrops cannot accumulate large numbers of granules of TSH because they are chronically stimulated.

(b) In other conditions where thyroid function is elevated, high levels of thyroxine cause a lack of releasing hormone production, which in turn results in a lack of stimulation of thyrotrops.

(4) Thyrotrops undergo regression when there is hyperthyroidism unrelated to excessive TSH secretion.

(5) The thyroxine-TRH-TSH feedback system is a simple negative feedback loop.

(a) High levels of thyroxine inhibit TRH production by the hypothalamus.

(b) Low levels of TRH then result in poor TSH production.

(c) When the thyroxine levels fall, TRH is secreted.

(d) This causes TSH secretion, stimulating the thyroid to produce thyroxine, which enters the circulation and decreases TRH production once again.

d. **Lactotrops and Somatotrops.**

(1) One type of acidophil in the adenohypophysis, the lactotrop, is responsible for the production of prolactin, a low molecular weight polypeptide that stimulates glandular development and milk secretion by the mammary gland.

(2) Another type of acidophil in the pars distalis is the somatotrop. This cell has a well-developed endoplasmic reticulum and a large number of granules that are 35 nm in diameter.

(3) Somatotrops secrete GH, which causes growth of various portions of the body.

(a) In the absence of GH, a patient is small in stature due to a failure of somatotrops to secrete GH, a condition called pituitary **dwarfism**.

(b) If GH is secreted to excess during childhood, a patient will grow to large size, a condition called **gigantism**.

(c) In the adult, hypersecretion of GH will result in **acromegaly**, a condition that involves enlargement of some of the facial bones as well as the hands and feet.

VI. RELEASING HORMONES AND THE PARS DISTALIS

A. FUNCTIONAL INTERACTION

1. The normal function of the pars distalis depends on its functional interconnection to the median eminence of the hypothalamus.

2. In animal studies when the pars distalis is removed from the sella turcica and transplanted elsewhere in the body, it loses its complement of basophils. When transplanted back to the region of the median eminence, however, the basophils in the pars distalis return, probably because they have redifferentiated from a dedifferentiated population of cells under the influence of **hypothalamic releasing hormones**.

B. THE PORTAL SYSTEM AND RELEASING HORMONES

1. The **hypophyseoportal system** drains the hypothalamus and immediately enters the adenohypophysis, specifically the pars distalis.

2. The neurons responsible for the secretion of releasing hormones have cell bodies in the supraoptic and paraventricular nuclei and have axons that project away from these nuclei and end on capillaries in the median eminence.

a. The neurons secrete releasing hormones into the capillaries from axonal terminations that end directly on the capillaries.

b. The releasing hormones then are carried directly to the capillaries that surround the cords and nests of cells within the pars distalis.

3. Thyrotropin releasing hormone (TRH), gonadotropin releasing hormone (Gn-RH), luteinizing hormone releasing hormone (LH-RH), and an inhibitor of GH secretion called **somatostatin**, all have been isolated and identified chemically.

4. **Birth Control Pills and LH-RH.**
   a. In the female, when estradiol levels are elevated, there is a block on LH-RH secretion and thus no LH formation. LH formation is required for ovulation. This is the basis for the action of most oral contraceptives.
   b. Basically, a high estrogen level results in inhibition of hypothalamic LH-RH production, a lack of LH secretion, and thus a lack of ovulation itself.

VII. NEUROHYPOPHYSIS

A. MICROSCOPIC ANATOMY

1. This portion of the pituitary gland actually is an extended lobe of the base of the brain.

2. The bulk of the pars nervosa of the pituitary gland is occupied by a large number of unmyelinated nerve fibers gathered together in a conspicuous **hypothalamicohypophyseal tract**.

3. In addition, there are large numbers of non-neuronal cells called **pituicytes** in the pars nervosa. Presumably these cells are homologous to the neuroglia.

4. It is believed that the neurons with cell bodies in the hypothalamic nuclei and axons in the hypothalamicohypophyseal tract may secrete the hormones of the pars nervosa from membrane-delimited vesicles that are in the axon terminations closely associated with capillaries (Fig. 20-4).

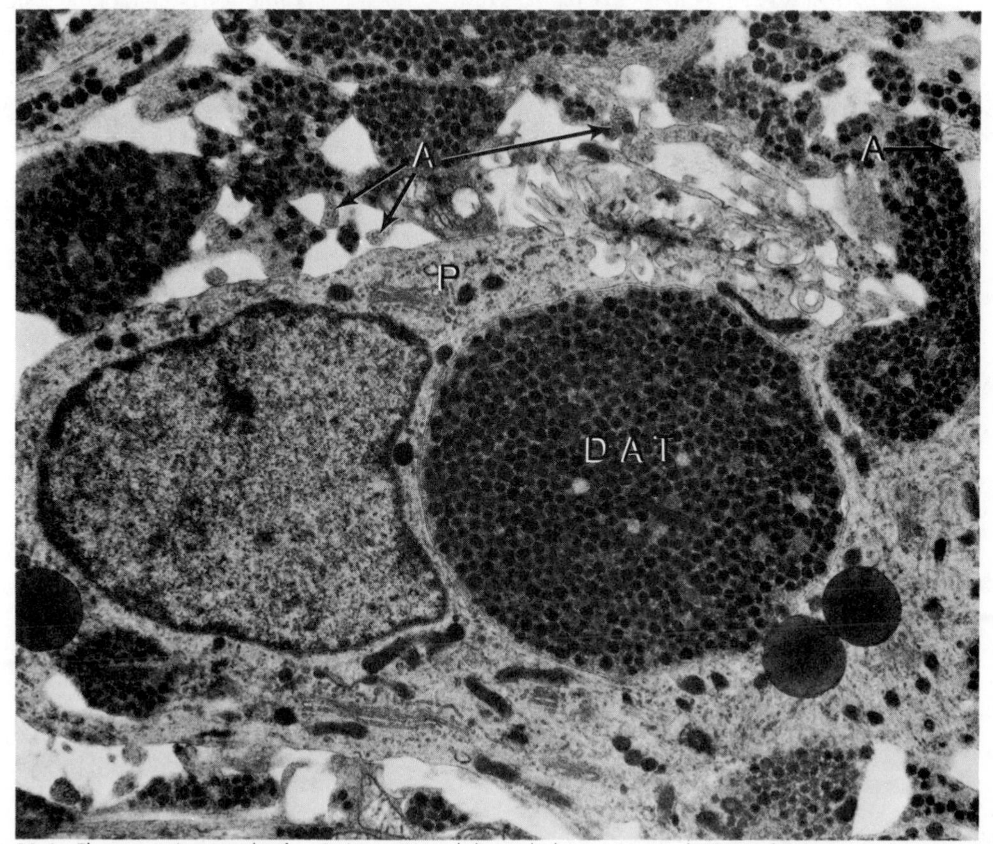

Figure 20-4. Electron micrograph of a pituicyte (*P*) and distended axon terminal (*DAT*) of an axon in the pars nervosa of the pituitary gland. Several small, unmyelinated axons (*A*) also are visible. The distended axon terminal is filled with electron-dense granules that represent secretion products of the pars nervosa coupled to neurophysins. (Courtesy of Dr. Raymond J. Walsh, Department of Anatomy, George Washington University.)

B. FUNCTION

1. Releasing hormones are secreted into the capillaries that drain into the portal system.

2. In addition to these releasing hormones, the pars nervosa produces two other important hormones.
   a. **Antidiuretic hormone** (ADH or vasopressin) causes water resorption in the renal collecting tubules, reducing the volume of the urinary filtrate considerably by resorbing most of the water from glomerular filtrate into the circulation.
   b. **Oxytocin** stimulates uterine contractions during labor and delivery and is also part of the "let-down" of milk reflex in lactating women.

3. ADH and oxytocin hormones are produced and probably secreted in close association with two **neurophysins** that are non-covalently coupled to the ADH and oxytocin but become dissociated from these hormones.

4. Fiber tracts running from brain nuclei may end either on capillary beds or in close association with processes from specialized ependymal cells called **tanycytes** located in the median eminence.
   a. These tanycytes may be involved in the transport of neurosecretions from nerve fibers, or even transport of the cerebrospinal fluid, to the hypophyseoportal circulation.
   b. Most substances are not allowed to pass from the blood into the brain parenchyma. The endothelial cells of the capillaries in the brain parenchyma (e.g., near the supraoptic nucleus) are tightly joined to one another. In contrast, the capillaries of the pars nervosa are distinctly leaky.
   c. The capillaries that pass through the supraoptic nucleus cannot pick up the releasing hormone or ADH-neurophysin complexes secreted by nerve fibers because there is a blood-brain barrier present here. In the pars nervosa, however, there is no such barrier.
   d. Thus, materials secreted from nerve endings, even if these materials have a substantial molecular weight, can enter the portal drainage of the pars nervosa and eventually make their way either to cells of the pars distalis (releasing hormones) or cells of the systemic circulation (ADH and oxytocin).

STUDY QUESTIONS

Directions: Each question below contains four suggested answers of which **one or more** is correct. Choose the answer

- **A** if **1, 2, and 3** are correct
- **B** if **1 and 3** are correct
- **C** if **2 and 4** are correct
- **D** if **4** is correct
- **E** if **1, 2, 3, and 4** are correct

1. The pars distalis contains which of the following components?

(1) Reticular fibers
(2) Connective tissue
(3) Collagen
(4) Tanycytes

2. Which of the following physiologic changes would be expected to occur following bilateral adrenalectomy?

(1) A decrease in glucocorticoid levels
(2) An increase in adrenocorticotropic hormone (ACTH) titers
(3) Hypertrophy of certain basophils
(4) Degranulation of corticotrops

Directions: The group of questions below consists of lettered choices followed by several numbered items. For each numbered item select the **one** lettered choice with which it is **most** closely associated. Each lettered choice may be used once, more than once, or not at all.

Questions 3–6

Match the following.

(A) Rathke's pouch
(B) Infundibulum
(C) Both
(D) Neither

3. Forms the pars nervosa and the stem of the infundibulum

4. Forms the pars distalis, pars tuberalis, and pars intermedia

5. Derived from ectoderm but not neuroectoderm

6. Can form ectopic endocrine tissue in the sphenoid bone

ANSWERS AND EXPLANATIONS

1. The answer is B (1, 3). (*V B 2*) Tanycytes are ependymal cells of the neurohyopophysis. They are not found in the pars distalis. There also is little if any connective tissue in the pars distalis.

2. The answer is E (all). (*V B 2*) Adrenalectomy would remove the negative feedback inhibition of adrenal hormones such as glucocorticoids. There would be a hypertrophy of basophils, degranulation of corticotrops, and an increase in adrenocorticotropic hormone (ACTH).

3–6. The answers are: 3-B, 4-A, 5-A, 6-A. (*III B, C*) The infundibulum forms from neuroectoderm (i.e., the floor of the diencephalon). The sphenoid bone forms after Rathke's pouch evaginates from the roof of the mouth, and it can contain fragments of ectopic pituitary tissue. Rathke's pouch is derived from ectoderm in the roof of the mouth and later gives rise to the pars distalis, pars tuberalis, and pars intermedia.

21
Female Reproductive System

I. INTRODUCTION

A. COMPONENTS

1. Ovaries, where **follicles** containing **oocytes** are stored and from which ova are shed gradually throughout reproductive life
2. Oviducts, which convey ova to the uterus
3. Uterus
4. Vagina
5. External genitalia
   a. Mons pubis
   b. Labia majora
   c. Labia minora
   d. Clitoris

B. FUNCTIONS

1. To produce and transport ova
2. To support a developing embryo

C. CYCLIC CHANGES—THE MENSTRUAL CYCLE. The female reproductive system, under the influence of the pituitary gland, undergoes cyclic changes that are related to reproductive function.

1. A complete menstrual cycle lasts about 29 days.
2. A new cycle, by convention, is considered to begin on the first day of menses.
3. At this point, the uterine epithelium begins to bleed and slough off. Next, it heals and starts to grow thicker as glands begin to form.
4. During this early phase of the cycle, several follicles grow larger in the ovary and secrete hormones that promote the proliferative development of the uterine epithelium.
5. At midcycle, one of these follicles usually sheds a single ovum from the ovary.
6. Follicular development.
   a. The developing follicle that has released the ovum then becomes a **corpus luteum**, a different kind of endocrine organ. The corpus luteum produces hormones that stimulate the uterine epithelial glands to mature and to secrete trophic substances for a potential embryo.
   b. If fertilization does not occur, the hormone-producing remnants of the follicle undergo a decrease in functional activity.
   c. If the ovum is fertilized, it implants in the wall of the uterus, becoming a hormone-producing embryo and preventing the sloughing of the uterine epithelium.
7. When fertilization does not occur, hormonal support of the uterine epithelium is withdrawn and is followed by menstruation (i.e., three to five days of flow of blood, mucus, and cellular debris).

D. HORMONAL CONTROL OF THE MENSTRUAL CYCLE is complex and involves endocrine organs scattered widely throughout the body, including the hypothalamus, pars distalis, ovaries, and the developing placenta, if one forms. This control basically is designed to ensure that the uterine epithelium is in a state to receive a developing embryo when that embryo is capable of implanting itself in the wall of the uterus.

E. DEVELOPMENT. Also discussed in this chapter is the development of

1. Ovaries
2. Reproductive ducts
3. External genitalia

II. OVARIES

A. GROSS ANATOMY

1. The two ovaries, as well as the oviducts, are attached to the body wall by the **suspensory ligament** of the ovary and are covered by peritoneal folds. The ovaries also are attached to the **broad ligament** by the **mesovarium**.
2. The mesothelium of the peritoneum covering the ovary is a cuboidal epithelium called the **germinal epithelium**. It has no relationship, however, to the production of gametes.
3. The ovaries are roughly ellipsoid and are about 4 cm long, 2 cm wide, and 1 cm thick, although these dimensions vary tremendously depending upon such factors as age, time during the menstrual cycle, and whether or not a woman is pregnant.
4. The ovary has a **hilus** near its attachment to the suspensory ligament, where it receives blood vessels, nerves, and lymphatics.

B. MICROSCOPIC ANATOMY

1. The ovary is divided roughly into a central **medulla** and a peripheral **cortex**.
2. The medulla is occupied largely by twisting blood vessels and a dense connective tissue stroma.
3. The stroma extends into the cortex all the way to the capsule of the ovary, which is known as the **tunica albuginea**. (The testes are encapsulated in a similar structure bearing the same name.)
4. The cortex also contains numerous **ovarian follicles** in various stages of development.
5. In each follicle there is an epithelium of variable thickness enclosed by a basement membrane.
6. The follicular epithelial cells, in turn, surround a single oocyte, which may undergo development that eventually leads to ovulation and the formation of an ovum.
7. In females who are reproductively competent (i.e., between approximately 14 and 50 years of age), the ovaries contain follicles in many different stages of development and regression; the presence of these different developmental stages is the dominant histologic feature of the ovarian cortex.

III. FOLLICULAR DEVELOPMENT

A. GENERAL INFORMATION

1. At birth, each ovary contains approximately 1,000,000 oocytes, each surrounded by **primordial follicles**. Many of these degenerate between birth and puberty.
2. Of the total 40,000 or so oocytes present at puberty, only about 400–500 eventually are ovulated; all the rest undergo a regression and absorption known as **atresia**.
3. During each menstrual cycle, commencing with the menarche (at about 14 years of age) and ceasing with menopause (at about 50 years of age), one ovum is ovulated from one of the two ovaries.
4. Which of the ovaries ovulates each month is determined randomly.
5. Not all cycles result in ovulation, and sometimes more than one ovum is shed during a single cycle, potentially resulting in **dizygotic twins**.
6. In most women, a single ovum is produced each month for 35–40 years.
7. During each cycle, many follicles undergo the late stages of follicular development, but normally only one follicle releases an ovum; the others undergo atresia.

B. PRIMORDIAL AND PRIMARY FOLLICLES (Fig. 21-1)

1. In the most immature follicles, called **primordial follicles**, a single oocyte is surrounded by a single layer of flattened follicular epithelial cells. These follicular cells sometimes are called **granulosa cells**.
2. The follicular border is demarcated by a basement membrane of the follicular epithelium.
3. In the immediate vicinity of the follicle, cortical stroma cells form a **theca interna** and a **theca externa**; these two layers are not distinctly divided from one another.
4. As follicles continue to develop, the follicular cells become active and grow into a single cuboidal layer; at this point, follicles are called **primary follicles**. (A large primary follicle is shown in Figure 21-2.)
5. A hyaline, amorphous extracellular material called the **zona pellucida** then becomes interposed between the oocyte and the follicular epithelium. Although the origin of this layer is not clearly understood, the layer is believed to be secreted by both the developing oocyte and the follicular epithelial cells. (The zona pellucida eventually disappears following fertilization but prior to implantation of the embryo.)
6. The follicular cells by this stage have become proliferative and are contributing to a continuously growing layer of follicular cells.
7. While the follicular epithelium is developing and becoming multilayered, the follicle is still called a primary follicle.
8. As the follicle enlarges, it begins to sink further and further into the medulla of the ovary, leaving more primitive follicles behind in the cortex.
9. While the follicular cells are proliferating, the cells of the theca interna also are increasing in number and beginning to differentiate into cells with lipid droplets in the cytoplasm and other characteristics typical of steroid-secreting cells.

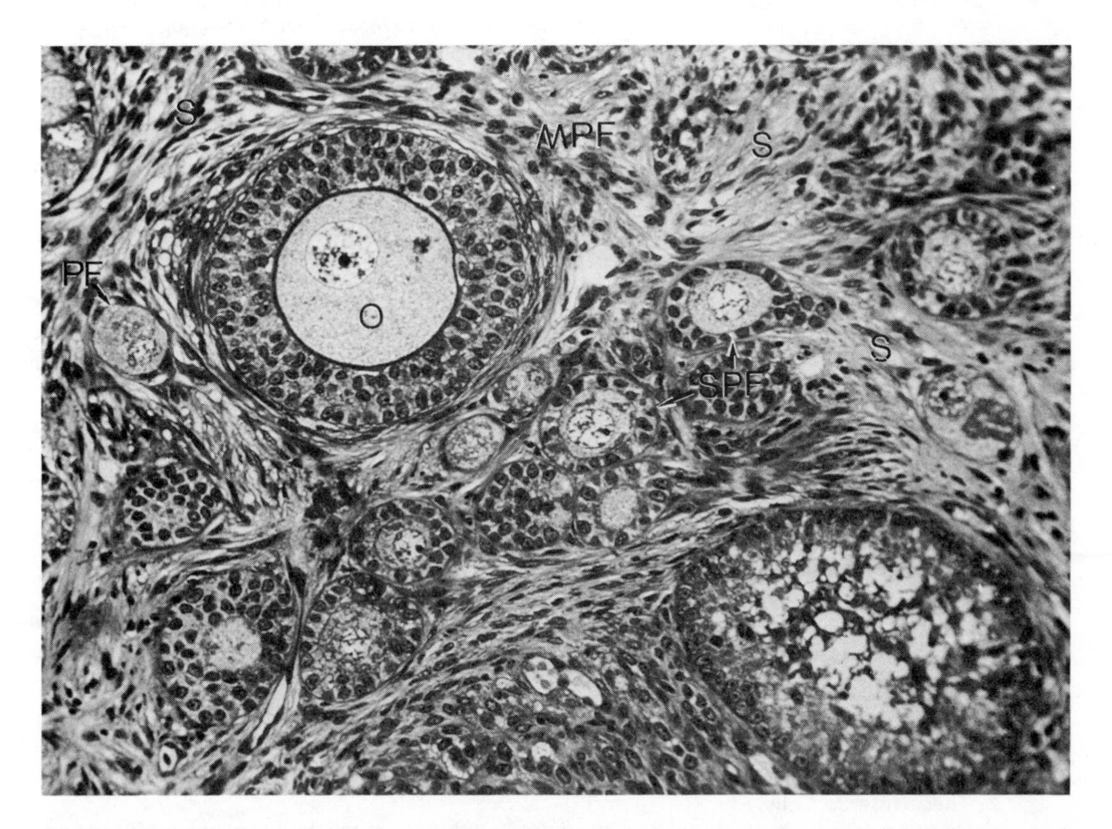

Figure 21-1. Light micrograph of a section through the ovarian cortex, showing primordial follicles (*PF*), single-layered primary follicles (*SPF*), and multilayered primary follicles (*MPF*). Each follicle contains an oocyte (*O*) and is embedded in connective tissue stroma (*S*).

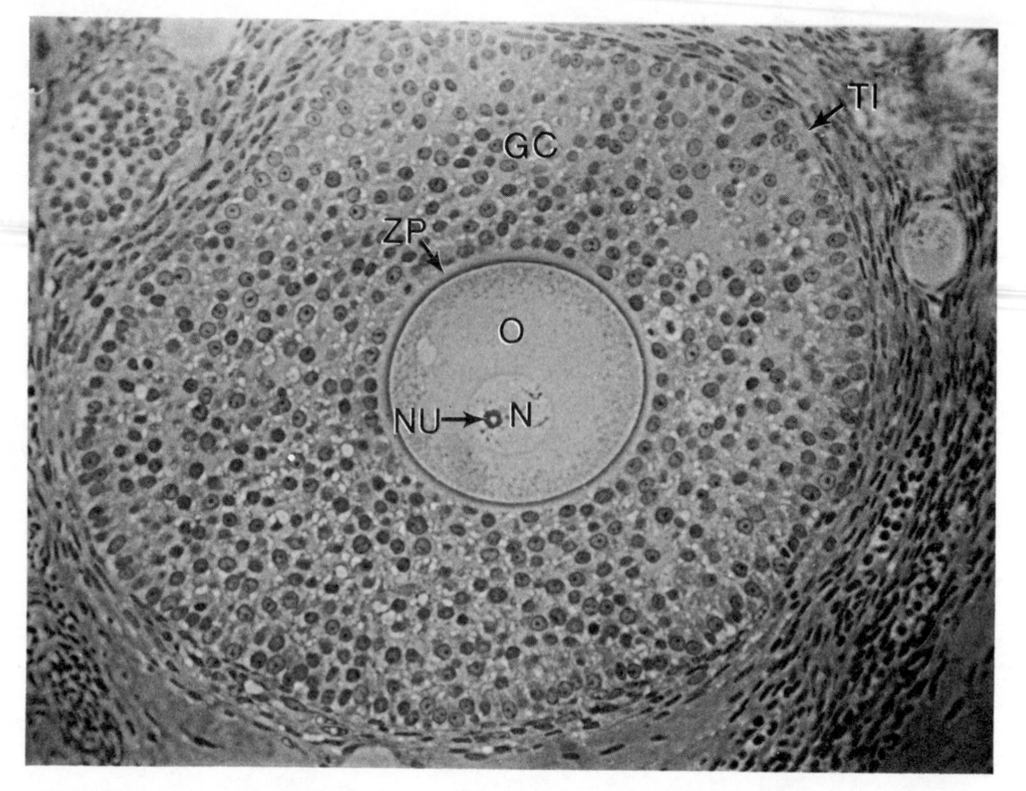

Figure 21-2. Light micrograph of a large primary ovarian follicle. The granulosa cells (*GC*) at this stage are many layers thick. The theca interna (*TI*) also is several layers thick. The granulosa cells surround an extracellular zona pellucida (*ZP*), which encloses an oocyte (*O*). The nucleus (*N*) and nucleolus (*NU*) of the oocyte can be seen. (Reprinted with permission from Johnson KE: *Histology: Microscopic Anatomy and Embryology*. New York, John Wiley, 1982, p 304.)

C. SECONDARY AND PREOVULATORY FOLLICLES

1. As a follicle continues to mature, the granulosa cells begin to secrete glycoproteins into the spaces surrounding them.
2. These glycoproteins coalesce into small lakes called **Call-Exner bodies**, and these in turn probably fuse to form the single, eccentrically placed cavity called the **antrum**.
3. As soon as the antrum appears among the follicular epithelial cells, the entire follicle is called a **secondary** or **antral follicle** (Fig. 21-3).
4. The fluid within the antrum is called the **liquor folliculi**. As this material accumulates, the antrum grows in volume.
5. At the same time, the follicular epithelial cells and theca cells are still increasing in number, causing a striking increase in the overall volume of the follicle. The follicle eventually becomes so large that it forms a blister on the surface of the ovary and occupies much of the entire width of the ovary.
6. This type of follicle is called a **mature** or **preovulatory follicle**.
7. The growth of the antrum pushes the oocyte to one side of the follicle where the oocyte remains surrounded by the **cumulus oophorus**, a mound composed of granulosa cells.

D. ULTRASTRUCTURE OF FOLLICULAR AND ASSOCIATED CELLS

1. **Granulosa Cells.**
   a. These cells have an abundance of rough endoplasmic reticulum and free ribosomes, many mitochondria, and a prominent Golgi apparatus.
   b. Granulosa cells play a role in secreting the liquor folliculi and in androgen aromatization.
2. **Theca Interna Cells.**
   a. These cells have an abundance of smooth endoplasmic reticulum, peculiar mitochondria with tubular cristae, and scattered lipid droplets.

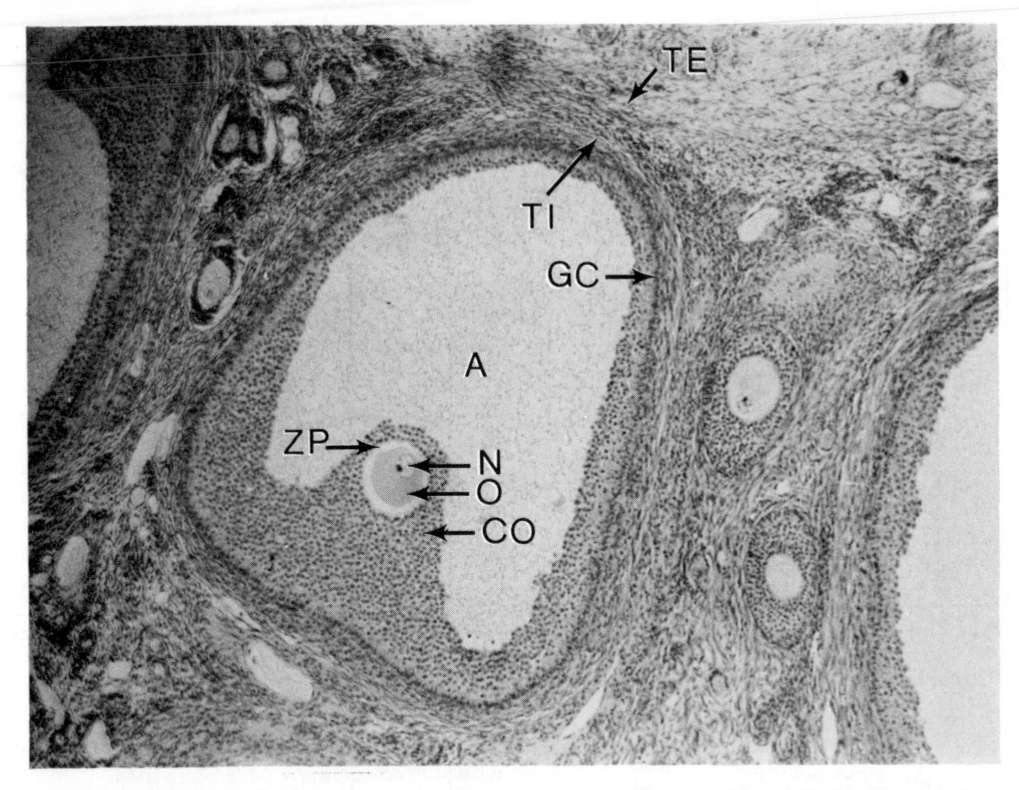

Figure 21-3. Light micrograph of an antral follicle. The antrum (*A*) is filled with liquor folliculi. The granulosa cells (*GC*) are thrown up into a heap surrounding the oocyte (*O*). This heap is known as the cumulus oophorus (*CO*). The ovum is surrounded by a zona pellucida (*ZP*). The theca interna (*TI*) and the theca externa (*TE*) are visible. (Reprinted with permission from Johnson KE: *Histology: Microscopic Anatomy and Embryology*. New York, John Wiley, 1982, p 305.)

b. Theca interna cells secrete androgens which, in the granulosa cells, convert to **estrogens**; the estrogens in turn stimulate the proliferative phase of the endometrium.

c. They also are intimately associated with a dense capillary plexus, whereas the follicular epithelium is avascular.

E. OVULATION

1. The preovulatory antral follicle continues to grow until it occupies the entire cortical thickness and bulges from the surface of the ovary.
2. In a short period immediately prior to ovulation, the follicle swells rapidly due to a sudden burst of follicular proliferation, an accumulation of follicular fluids, and a folding of the thecal and granulosa layers.
3. The cells of the cumulus oophorus become less firmly attached to the other follicular cells.
4. For unknown reasons, the follicle then ruptures through the wall of the ovary, expelling the oocyte and an attached layer of granulosa cells called the **corona radiata**, as well as the viscous liquor folliculi.
5. This mucoid material and the attached **gamete** are then swept rapidly into the oviducts by the beating of cilia on the fimbriae.

F. FORMATION OF THE CORPUS LUTEUM

1. Following ovulation, the follicle undergoes a process known as **luteinization**, whereby a new hormone-producing body, the corpus luteum, arises (Fig. 21-4).
2. The corpus luteum is formed from both granulosa cells and theca interna cells.

a. Granulosa-Lutein Cells.

(1) The granulosa cells undergo a fundamental morphologic change.

(2) They are converted rapidly from glycoprotein-secreting cells to progesterone-secreting

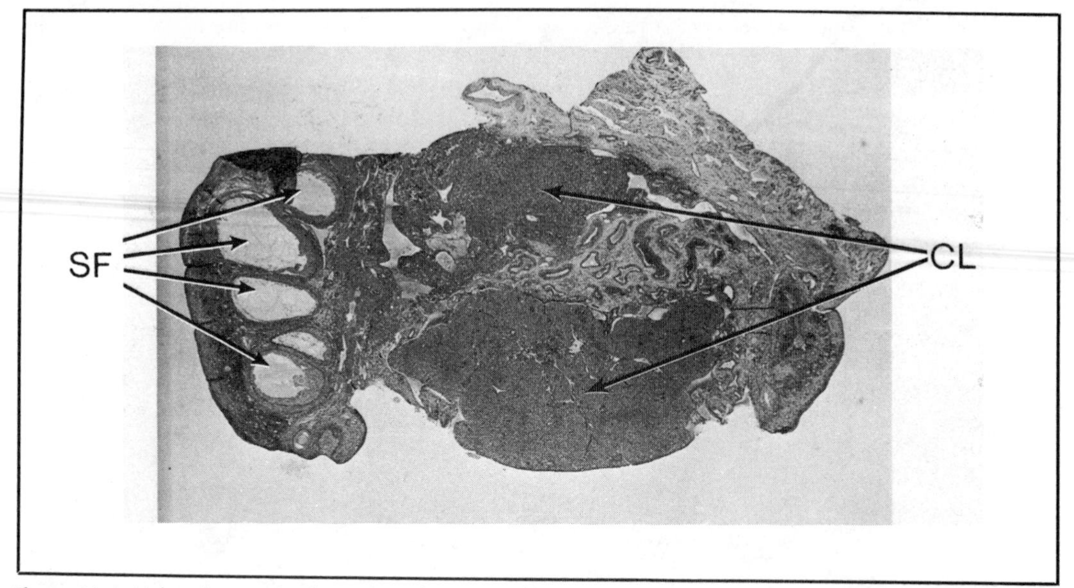

Figure 21-4. Light micrograph of a section of an ovary showing numerous cortical secondary follicles (*SF*) and two corpora lutea (*CL*). The upper corpus luteum is left over from a previous cycle; the lower one is from the most recent cycle. (Courtesy of Dr. Bela Gulyas, National Institute of Child Health and Human Development.)

cells, complete with the masses of lipid droplets, smooth endoplasmic reticulum, and peculiar mitochondria associated with steroid-secreting cells.

(3) Granulosa-lutein cells secrete the bulk of the **progesterone** produced in the corpus luteum and become, by far, the dominant cell type.

b. Theca-Lutein Cells. The theca interna cells contribute to the corpus luteum as theca-lutein cells, which probably continue to produce steroids other than progesterone.

3. The progesterone produced in the corpus luteum stimulates the secretory phase of the endometrium.

4. If conception occurs, chorionic gonadotropins are produced by the developing trophoblast. Human chorionic gonadotropin (HCG) supports the corpus luteum and allows it to continue to secrete progesterones early in pregnancy.

5. Without chorionic gonadotropins, the corpus luteum degenerates into a scar-like **corpus albicans**.

IV. OOGENESIS

A. PRIMORDIAL GERM CELLS

1. In human embryos, primordial germ cells originate in the wall of the **yolk sac** (i.e., endoderm) and migrate to the genital ridges.

2. In the primitive gonad, primordial germ cells differentiate into **oogonia**, a population of diploid proliferative cells.

3. By the third month gestation, the embryonic ovary contains nests of oogonia surrounded by a flattened layer of primitive follicular cells derived from the lining of the coelomic cavity over the ovary (i.e., mesoderm).

B. PRIMARY OOCYTES

1. By the seventh month gestation, most oogonia have degenerated.

2. Of the remaining oogonia, most are surrounded by a few flattened follicular cells and have begun their first meiotic division; at this stage they have become primary oocytes.

3. A primary oocyte, together with its surrounding cells, constitutes a **primordial follicle**.

4. At birth, all follicles present in the ovaries are primordial follicles, with primary oocytes arrested at the **dictyotene** stage of the first meiotic prophase.

5. Primary oocytes **do not complete their first meiotic division until after puberty**, and only one divides during each ovulation.

6. During childhood, about 90 percent of the primordial follicles that are present in the ovaries undergo atresia.

7. The rest remain with the potential to undergo the developmental sequence leading tofollicular growth and, in a few cases, ovulation.

8. During the entire growth process of the primary and secondary follicles, which commences at puberty, the follicles contain primary oocytes.

C. SECONDARY OOCYTES

1. At the end of the maturation period of the secondary follicle, immediately prior to or during ovulation, the long-arrested first meiotic prophase is interrupted, the first meiotic division is completed, and a **secondary oocyte** and a **first polar body** are formed.

2. The first polar body is a small fragment of the egg cell containing 23 chromosomes, with 2 chromatids to each chromosome. The first polar body may degenerate or it may divide into two polar bodies.

3. Similarly, the secondary oocyte contains 23 chromosomes with 2 chromatids in each. Unlike the first polar body, however, the secondary oocyte carries the bulk of the cytoplasm that was present in the primary oocyte.

D. EVENTS AT FERTILIZATION

1. Following fertilization, the second meiotic division is completed, and a second polar body is shed.

2. Thus, a fertilized human ovum usually has two or three small haploid polar bodies and a diploid **zygote**, formed by the union of the haploid female and male **pronuclei**.

3. It is important to note that the first meiotic prophase extends from approximately the time of birth until the time immediately preceding ovulation and, thus, may last as long as 50 years.

4. Some believe that this extended meiotic prophase may account for the increasing frequency of some congenital birth defects with increasing maternal age. There is no doubt that birth defects such as trisomy 21 (or Downs syndrome) occur more commonly as maternal age advances, but there is controversy over the mechanism of this effect.

V. OVIDUCTS

A. GROSS ANATOMY

1. The oviducts, also known as the fallopian or uterine tubes, are paired structures that measure about 12 cm long and that extend away from the uterus toward each ovary.

2. They are supplied with their own mesentery, the **mesosalpinx**, which carries a blood supply to them.

3. The gross anatomic and the histologic structure of the oviducts varies in different areas.

4. Segments of the Oviduct.
   a. An **interstitial portion** lies within the wall of the uterus and attaches the oviduct to the uterus.
   b. The **isthmus** connects the interstitial portion and a dilated portion called the **ampulla**.
   c. The ampulla, in turn, opens into a funnel-shaped **infundibulum**. The infundibulum has a large number of projections from its surface, known as the **fimbriae**.

B. OVIDUCTS AND OVUM TRANSPORT

1. In some animals at about the time of ovulation, the infundibulum essentially engulfs the ovary, ensuring that the ovum, released from the surface of the ovary, will be swept into the oviduct rather than into the peritoneal cavity.

2. It is not known if this phenomenon occurs in humans as well.

3. It is known, however, that fertilized ova can enter the peritoneal cavity and either implant ectopically or, after considerable migration, reenter the reproductive tract.
   a. Cases have been reported of conception in women who have an ovary on one side of the body associated with a damaged or atrophic oviduct on that side, and no ovary on the opposite side of the body but a normal oviduct.
   b. For such women to have conceived indicates that an ovum must have traveled extensively

through the peritoneal cavity, either before or after fertilization, later to implant a considerable distance from the site of ovulation.

C. OVIDUCTAL MUCOSA (Fig. 21-5)

1. The mucosa of the oviduct is lined by a simple columnar epithelium with a mixture of glandular cells and ciliated cells. The oviductal mucosa is relatively smooth everywhere except in the ampulla and the infundibulum, where it is thrown into complex folds.
2. The glandular cells are believed to secrete trophic (i.e., nutritive) substances for the developing ovum. (The ovum usually spends the first three days or so following fertilization in the oviduct, and the developing morula probably needs nutrients until it can establish an implantation site in the uterine epithelium.)
3. The ciliated cells are involved in ovum transport. They are quite numerous on the fimbriae and in the ampulla but are sparse in the isthmus and interstitial portions.
4. The functioning of the cilia is related to the activity of the ovaries.
   a. Following castration in animals, ciliated cells are lost, but they return following administration of estrogens.
   b. Following menopause in women, ciliated cells disappear, and the histologic differences among parts of the oviducts become less pronounced.
   c. Cyclic changes occur in ciliary activity and in other functions of the oviducts.
5. The cilia of the fimbriae beat into the oviducts and clearly are designed to propel the ovum into the reproductive tract.

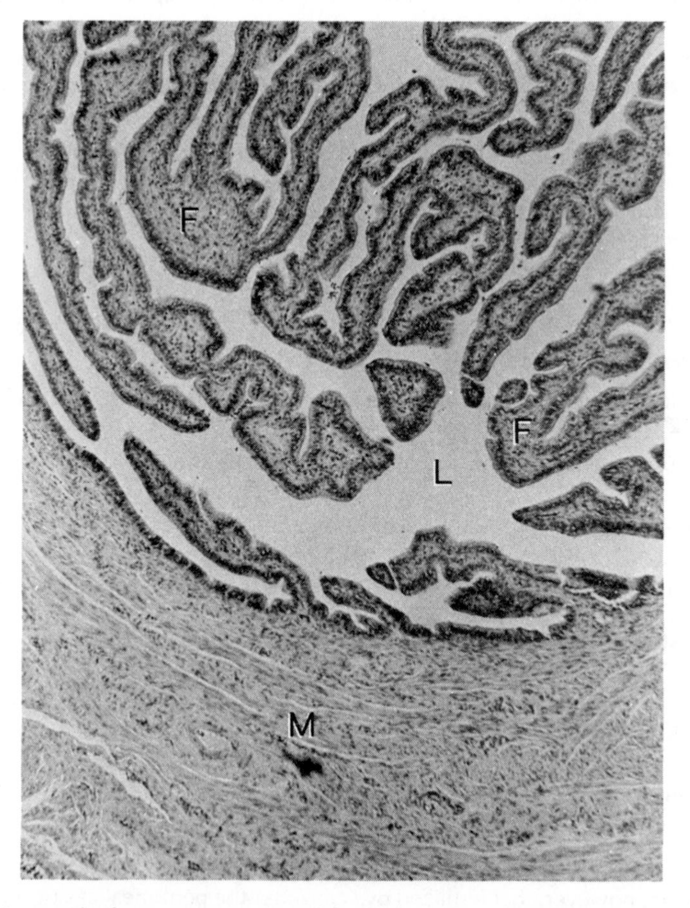

Figure 21-5. Light micrograph of a transverse section of the oviduct. This tubular organ has a lumen (*L*) with folded projections (*F*) of the mucosa into the lumen. These folds are coated by a ciliated columnar epithelium. The oviduct also has a thick muscular coat (*M*). (Reprinted with permission from Johnson KE: *Histology: Microscopic Anatomy and Embryology*. New York, John Wiley, 1982, p 308.)

6. The epithelium of the oviductal mucosa rests on a typical basement membrane. There is a thin lamina propria with connective tissue fibers, fibroblasts, and possibly some smooth muscle cells under the epithelium.

D. OVIDUCTAL MUSCULARIS AND SEROSA (see Fig. 21-5)

1. There is a good deal of smooth muscle in the muscularis, with an inner circular layer and an outer longitudinal layer; the layering, however, is not striking in humans.
2. Toward the uterus, the muscularis is quite thick.
3. Presumably, peristaltic contractions of the muscularis propel a developing embryo into the uterus.
4. The serosa of the oviducts is a loose connective tissue layer of little importance; it is coated by a mesothelium continuous with the lining of the rest of the peritoneal cavity.

VI. UTERUS

A. GROSS ANATOMY

1. The uterus is a remarkable organ specialized for the nurturing and carrying of a fetus.
2. The uterus in a nonpregnant woman is approximately 6 cm long, 5 cm wide, and 3 cm thick. This peritoneal organ lies between the rectum and the urinary bladder.
3. The uterus has a **body**, a **fundus**, and a **cervix**; the cervix opens directly into the vagina.
4. The wall of the uterus is divided into an endometrium, a myometrium, and a perimetrium.

B. UTERINE ENDOMETRIUM

1. The endometrium (or luminal layer of the uterus) consists of a complex epithelium and a connective tissue stroma that represents the lamina propria of the mucosa.
2. The endometrium has two layers.
   a. One is a deep basal layer that is not sloughed during menstruation.
   b. The other is a more superficial layer that grows, sloughs, and reforms over and over again from the residual basal layer.
3. The endometrium also has long glands extending over its entire thickness and even into the superficial portions of the myometrium.
4. **Cyclic Changes in the Endometrium.** The luminal layer undergoes extensive changes during the menstrual cycle, varying in height from as little as 1 mm immediately following menstruation to as much as 7 mm immediately before menstruation.
   a. **The Follicular or Proliferative Phase.**
      (1) During the proliferative phase, which begins at the end of menstrual flow, the glands are straight and their lumina are narrow.
      (2) During this phase, endometrial epithelial cells are tall columnar cells with numerous microvilli, no glycogen, and an otherwise unremarkable fine structure except for a prominent apical Golgi apparatus.
      (3) The proliferative phase of the endometrium is controlled by estrogen secretion from the theca interna and granulosa cells of the developing follicle. It extends until midcycle.
   b. **The Secretory or Luteal Phase.**
      (1) Under the influence of a growing corpus luteum and its progesterone, the endometrial glands begin to secrete materials.
      (2) The endometrial epithelial cells accumulate glycogen, which presumably contributes sugar precursors for the glycosylated secretions of the endometrial glands. In the late secretory phase, glycogen migrates apically and may even be released into the secretion product.
      (3) During the secretory phase, the glands continue to elongate; they also become quite wavy and may branch.
      (4) Their lumina become engorged with secretion products.
      (5) During the late secretory phase, endometrial epithelial cells become filled with lysosomes and myelin figures.
   c. **Endometrial Blood Vessels.**
      (1) The endometrium is supplied with arterial blood via **coiled arteries**.
      (2) These arteries increase in length and complexity during the proliferative and secretory phases.
      (3) About two weeks after ovulation, if fertilization does not occur, coiled arteries become

alternately engorged with blood and constricted. Presumably, the intermittent blood supply causes degenerative changes, and finally the epithelium begins to break down.

d. Menstrual Phase.

(1) If fertilization does not occur, chorionic gonadotropins are not produced to support the corpus luteum. Consequently, the corpus luteum begins to regress and ceases to provide progesterone for stimulation of the endometrium.

(2) The entire functional endometrial layer thus degenerates and is lost, along with mucus and blood, during menstruation.

(3) The basal layer remains and replaces the superficial layer.

(a) After menstruation is completed and bleeding has stopped, the basal layer epithelium heals over the endometrial stroma.

(b) Basal layer cells become proliferative and, together with the growing stromal cells, contribute to a thickening of the endometrium.

(c) The endometrium thickens, endometrial glands begin to grow in length, and the cycle begins again.

e. Conception.

(1) If conception occurs, a trophoblast that secretes chorionic gonadotropin is rapidly formed by the developing embryo.

(2) Chorionic gonadotropin stimulates the corpus luteum of pregnancy to grow and to continue to produce progesterone.

(3) The endometrium is not sloughed; instead, an environment suitable for the early development and implantation of the embryo is provided.

C. UTERINE MYOMETRIUM

1. The myometrium is a thick muscular coat with three indistinct layers.
2. The middle layer is roughly circular and richly vascularized; it blends gradually into inner and outer layers of longitudinal or oblique fibers.
3. In the cervix, the myometrium is more regular than throughout the rest of the uterus; it has a middle circular layer and inner and outer longitudinal layers.
4. Elastic fibers are relatively sparse among the uterine smooth muscle layers but considerably more abundant in the cervix.
5. In the nonpregnant uterus, the smooth muscle cells are relatively small; during pregnancy, they can grow enormously, reaching lengths in excess of 5 mm by term.

D. THE PERIMETRIUM is an ordinary connective tissue layer coated in most places by the peritoneum.

E. CERVIX

1. The cervix connects the uterus to the vagina.
2. The cervical mucosa is folded elaborately into deep crypts and clefts—not glands—which are lined by an abundance of mucus-producing columnar cells and a few ciliated cells.
3. The secretions of the cervical mucosal cells vary during different parts of the menstrual cycle.
4. The cervical canal is filled with a mucous plug, which varies in its chemical and physical properties depending on the time of the cycle and whether or not there is a pregnancy.

VII. VAGINA

A. VAGINAL EPITHELIUM. The vagina is lined by a stratified squamous epithelium. It is not keratinized in humans, although scattered keratohyalin granules are seen occasionally in the outer layers of cells. Many animals, however, have a keratinized vaginal epithelium.

B. VAGINAL pH AND GLYCOGEN

1. The vaginal epithelial cells contain a good deal of glycogen, which is released most actively at about the time of ovulation.
2. Vaginal bacteria ferment this glycogen to lactic acid, lowering the vaginal pH and thus providing an environment that is thought to be more hospitable to spermatozoa and less hospitable to harmful bacteria.
3. The production of glycogen is controlled partially by estrogen secretion.

4. The withdrawal of estrogen later in the cycle decreases glycogen secretion, increases vaginal pH, and increases the likelihood that various infectious agents will invade the vagina.

C. VAGINAL LEUKOCYTES. The **lamina propria** of the vagina is unremarkable except for its scattered lymphoid nodules. Also, there is a massive influx of lymphocytes and polymorphonuclear leukocytes through the lamina propria into the vagina itself at the end of the menstrual cycle. This influx of lymphocytes and polymorphonuclear leukocytes may function to combat the increased risk of infection that occurs late in the cycle.

D. VAGINAL LUBRICATION

1. The vagina receives lubricating mucus from the cervix.
2. A rich blood supply in the lamina propria becomes engorged during sexual arousal, increasing the coloration of the vaginal mucosa and producing a watery transudate which, along with the secretions of the **vestibular glands** (or glands of Bartholin), serves as a lubricant.

VIII. EXTERNAL GENITALIA

A. LABIA

1. The **labia majora** are folds of body skin which contain some smooth muscle. They are covered with stratified squamous keratinized epithelium and after puberty are equipped with coarse pubic hairs, apocrine sweat glands, and sebaceous glands.
2. The **labia minora** lie inside the labia majora. They are covered by a nonkeratinized, hairless, stratified squamous epithelium that is usually more pigmented than the surrounding epithelia. Sebaceous glands lie on both sides of the labia minora.

B. CLITORIS AND VESTIBULAR GLANDS

1. The clitoris is an erectile organ similar morphologically to the penis, but smaller.
2. The clitoris is covered by a stratified squamous epithelium and is richly endowed with various specialized sensory nerve endings.
3. In the area between the opening of the vagina and the labia minora, there are several types of vestibular glands that secrete lubricating substances.

IX. HORMONAL CONTROL OF THE REPRODUCTIVE CYCLE

A. GONADOTROPINS

1. The cyclic changes that occur in the ovary are under the control of the adenohypophysis.
2. Gonadotrops secrete **follicle-stimulating hormone (FSH)**, which stimulates growth of ovarian follicles up to the point of ovulation. FSH also stimulates the secretion of estrogens by the theca interna and granulosa cells.
3. Later in the proliferative phase of the menstrual cycle, the adenohypophysis continues to produce FSH but also produces a sudden burst of **luteinizing hormone** (LH), which is required for ovulation and for the early development of the corpus luteum.

B. NEUROHYPOPHYSIS

1. Both FSH and LH are, in turn, controlled by the secretion of **gonadotropin-releasing hormone (Gn-RH)**, which is elaborated by the hypothalamus.
2. The estrogen and progesterone produced by growing follicles and by the corpus luteum in turn affect the control of Gn-RH secretion, completing a feedback loop.
3. The levels of estrogen and progesterone are related to the cyclic changes that occur in the endometrium (see VI B 4).

X. DEVELOPMENT OF THE FEMALE REPRODUCTIVE SYSTEM

A. THE OVARIES develop from three rudiments.

1. **Primordial Germ Cells.**
   a. These cells migrate from the endodermal lining of the yolk sac along the hindgut mesentery into the genital ridges.
   b. They give rise to the oogonia and oocytes.

2. Coelomic Epithelial Lining Cells.
a. These cells migrate from the lining of the coelom into the genital ridges.
b. They surround primordial germ cells and become follicular epithelial cells.

3. Genital Ridge Mesenchyme.
a. These cells arise in situ from mesenchymal cells in the body wall associated with the developing mesonephros.
b. They form the connective tissue stroma of the ovaries, as well as of the theca externa and theca interna.

B. REPRODUCTIVE DUCTS

1. The mesonephric (wolffian) ducts degenerate almost completely in the female.
2. The epoophoron, paroophoron, and cystic structures in the broad ligament (called Gartner's cysts) are formed from the mesonephric tubules and ducts.
3. The paramesonephric (müllerian) ducts undergo extensive development and, along with surrounding mesenchyme, form into the fallopian tubes, the uterus, and the upper part of the vagina.
4. The lower part of the vagina develops from the urogenital sinus.

C. EXTERNAL GENITALIA

1. The genital tubercle forms the clitoris.
2. The urethral folds form the labia minora.

STUDY QUESTIONS

Directions: Each question below contains four suggested answers of which **one or more** is correct. Choose the answer

A if **1, 2, and 3** are correct
B if **1 and 3** are correct
C if **2 and 4** are correct
D if **4** is correct
E if **1, 2, 3, and 4** are correct

1. True statements concerning the uterine myometrium include which of the following?

(1) It may have glandular elements of the endometrium penetrating it
(2) It may contain abundant skeletal muscle during pregnancy
(3) It has multiple layers of smooth muscle
(4) It usually undergoes little change during pregnancy

2. True statements concerning the zona pellucida include which of the following?

(1) It is produced by theca interna cells
(2) It is secreted by granulosa cells
(3) It is composed of a simple squamous epithelium
(4) It is shed from the embryo before implantation

Continued on next page

Directions: The group of questions below consists of lettered choices followed by several numbered items. For each numbered item select the **one** lettered choice with which it is **most** closely associated. Each lettered choice may be used once, more than once, or not at all.

Questions 3–7

The electron micrograph below shows a granulosa-lutein cell in a functional corpus luteum. For each description that follows, select the lettered area in the electron micrograph with which it is most closely associated.

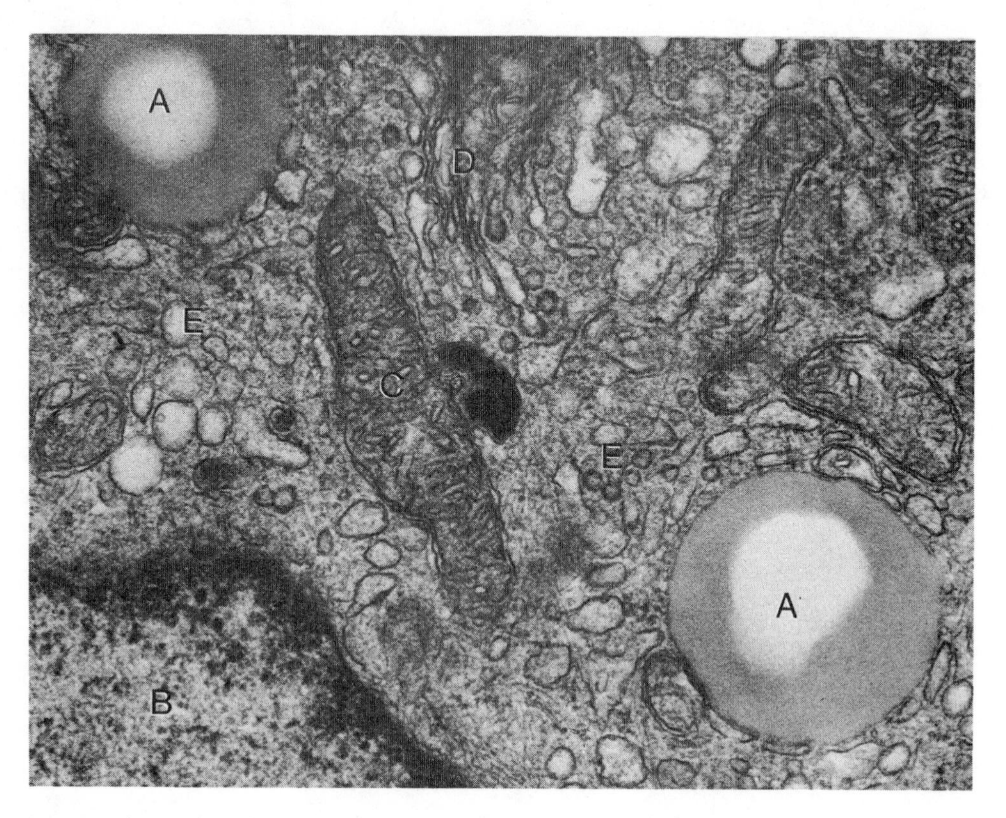

Courtesy of Dr. Bela Gulyas, National Institute of Child Health and Human Development.

3. Source of energy-rich compounds used in steroid biosynthesis

4. Vacuole-containing stored cholesterol esters for steroid biosynthesis

5. This structure is bounded by a double unit membrane and contains the bulk of genetic material in a cell

6. Golgi apparatus

7. Membranous labyrinth, the location of many enzymes for steroid biosynthesis

ANSWERS AND EXPLANATIONS

1. The answer is B (1, 3). (*VI B 3; VI C 1, 5*) The uterine myometrium has three indistinct muscular layers, and long glands from the endometrium may extend into its superficial portions. It contains no skeletal muscle during pregnancy but undergoes a dramatic increase in the size of its smooth muscle cells at this time.

2. The answer is C (2, 4). (*III B 6*) The zona pellucida is an acellular secretion product of the granulosa cells and the developing oocyte. It is shed during the early blastocyst stage of embryonic development, before implantation.

3–7. The answers are: 3-C, 4-A, 5-B, 6-D, 7-E. [*III F 1 a (2)*] In the electron micrograph, (A) is a lipid droplet, (B) is the nucleus, and (C) is a mitochondrion. Notice that the cristae of the mitochondrion have a peculiar tubular arrangement. Although the functional significance of this pattern is not known, this particular arrangement is a feature common to mitochondria in many steroidogenic cells. (D) is the Golgi apparatus. (E), the smooth endoplasmic reticulum, contains several enzymes for steroid biosynthesis and, thus, is always prominent in steroidogenic cells such as granulosa-lutein cells.

ANSWERS AND EXPLANATIONS

22
Male Reproductive System

I. INTRODUCTION

A. COMPONENTS (Fig. 22-1)

1. Testes
2. Excretory ducts
3. Epididymis
4. Ductus deferens
5. Seminal vesicles and prostate
6. Urethra and penis

B. FUNCTIONS

1. Production of gametes
2. Production of fluid medium for gametes
3. Production of male sex hormones
4. Sexual intercourse
5. Elimination of urine

C. DEVELOPMENT. Also discussed in this chapter is the development of

1. Gonads
2. Mesonephric ducts
3. External genitalia

II. MICROSCOPIC ANATOMY OF TESTES

A. These paired organs are located in the scrotum and are partially surrounded by an outpocketing of the peritoneal cavity known as the **tunica vaginalis**.

B. DESCENT OF TESTES

1. Testes develop retroperitoneally within the abdominal cavity and descend into the scrotum, where the cooler temperatures are more hospitable for spermatogenesis.
2. During its descent, the testis carries an evagination of the peritoneal cavity into the scrotum.
3. Usually, the connection between this **processus vaginalis** and the peritoneal cavity proper is obliterated later in life, leaving a tunica vaginalis.
4. Thus, the anterior and lateral aspects of the testis are coated with a mesothelium that once was continuous with the mesothelium lining the peritoneal cavity.

C. TUNICA ALBUGINEA

1. The tunica albuginea is a thick, fibrous connective tissue capsule that lies beneath the mesothelium.
2. It penetrates deeply into the posterior aspect of the testis, forming the **mediastinum testis**.
3. The hilus of the testis is located in the mediastinum testis, and it is through here that ducts, blood vessels, nerves, and lymphatic vessels enter and leave.

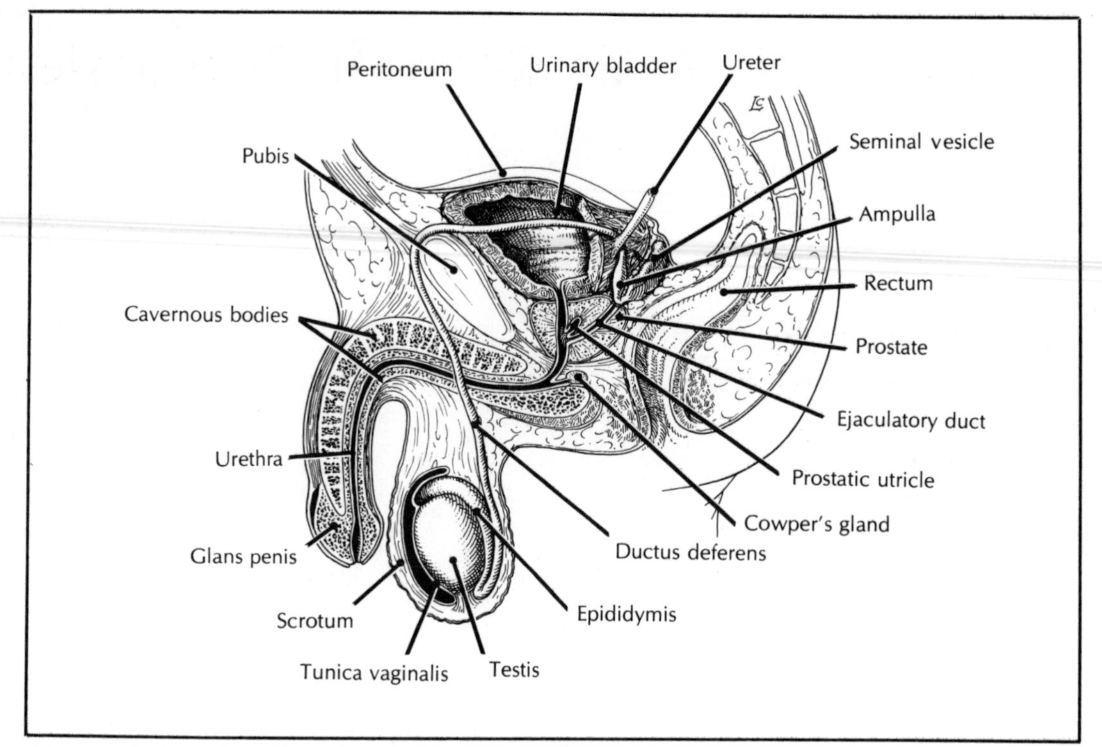

Figure 22-1. Illustration of the male reproductive system. (Reprinted with permission from Bloom W, Fawcett DW: *A Textbook of Histology.* Philadelphia, WB Saunders, 1975.)

D. Connective tissue trabeculae project from the mediastinum into the interior of the testis, dividing it into several hundred lobules.

E. Within each lobule are several long (60 cm), thin (200 μm), coiled seminiferous tubules. Typically, these form complex loops that begin and end at the mediastinum; occasionally, however, blind-ending tubules or anastomoses from one loop to another are seen.

F. Because each ejaculate contains millions of spermatozoa, it is not surprising that there are over 500 m of seminiferous tubules in both testes combined.

G. INTERSTITIAL CELLS

1. The interstices between the seminiferous tubules are filled with small vessels and lymphatics.
2. Connective tissue fibroblasts and, possibly, macrophages and lymphocytes occur here. Most importantly, the **Leydig** cells, specialized for the secretion of the male sex hormone **testosterone**, are found here.

III. SEMINIFEROUS EPITHELIUM (Fig. 22-2)

A. SPERMATOGENIC CELLS

1. Spermatogonia proliferate constantly after puberty.
2. They produce spermatocytes, which divide into spermatids; spermatids then differentiate into spermatozoa.

B. SERTOLI CELLS (Fig. 22-3)

1. **Structure.**
   a. Sertoli cells comprise a continuous epithelium in the seminiferous tubule.
   b. They are irregularly shaped cells which extend from the basement membrane of the epithelium all the way to the lumen of the tubule.

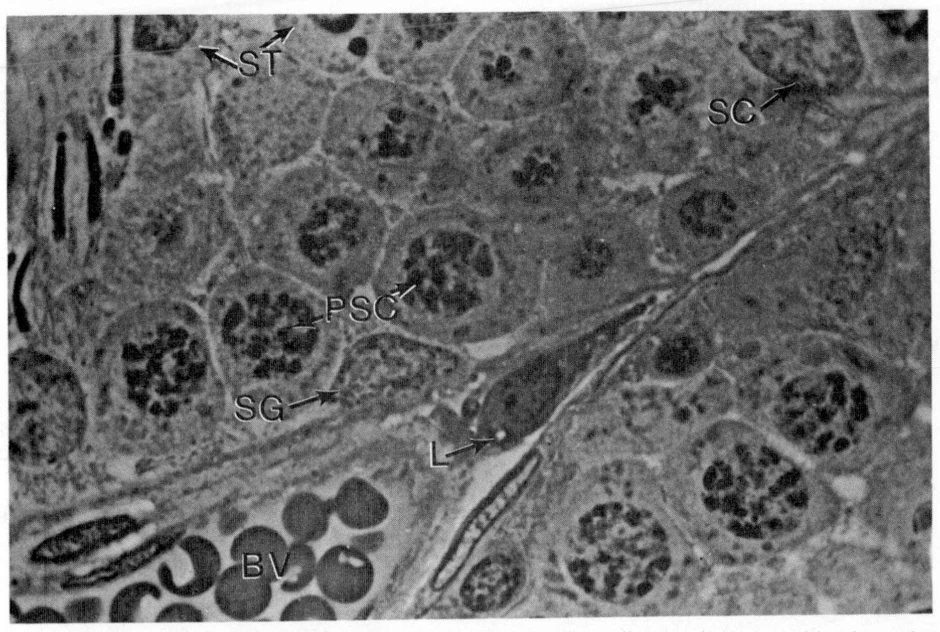

Figure 22-2. Light micrograph of the testis. A solitary interstitial (Leydig) cell (*L*) is adjacent to a blood vessel (*BV*). In the seminiferous epithelium, spermatogonia (*SG*), primary spermatocytes (*PSC*), and spermatids (*ST*) are visible. In the extreme upper right, part of an irregular Sertoli cell nucleus (*SC*) is visible. (Reprinted with permission from Johnson KE: *Histology: Microscopic Anatomy and Embryology*. New York, John Wiley, 1982, p 336.)

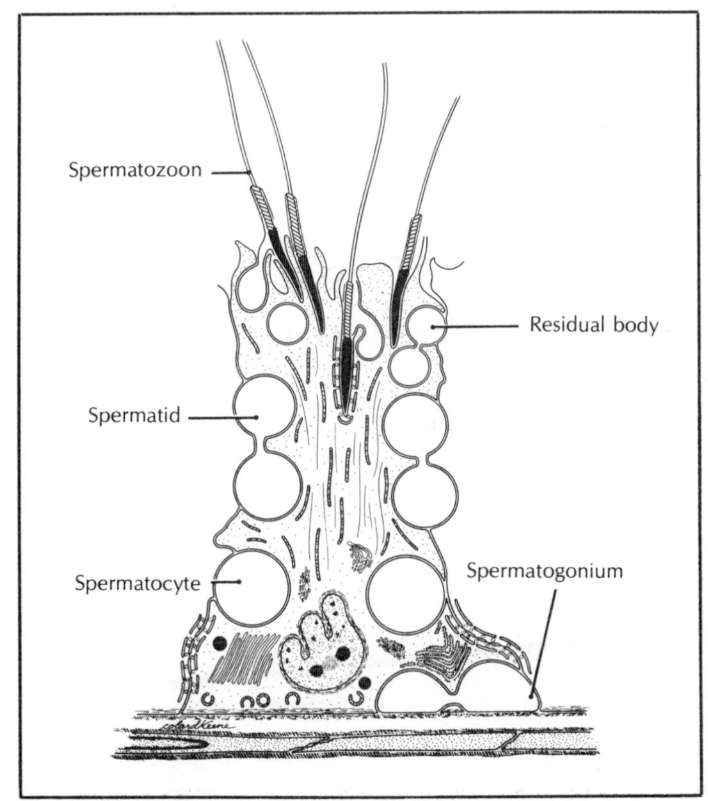

Figure 22-3. Diagram illustrating the Sertoli cell and its relationship to other cells in the seminiferous epithelium. The spermatogonia reside in a basal compartment that is separated from an adluminal compartment by an extensive network of tight junctions. (Reprinted with permission from Dym M, Fawcett DW: The blood-testis barrier in the rat and the physiological compartmentation of the seminiferous epithelium. *Biol Reprod* 3:308, 1970.)

2. **Compartmentalization.**
   a. Sertoli cells have complex tight junctions between them, which define an adluminal and a basal compartment.
   b. Spermatogonia reside in the basal compartment where they are stimulated by hormones in the blood.
   c. Primary spermatocytes and germ cells of later stages are found in successively higher levels, above the zone of tight junctions.

3. **Functions** of Sertoli cells include
   a. Provision of nutritive support for developing spermatozoa (a presumed function)
   b. Phagocytosis of residual bodies (bits of spermatid cytoplasm shed during the final phases of spermiogenesis)
   c. Elaboration of androgen-binding proteins

IV. SPERMATOGENESIS AND SPERMIOGENESIS

A. SPERMATOGENESIS refers to the entire process of converting diploid spermatogonia into haploid spermatozoa. The spermatogenic cycle, from spermatogonia to spermatozoa, is initiated at puberty and lasts about 64 days. This process requires testosterone secretion by the Leydig cells.

1. Spermatogonia begin to proliferate rapidly at puberty and, by division, produce primary spermatocytes.

2. Primary spermatocytes enter meiosis I, the first meiotic division, and produce secondary spermatocytes.

3. In meiosis II, secondary spermatocytes divide and form haploid spermatids.

B. SPERMIOGENESIS refers to the process of converting spermatids into spermatozoa. There are three events during spermiogenesis.

1. Acrosome Formation.
   a. Golgi membranes coalesce to form an acrosomal granule.
   b. The acrosomal granule flattens around the nucleus to form an acrosomal cap.

2. Flagellum and Midpiece Formation.
   a. Centrioles migrate to the spermatid pole, opposite the acrosome.
   b. One centriole forms a flagellum.
   c. Mitochondria group around the base of the flagellum to form the midpiece.

3. Streamlining of Spermatozoa.
   a. Nuclear condensation results in a pointed, streamlined nucleus.
   b. Excess cytoplasm is shed as a residual body which is engulfed by the Sertoli cells.

V. CONSTITUENTS OF THE INTERSTITIUM

A. LEYDIG CELLS

1. The tissue between the seminiferous tubules, called the interstitial tissue, contains the Leydig cells.

2. Leydig cells are about 20 μm in diameter and usually are found in close association with small blood vessels; this is not surprising since they are responsible for the secretion of testosterone into the blood.

3. Each Leydid cell has an acidophilic cytoplasm and a prominent nucleus with a nucleolus, when seen in the light microscope.

4. With the electron microscope they exhibit an abundance of smooth endoplasmic reticulum and a prominent Golgi apparatus.

5. Leydig cells contain cholesterol which they use to synthesize testosterone on membranes of the smooth endoplasmic reticulum.

6. Androgen Secretion.
   a. The production of androgen by Leydig cells is under the control of the pituitary gland.
   b. Cells of the pars distalis secrete luteinizing hormone (LH), which sometimes is called interstitial cell-stimulating hormone (ICSH). LH, a gonadotropic hormone, stimulates the Leydig cells to release testosterone.

c. High levels of testosterone then feed back on the hypothalamus and inhibit the production of luteinizing hormone releasing hormone (LH-RH) in a classic negative-feedback loop.
d. Following hypophysectomy, there is a complete degeneration of the interstitial Leydig cells, no testosterone production, and a cessation of spermatogenesis.
e. Following castration, testosterone levels fall rapidly and the gonadotrops are under chronic hypothalamic stimulation by LH-RH, resulting in the appearance of **castration cells** in the pars distalis.
f. The role of follicle-stimulating hormone (FSH) in the male reproductive system is still somewhat controversial, although there is some evidence that FSH alters the function of Sertoli cells.

B. BLOOD VESSELS

1. The interstitial tissue has a rich capillary network to supply endocrine cells.
2. These capillaries are supplied by arterioles and are drained by venules.

C. CONNECTIVE TISSUE

1. Fibroblasts and collagen fibers hold seminiferous tubules together.
2. Mast cells, macrophages, and lymphocytes occur here as well.

VI. TESTICULAR DRAINAGE DUCTS

A. TUBULI RECTI are straight portions of loops of seminiferous tubules. They are lined by Sertoli cells, and they join the **rete testis**.

B. RETE TESTIS AND EFFERENT DUCTULES

1. The rete testis conveys gametes from the tubuli recti to the **efferent ductules**. The efferent ductules are continuous with the rete testis.
2. The rete testis is lined by cuboidal ciliated cells.
3. The efferent ductules are lined by patches of tall and low columnar cells with or without cilia, they are surrounded by smooth muscle cells, and they empty into the epididymis.

C. EPIDIDYMIS

1. **Epithelial Structure.** The epididymis is lined by a tall pseudostratified epithelium.
   a. Tall cells have long, apical nonmotile stereocilia. These cells also are secretory, producing certain constituents of ejaculate.
   b. Short cells rest on a basement membrane and probably serve as a stem cell population to replace effete tall cells.
2. **Function.**
   a. Although its exact function still is not clear, the epididymis probably modifies the contents of ejaculate by removal and addition of various materials (i.e., stereocilia absorb and tall columnar cells secrete).
   b. **Maturation.** Spermatozoa from the head of the epididymis are poorly motile and incapable of fertilizing ova in vitro, whereas spermatozoa from the tail of the epididymis are highly motile and capable of fertilizing ova in vitro. This change in the spermatozoa is called maturation.

D. DUCTUS DEFERENS

1. **Gross Anatomy.**
   a. The tubular ductus deferens (vas deferens) receives spermatozoa and secretions from the epididymis and conveys them to the prostate. Near the epididymis the ductus deferens is relatively coiled, but close to the prostate it is rather straight.
   b. Proximal to the prostate gland, the ductus deferens is dilated into an **ampullary portion**, where it is joined by a sac-like diverticulum called the **seminal vesicle**.
   c. After they join, the ductus deferens and seminal vesicle enter the prostate as an **ejaculatory duct**.
2. **Microscopic Anatomy.**
   a. The ductus deferens is lined by a pseudostratified epithelium that is supported by testosterone secretion from Leydig cells. The tall columnar cells of this epithelium bear

long, nonmotile microvilli called stereocilia. The ductus deferens has a **mucosa**, a **muscularis**, and an **adventitia** (Fig. 22-4).

(1) The mucosa is thrown into gentle undulations, and it has a thin lamina propria.

(2) The muscularis is composed of three concentrically arranged layers—an inner and outer layer arranged longitudinally and a middle circular layer.

(3) The adventitia contains the usual small amounts of connective tissue, blood vessels, and nerves. It lacks a mesothelium.

b. In the ampullary portion of the ductus deferens, the mucosal folds are considerably higher and more complicated. The ampulla, like the seminal vesicle, looks like a gland.

E. EJACULATORY DUCTS

1. There are two ejaculatory ducts, each formed at the junction of a ductus deferens and seminal vesicle. The ejaculatory ducts penetrate the prostate and join the prostatic urethra in a single midline thickening of the mucosa known as the **colliculus seminalis**.

2. Each ejaculatory duct usually is lined by a pseudostratified epithelium, although columnar epithelium may be found here as well.

VIII. GLANDS OF THE MALE REPRODUCTIVE SYSTEM

A. SEMINAL VESICLES

1. **Structure.**

a. Like the ductus deferens, each seminal vesicle is lined by a pseudostratified epithelium that is supported by testosterone secretion from Leydig cells.

b. With the electron microscope, the tall columnar secretory cells of the epithelium show a basal rough endoplasmic reticulum, a prominent apical Golgi apparatus, and secretion granules.

c. The smooth muscle layers in the seminal vesicles are not neatly arranged; rather, they

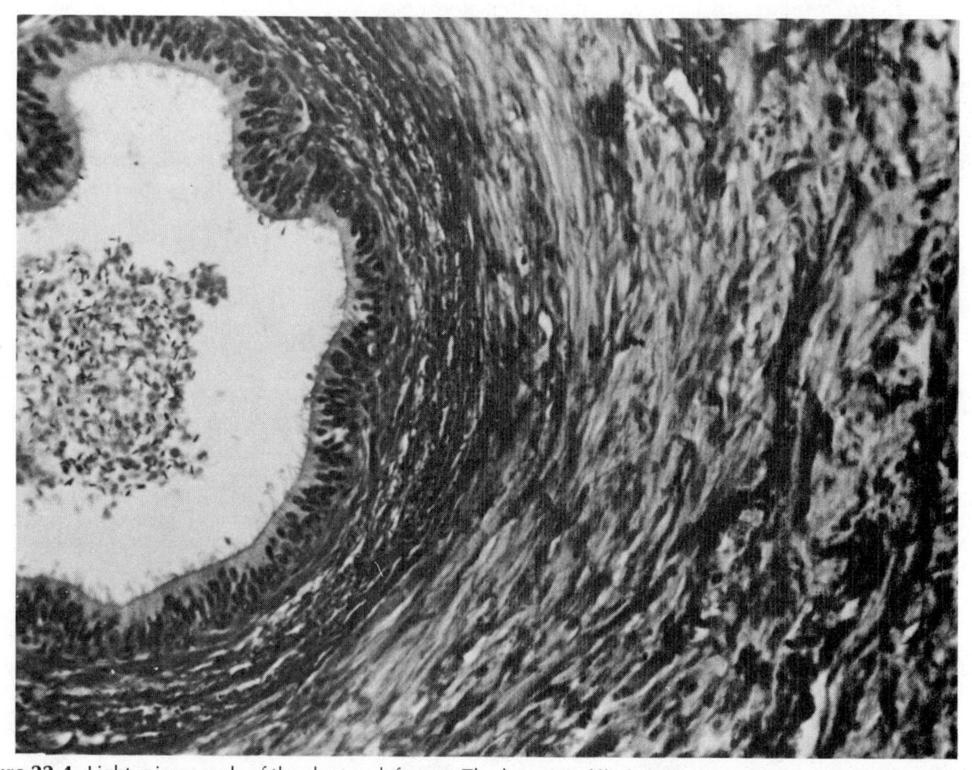

Figure 22-4. Light micrograph of the ductus deferens. The lumen is filled with numerous spermatozoa and is lined by a pseudostratified columnar epithelium with stereocilia. The lamina propria contains numerous elastic fibers, and the wall contains many smooth muscle cells. (Reprinted with permission from Johnson KE: *Histology: Microscopic Anatomy and Embryology*. New York, John Wiley, 1982, p 342.)

merge indistinctly into the smooth muscle surrounding the ejaculatory duct and into the prostate.

2. **Function.**
   a. The seminal vesicles produce a substantial volume of the ejaculate.
   b. The secretion of the seminal vesicles is viscous and yellowish and contains materials that are probably nutritive for the spermatozoa; it contains prostaglandins as well.

B. PROSTATE GLAND

1. **Structure.**
   a. The prostate gland lies just inferior to the bladder and receives ducts draining the urinary system (urethra) and reproductive system (paired ejaculatory ducts). The urinary and reproductive systems join at the prostatic urethra and exit the body through the penile urethra.
   b. The prostate is a compact mass of smooth muscle and elaborately branched glands. More detailed inspection reveals two types of glands.
      (1) The first is a relatively small set of periurethral mucosal glands.
         (a) Although small, these glands commonly undergo hyperplastic growth with advancing age. As a result, **benign prostatic hyperplasia** occurs, constricting the prostatic urethra.
         (b) Benign prostatic hyperplasia can be corrected surgically by a transurethral resection.
      (2) The second and more important are the main secretory glands of the prostate proper.
         (a) The epithelium of this gland is columnar or pseudostratified at the gland's main segment but grades into transitional epithelium where the gland empties into the prostatic urethra.
         (b) When viewed in the electron microscope the epithelial cells appear typical for protein-secreting cells.

2. **Function.**
   a. The prostate produces the majority of the volume of seminal fluid.
   b. The secretions of the prostate are rich in proteolytic enzymes (especially **fibrinolysin**), which probably function to liquefy the semen after it is deposited in the vagina.
   c. Presumably, this liquefaction allows spermatozoa to free themselves from the viscous ejaculate and begin traveling through the cervix and uterus to the fallopian tubes.
   d. These secretions solidify sometimes in the gland and can be seen, especially in older men, as **prostatic concretions**.

C. BULBOURETHRAL GLANDS

1. Just after the urethra leaves the prostate and enters the penis, it is joined by the ducts of the two paired bulbourethral glands (Cowper's glands).

2. These glands are typical compound tubuloalveolar glands with columnar epithelial cells lining the glandular lumina.

3. During sexual arousal, the bulbourethral glands secrete a mucoprotein which probably serves as a lubricant.

IX. PENIS.

The penis functions both as a sex organ and as an organ for urine excretion.

A. ANATOMY

1. The penis has paired dorsal **corpora cavernosa** and a single midline **corpus spongiosum**, which contains the urethra. These three bodies are bound together by a tough fibrous connective tissue capsule known as the **tunica albuginea**, and they are covered by a thin layer of skin that has an unusually rich sensory innervation.

2. The cavernous bodies of the penis consist of erectile tissue that grows and becomes rigid when filled with blood.

3. Helical arteries that drain into a venous plexus bring the cavernous bodies a rich supply of blood.

B. SEXUAL FUNCTION

1. During erection, the blood flow into the venous sinuses is greater than the blood flow out, causing the sinuses to fill with blood and causing tumescence of the organ.

2. Following a sexual encounter, the blood flow to these sinuses slows and the penis becomes flaccid again.

X. DEVELOPMENT OF THE MALE REPRODUCTIVE SYSTEM

A. TESTES develop from three rudiments.

1. **Primordial germ cells** migrate from the endodermal lining of the yolk sac along the hindgut mesentery into the genital ridges. These cells give rise to spermatogonia, spermatocytes, and eventually to spermatids and spermatozoa.

2. **Coelomic epithelial cells** migrate from the lining of the coelom into the genital ridges, surround primordial germ cells, and become Sertoli cells.

3. **Genital Ridge Mesenchyme.**
   a. This mesenchyme arises in situ from mesenchymal cells in the body wall that are associated with the mesonephros in the urogenital ridge.
   b. The cells here form the interstitial connective tissue elements, the blood vessels of the testis, and the hormone-producing Leydig cells.
      (1) Leydig cells are well-developed in the male fetus prior to birth.
      (2) They probably produce androgens to counteract maternal estrogens.

B. REPRODUCTIVE DUCTS

1. The mesonephric ducts (wolffian ducts) form the bulk of the epididymis, ductus deferens, and seminal vesicles.

2. The prostate forms from small glands that branch from the pelvic portion of the urogenital sinus into surrounding mesenchymal elements.

3. The penile urethra develops from the phallic portion of the urogenital sinus.

4. The paramesonephric ducts (mullerian ducts) almost completely degenerate but persist as the appendix testis and the prostatic utricle.

C. EXTERNAL GENITALIA

1. The genital tubercle forms the penis.

2. The genital tubercle forms the shaft of the penis, and the urethral folds unite to form the majority of the penile urethra.

3. The scrotal swellings form the scrotum.

STUDY QUESTIONS

Directions: Each question below contains five suggested answers. Choose the **one best** response to each question.

1. Which statement best describes the testis?

(A) It produces few mature spermatozoa before puberty
(B) Only proliferative cells are found in the seminiferous epithelium
(C) There is a large number of capillaries throughout the seminiferous epithelium
(D) Cells derived from primordial germ cells are not formed after puberty
(E) Spermatogenesis increases at elevated temperatures

2. Which of the following statements is true concerning the ductus deferens?

(A) Smooth muscle is absent in the wall
(B) It conveys spermatozoa to the prostate by ciliary currents
(C) It has an adventitia covered by a mesothelium
(D) It has a lumen lined by tall columnar cells with long apical stereocilia
(E) It carries mature spermatozoa to the epididymis

Directions: The question below contains four suggested answers of which **one or more** is correct. Choose the answer

A if **1, 2, and 3** are correct
B if **1 and 3** are correct
C if **2 and 4** are correct
D if **4** is correct
E if **1, 2, 3, and 4** are correct

3. Components of a spermatozoon include

(1) a streamlined nucleus
(2) microtubules
(3) a plasma membrane
(4) mitochondria

Directions: The group of questions below consists of lettered choices followed by several numbered items. For each numbered item select the **one** lettered choice with which it is **most** closely associated. Each lettered choice may be used once, more than once, or not at all.

Questions 4–7

For each description of characteristics of testicular ducts, choose the appropriate response.

(A) Epididymis
(B) Ductus deferens
(C) Both
(D) Neither

4. Lined by pseudostratified epithelium with stereocilia
5. All parts contain mature spermatozoa
6. Stimulated by testosterone
7. Derived from the mesonephric duct

ANSWERS AND EXPLANATIONS

1. The answer is A. (*III A, B*) The testis begins to produce mature spermatozoa after puberty, not before. The seminiferous epithelium contains some nonproliferative cells (e.g., Sertoli cells). The seminiferous epithelium, like all other epithelia in the body, is avascular. Spermatogenesis actually decreases at elevated temperatures, and the seminiferous epithelium will be damaged at body temperature.

2. The answer is D. (*VI D 2*) The ductus deferens has a good deal of smooth muscle in its wall. Peristaltic contractions of this smooth muscle are important for propelling spermatozoa; ciliary currents are not involved. Mesothelium does not cover the adventitia of this tube, and the tube carries capacitated sperm away from the epididymis. The ductus deferens is lined by a pseudostratified epithelium, and some of the cells in this epithelium are tall columnar cells with stereocilia.

3. The answer is E (all). (*IV B*) A spermatozoon is a complete cell, containing a streamlined nucleus and microtubules in the flagellum for propelling the cell. A spermatozoon also has mitochondria in the midpiece for adenosine triphosphate (ATP) production. Like all cells, a spermatozoon is surrounded by a plasma membrane.

4–7. The answers are: 4-C, 5-B, 6-C, 7-C. (*VI C, D; X B*) The epididymis and the ductus deferens both are stimulated by testosterone, are lined by a pseudostratified epithelium, and are derived from the mesonephric duct. Spermatozoa have matured by the time they enter the tail of the epididymis; however, they are not mature in the head of the epididymis.

23 Urinary System

I. INTRODUCTION

A. COMPONENTS

1. Kidneys
2. Ureters
3. Urinary bladder
4. Urethra

B. FUNCTIONS

1. Formation of urine, which contains wastes
2. Regulation of urine composition, which is essential for maintenance of the osmotic character of the body's internal milieu
3. Elimination of urine
4. Conveyance of gametes (in the male urethra)
5. Production of **erythropoietin**,which stimulates erythrocyte formation, and **renin**, which affects blood pressure

C. DEVELOPMENT

1. Three kidney-like structures take form.
   a. **Pronephros** never becomes functional.
   b. **Mesonephros** becomes functional and then degenerates.
   c. **Metanephros** becomes the definitive kidney.
2. Metanephric kidneys and ureters form from the ureteric bud and the metanephrogenic blastema.
3. The urinary bladder forms from the urogenital sinus.

II. MACROSCOPIC ANATOMY OF THE KIDNEY (Fig. 23-1)

A. GENERAL CONSIDERATIONS

1. The kidneys are paired bean-shaped organs located retroperitoneally, one on either side of the vertebral column.
2. They are 10–12 cm long, 5–6 cm wide, and 3–4 cm thick.
3. The kidneys are concave medially, with a **hilus** where the ureter, veins, and lymphatics exit and the arteries and nerves enter.

B. RENAL SINUS

1. Inward from the hilus and surrounded by the parenchyma of the kidney is a cavity called the **renal sinus**, which contains the **renal pelvis**.
2. At its upper portion the ureter expands into the renal pelvis, which in turn branches into two or three **major calyces**.
3. Several **minor calyces** branch from each major calyx.

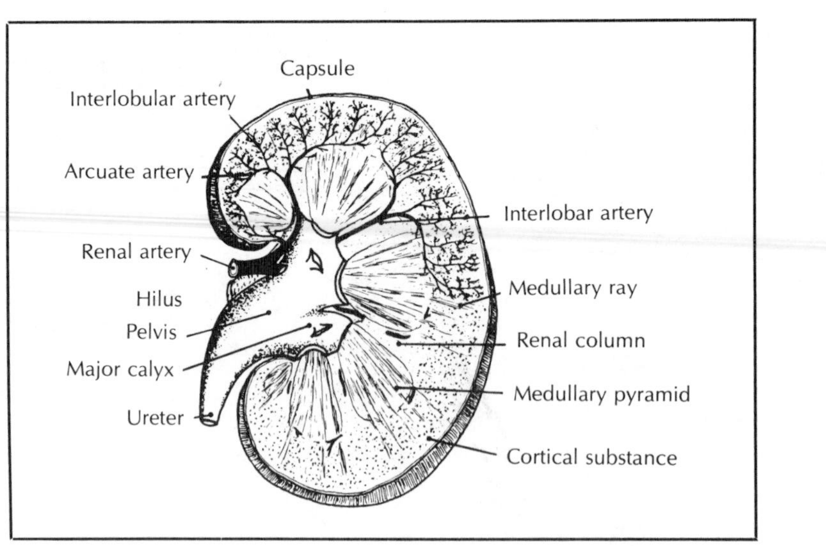

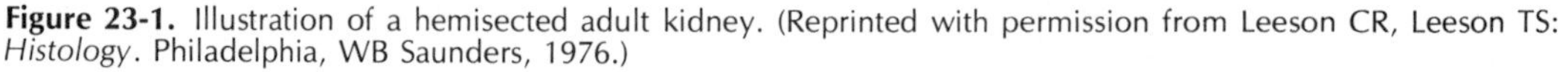

Figure 23-1. Illustration of a hemisected adult kidney. (Reprinted with permission from Leeson CR, Leeson TS: *Histology*. Philadelphia, WB Saunders, 1976.)

C. RENAL CORTEX AND RENAL MEDULLA

1. When a kidney is hemisected, a dark **cortex** and a light **medulla** are visible.
2. The cortex contains glomeruli and tubules. (The histologic details of the renal cortex are shown in Figure 23-2.)
3. The medulla is made up of many **renal pyramids**. The bases of the pyramids point toward the cortex, and the apices of the pyramids end in each of the minor calyces. The apex of a renal pyramid often is called the **papilla**.
   a. Each pyramid has uriniferous tubules and blood vessels.
   b. The tip of each papilla is perforated by many **papillary ducts**, which empty into minor calyces at the **area cribrosa**.
4. Between the renal pyramids, bits of cortical tissue called **renal columns** project toward the hilus.
5. Striations called **medullary rays** project from the bases of the renal pyramids into the cortex. Each medullary ray and its surrounding cortical tissue comprise a **renal lobule**. (Renal lobules are not demarcated by connective tissue septa as in some other glands.)

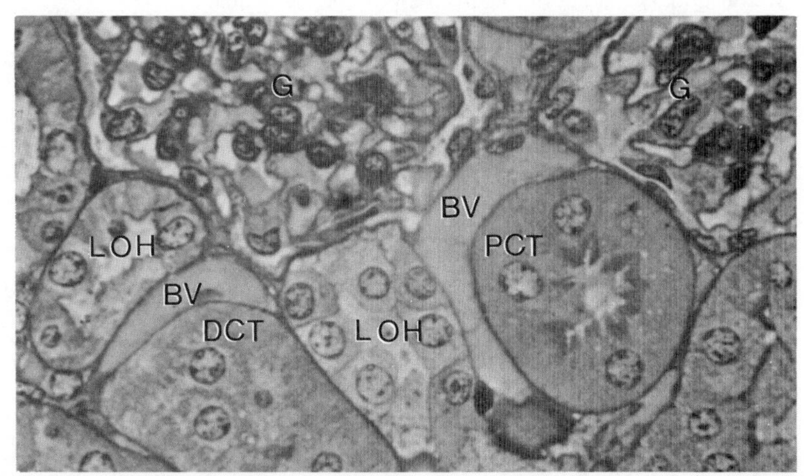

Figure 23-2. Light micrograph of the renal cortex showing proximal convoluted tubules (*PCT*), distal convoluted tubules (*DCT*), segments of the loop of Henle (*LOH*), blood vessels (*BV*), and glomeruli (*G*). (Reprinted with permission from Johnson KE: *Histology: Microscopic Anatomy and Embryology*. New York, John Wiley, 1982, p 355.)

III. HISTOLOGY OF URINIFEROUS TUBULES. These tubules are composed of **nephrons** for producing urine and **collecting tubules** for concentrating and conducting urine.

A. THE NEPHRON (Fig. 23-3) is the most complex element of the uriniferous tubule, both structurally and functionally.

1. **Bowman's capsule** is a distended epithelial structure with a deep indentation that is occupied by a glomerulus of fenestrated capillaries. Like all epithelia, Bowman's capsule rests on a basement membrane.
   - **a.** The parietal portion of Bowman's capsule is a simple squamous epithelium, whereas the visceral portion is composed of complex and highly modified epithelial cells called **podocytes** (Fig. 23-4).
   - **b.** The basement membrane separating the podocytes from the fenestrated glomerular capillaries is extremely thick and serves as one site of blood-protein retention.
   - **c.** Each podocyte has a cell body from which 5 to 10 **primary processes** radiate.
   - **d.** Each primary process branches extensively into **secondary processes** (sometimes called **pedicels** or **foot processes**). Pedicels of adjacent primary processes—from the same podocyte as well from different podocytes—interdigitate into a layer of processes with slits between them.
   - **e.** Across each slit is a thin **diaphragm** that aids in excluding certain blood constituents from urine. Together the slit diaphragm and basement membrane form the only barrier between the vascular space of the capillary and the urinary space.

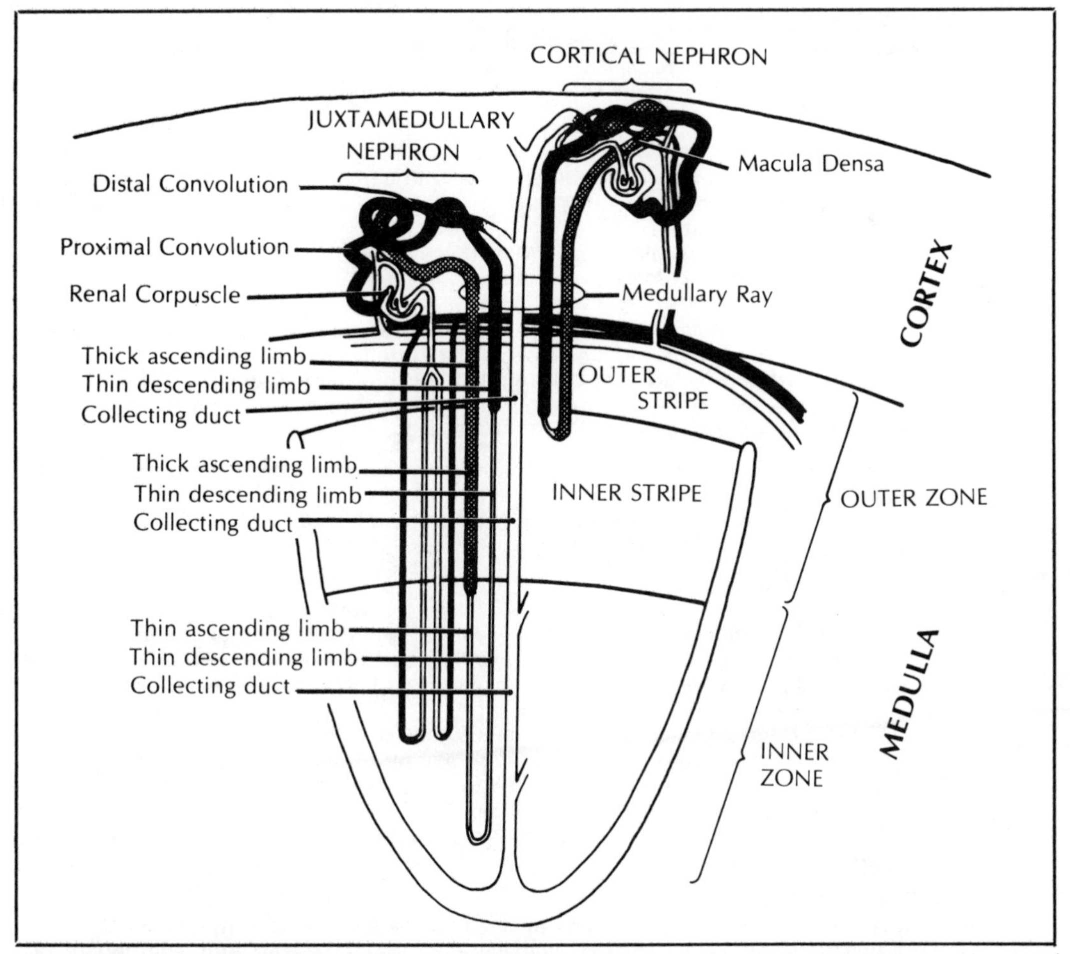

Figure 23-3. Diagram illustrating the difference between the anatomy of cortical and juxtamedullary nephrons and the relationship between nephrons and collecting tubules. (Reprinted with permission from Weiss L, Greep RO: *Histology*. New York, McGraw-Hill, 1977.)

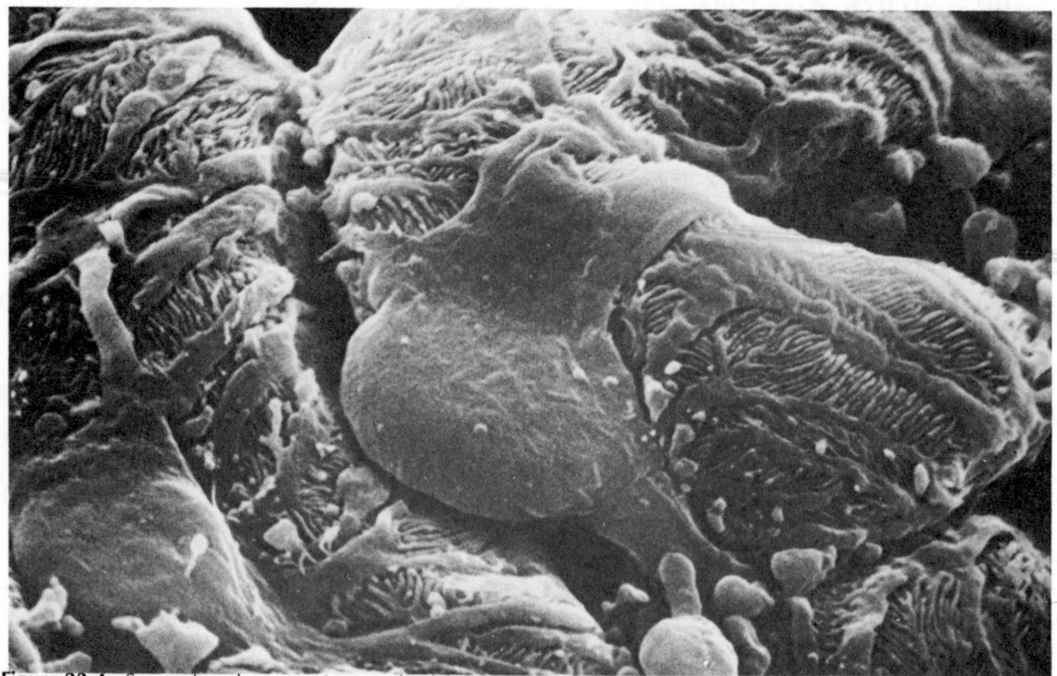

Figure 23-4. Scannning electron micrograph of the podocytes comprising the visceral layer of Bowman's capsule, viewed from the urinary space. (Courtesy of Dr. Marilyn J. Koering, Department of Anatomy, George Washington University.)

(1) A molecule of urea in the blood passes through the pore of the capillary, through the basement membrane, and through the slit diaphragm into the urinary space of Bowman's capsule.

(2) When large molecular weight (MW) tracer molecules are injected into the afferent vessels of a glomerulus, their different paths can be followed with the electron microscope (Fig. 23-5).

(a) Ferritin (MW = 400,000) escapes the fenestrated capillary but accumulates at the basement membrane on the capillary side.

(b) Myeloperoxidase (MW = 160,000) escapes the fenestrated capillary, crosses the basement membrane, and accumulates at the slit diaphragm.

(c) Horseradish peroxidase (MW = 40,000) enters the urine.

2. Renal Corpuscle. The glomerulus and Bowman's capsule together comprise a **renal corpuscle** which has a **vascular pole**, where afferent vessels enter the glomerulus and efferent vessels leave it, as well as a **urinary pole**, where the epithelium of Bowman's capsule is continuous with the **proximal tubule**.

3. Proximal Tubule.

a. The proximal tubule is convoluted as it leaves the Bowman's capsule and straight as it approaches the **thin segment** of the descending limb of the **loop of Henle**.

b. When viewed in cross section, the proximal tubule appears to have a cuboidal epithelium. Actually the cells are shaped more like pieces of a jigsaw puzzle, and their apical surface is covered with microvilli.

c. There is clear evidence that active pinocytosis of markers from the urine occurs in the proximal tubule. Horseradish peroxidase that leaves the blood and enters the urine also appears in proximal tubule cells that have recaptured the marker protein.

d. The cytoplasm of a proximal tubule cell is eosinophilic and loaded with mitochondria. These mitochondria are elongated on an axis perpendicular to the basement membrane, and in well-fixed specimens, the basal striations of proximal tubule cells are evident.

4. Distal Tubule.

a. The straight portion of the proximal tubule passes into the **thin segment**, which is a squamous epithelium with no microvilli. After forming the **ascending limb** of the loop of Henle, the thin segment becomes the epithelium of the **distal tubule**.

b. The distal tubule is straight at the thin segment and convoluted at its distal portion.

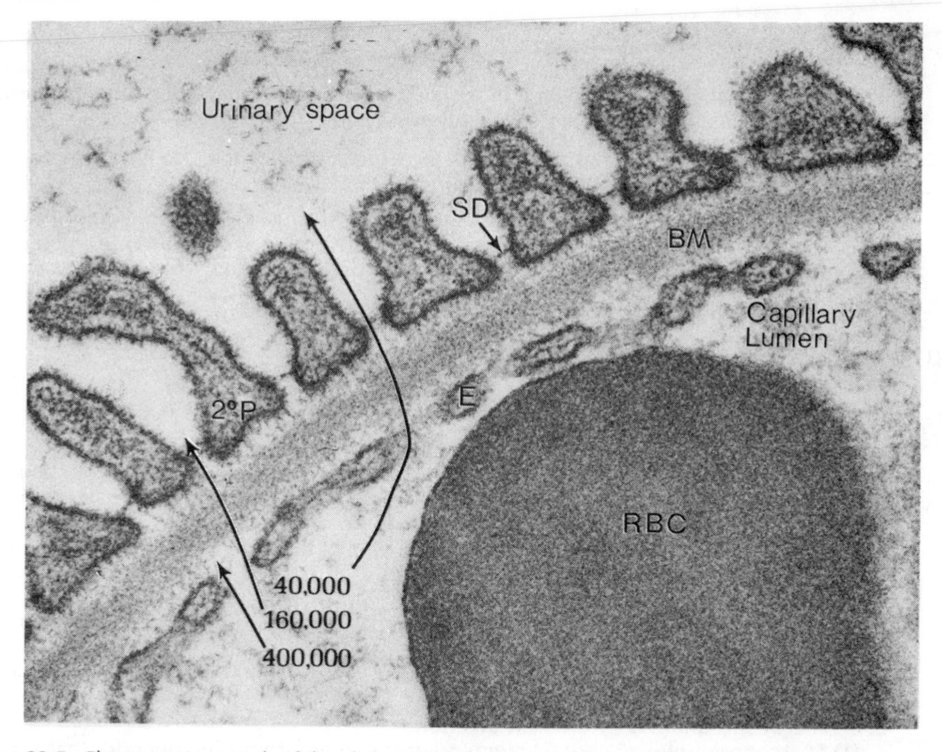

Figure 23-5. Electron micrograph of the glomerular basement membrane (*BM*). The capillary lumen is lined by a fenestrated endothelium (*E*), and the urinary space is lined by a complex barrier formed by podocyte secondary processes (*2°P*) and slit diaphragms (*SD*). Note that 400,000 molecular weight (MW) substances do not cross the glomerular basement membrane; 160,000 MW substances do not cross the slit diaphragm; and 40,000 MW substances enter the glomerular filtrate. (Reprinted with permission from Weiss L, Greep RO: *Histology*. New York, McGraw-Hill, 1977.)

c. Unlike proximal tubules, distal tubules have cells that are pale staining, have few microvilli, and have prominent basal mitochondria.

d. Ions are actively pumped from the distal tubule into the interstitial fluid compartment. The basal mitochondria are involved in this ion pumping.

5. Loop of Henle. Variations in the lengths of different parts of the loop of Henle depend on whether nephrons are cortical or juxtamedullary (see Fig. 23-3).

B. COLLECTING TUBULES

1. In the collecting tubules, the basal mitochondria disappear and the cells grow in height as the diameter of the collecting tubules increases until the point where they join papillary ducts.

a. Papillary ducts empty into minor calyces at the area cribrosa.

b. Papillary ducts are lined by simple columnar epithelium.

2. Antidiuretic hormone (ADH) increases the water permeability of the collecting tubules, causing the production of a smaller volume of more concentrated urine.

IV. JUXTAGLOMERULAR APPARATUS.

In addition to its function in the production of urine, the kidney also secretes a hormone that is involved in the regulation of blood presssure and blood volume. This hormone is produced in the juxtaglomerular apparatus (JGA).

A. STRUCTURE.

The JGA is composed of three types of cells.

1. Specialized cells in the wall of the distal convoluted tubule at the vascular pole of the renal corpuscle comprise the **macula densa.**

2. Modified smooth muscle cells in the wall of the afferent arteriole, called the **juxtaglomerular cells**, contain granules of a protein called **renin**.

3. Mesangial cells (also called **lacis cells**) are the third component of the JGA. They are closely associated with glomerular capillaries near the vascular pole.

B. FUNCTION

1. Renin is an enzyme that is produced in the JGA. It acts on blood protein **angiotensinogen** to cleave off the inactive decapeptide, **angiotensin I**.
2. Another enzyme in the lungs cleaves off two more amino acids to produce the potent vasoconstrictor octapeptide, **angiotensin II** (Fig. 23-6).
3. In addition to its vasoconstrictor activity, angiotensin II causes the **zona glomerulosa** of the adrenal gland to secrete the mineralocorticoid, **aldosterone**. This hormone influences salt resorption in the kidney, thereby regulating blood and tissue fluid volume.
4. There is some evidence that the JGA produces erythropoietin.

V. BLOOD, LYMPH, AND NERVE SUPPLY OF THE KIDNEY

A. BLOOD. A large volume of blood—1.2 liters—flows through the kidneys each minute.

1. **Arteries.**
   a. The **renal artery** branches from the abdominal aorta to supply each kidney. It then divides into a dorsal branch, a ventral branch, and several **interlobar arteries**, all of which are located in the connective tissue at the hilus of the organ.
   b. The interlobar arteries then enter the parenchyma of the kidney. At the junction between the cortex and medulla, the interlobar arteries arch parallel to the convex surface of the kidney and form the **arcuate arteries**.
   c. Small **interlobular arteries** branch from the arcuate arteries and immediately penetrate the cortex, extending all the way to the capsule.
   d. Numerous small branches of the interlobular arteries enter the glomeruli as **afferent arterioles**.
   e. Glomeruli are drained by **efferent arterioles.**
      (1) Efferent arterioles of cortical glomeruli form an anastomosing network that surrounds tubules.
      (2) Efferent arterioles of juxtamedullary glomeruli descend into the medulla to form the **vasa recta**, which form hairpin loops at different levels in the medulla and then return to the cortex to become veins (see Fig. 23-2).
2. **Veins.**
   a. Superficial capillaries of the cortex drain into **superficial cortical veins**, then into **interlobular veins**, next into **arcuate veins**, and finally into the **renal veins**.
   b. Capillaries deep in the cortex drain into **deep cortical veins**, which run parallel to interlobular arteries. The deep cortical veins then drain into **interlobular veins** and finally into the **renal veins**.

B. LYMPH AND NERVES

1. Lymphatic vessels are found in both the capsules and the parenchyma of the kidney. Lymphatic capillaries form dense networks around the cortical tubules, which drain into larger lymphatics that leave the organ with the blood vessels.
2. There is debate over whether or not uriniferous tubules have a nerve supply.

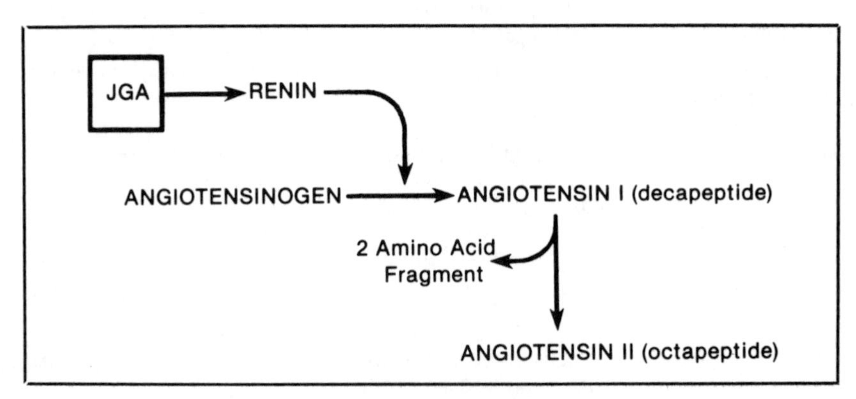

Figure 23-6. Diagram illustrating the role of the juxtaglomerular apparatus and renin in the production of angiotensin II. (Reprinted with permission from Johnson KE: *Histology: Microscopic Anatomy and Embryology*. New York, John Wiley, 1982, p 358.)

VI. URETERS, URINARY BLADDER, AND URETHRA

A. TRANSITIONAL EPITHELIUM

1. The passages for urine excretion are lined with a peculiar transitional epithelium. When the urinary passages are distended, the transitional epithelium is thinner and has fewer cell layers from the basement membrane to the lumen than when the bladder is empty.
2. With the electron microscope, the superficial epithelial cells show a peculiar modification. The surfaces of the cells are elaborately **scalloped**, and their apices are loaded with flattened **fusiform vesicles** that can insert their membranes into the surfaces of the superficial epithelial cells. The fusiform vesicles presumably represent a reserve of cell surface for when the bladder fills and empties.
3. The membrane facing the free surface of the urinary tract also is unusual in that it contains membrane particles in a hexagonal array that is distinct from the hexagonal array for particles in the gap junction. These membrane subunits may be related to the unique permeability characteristics of the urinary passages.
4. When urine reaches the urinary passages it is **hypertonic** to the blood, and the transitional epithelium presumably serves as an effective barrier to diffusion.
5. The epithelium rests on a relatively thin basement membrane.

B. MUSCULAR COATS

1. The sparse lamina propria grades into **smooth muscular coats** that do not form distinct layers as they do in the gastrointestinal tract.
2. The arrangement of these smooth muscle fibers, when evident, is as follows.
   a. The upper two-thirds of the ureter have an inner longitudinal layer and an outer circular layer.
   b. The remaining lower portion of the ureter has an inner and an outer longitudinal layer and a middle circular layer.
   c. The smooth muscle fibers in the urinary bladder have the same arrangement as in the distal portion of the ureters. (The microscopic anatomy of the urinary bladder is shown in Figure 23-7.)

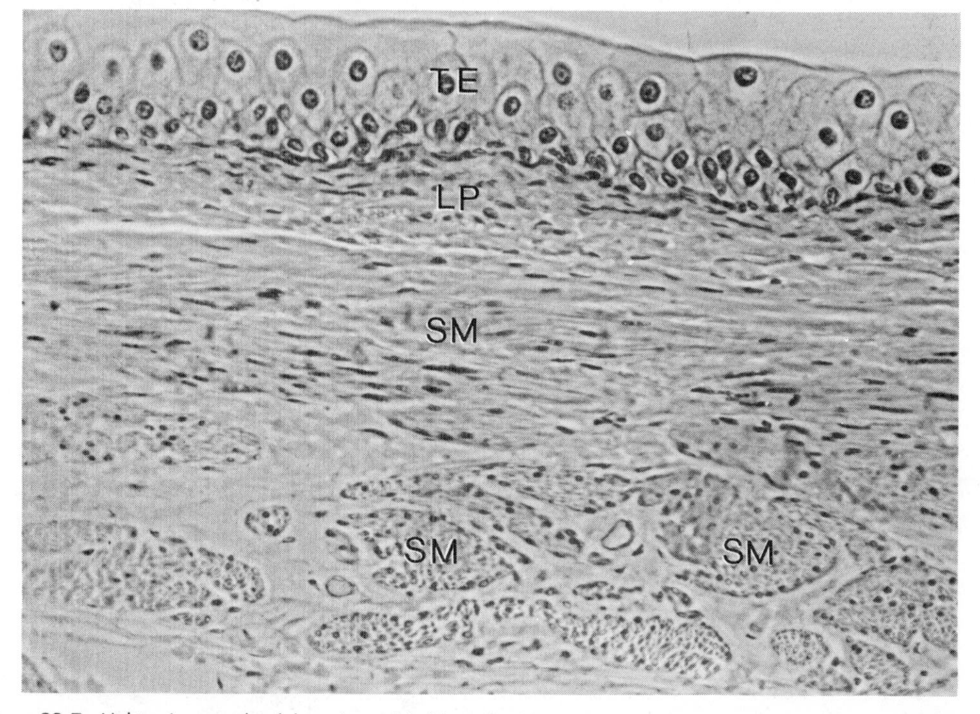

Figure 23-7. Light micrograph of the urinary bladder. The lumen of the bladder, at the top of the photograph, is lined by a transitional epithelium (*TE*) that rests on a thin lamina propria (*LP*). Several different arrangements of smooth muscle (*SM*) masses are visible.

C. VARIATIONS WITH SEX

1. **Male Urethra.**
   - a. The male urethra leaves the bladder, penetrates the prostate and the urogenital diaphragm, and exits through the penis.
   - b. The **pars membranacea** is lined with stratified columnar or pseudostratified columnar epithelium.
   - c. The **pars spongiosa** has patches of stratified squamous epithelium in the fossa navicularis.
2. **Female Urethra.** The female urethra usually is lined with stratified squamous epithelium but may be lined with pseudostratified columnar epithelium.

VII. DEVELOPMENT OF THE URINARY SYSTEM

A. PRONEPHROS.
This rudiment forms and degenerates rapidly in humans. It never forms functional nephrons.

B. MESONEPHROS

1. This structure probably forms functional nephrons and hypotonic urine before it degenerates.
2. The mesonephric duct forms a branch called the **ureteric bud**, which forms the collecting system in the definitive kidney.
3. The mesonephric duct also contributes to the development of the male reproductive tract.

C. METANEPHROS

1. The ureteric bud, a branch of the mesonephric duct, forms collecting tubules, papillary ducts, minor and major calyces, the lining of the renal pelvis, and the ureters.
2. The metanephrogenic blastema forms from the mass of mesenchyme at the tips of the ureteric buds and gives rise to the glomerulus, Bowman's capsule, proximal and distal convoluted tubules, and the loop of Henle.

D. UROGENITAL SINUS

1. This sac forms from the hindgut. The urorectal septum divides the cloaca into the anterior urogenital sinus and the posterior rectum.
2. The urogenital sinus forms the bladder and part of the urethra.

STUDY QUESTIONS

Directions: The question below contains five suggested answers. Choose the **one best** response to the question.

1. All of the following statements concerning the kidney are true EXCEPT

(A) renal pyramids empty into the major calyces
(B) papillary ducts empty into the minor calyces through the area cribrosa
(C) each major calyx connects with more than one minor calyx
(D) cortical tissue occurs in the renal columns between pyramids
(E) some nephrons have medullary components

Directions: Each question below contains four suggested answers of which **one or more** is correct. Choose the answer

A if **1, 2, and 3** are correct
B if **1 and 3** are correct
C if **2 and 4** are correct
D if **4** is correct
E if **1, 2, 3, and 4** are correct

2. True statements concerning cells in the proximal convoluted tubule include which of the following?

(1) They are involved in modifying the composition of the filtrate
(2) They can phagocytose proteins from the glomerular filtrate
(3) They have a prominent brush border composed of microvilli
(4) They contain few mitochondria

3. True statements concerning the juxtaglomerular apparatus include which of the following?

(1) Macula densa cells are part of the proximal convoluted tubule
(2) Renin granules occur in juxtaglomerular cells
(3) Mesangial cells are part of the distal convoluted tubule
(4) Juxtaglomerular cells are part of the wall of the afferent arteriole

Directions: The group of questions below consists of lettered choices followed by several numbered items. For each numbered item select the **one** lettered choice with which it is **most** closely associated. Each lettered choice may be used once, more than once, or not at all.

Questions 4–8

For each description of components of the renal glomerulus and Bowman's capsule, choose the appropriate lettered structure shown in the micrograph below. (In this micrograph, there are several examples of some options.)

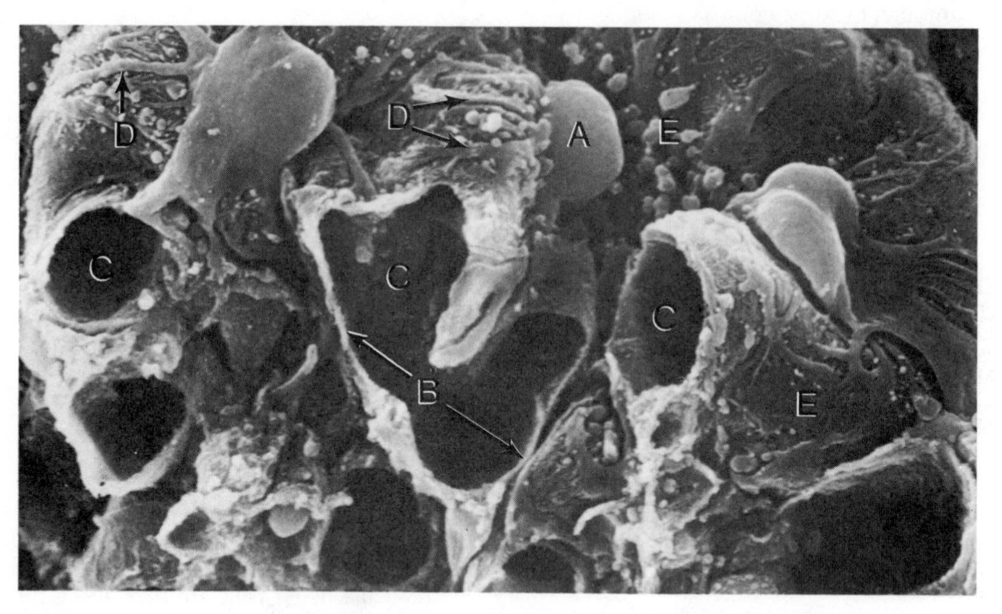

Courtesy of Dr. Marilyn J. Koering, Department of Anatomy, George Washington University.

4. Red blood cells move through this vascular channel

5. The initial portion of the protein sieving mechanism of the kidney

6. Podocyte primary process

7. Component characterized by transcellular fenestrations without diaphragms

8. The visceral portion of Bowman's capsule

ANSWERS AND EXPLANATIONS

1. The answer is A. (*II B, C*) Renal pyramids empty into minor calyces, which in turn join with major calyces. Minor calyces and renal pyramids are far more numerous than major calyces. The papillary ducts drain into minor calyces at the area cribrosa. Some nephrons have long loops of Henle that reach into the renal medulla.

2. The answer is A (1, 2, 3). (*III A 3*) The cells of the proximal convoluted tubule modify the composition of the glomerular filtrate by active transport and phagocytosis of proteins. These functions require adenosine triphosphate (ATP), which is produced in the numerous mitochondria in these epithelial cells. These cells have a prominent apical brush border.

3. The answer is C (2, 4). (*IV A*) The macula densa is part of the distal convoluted tubule. The juxtaglomerular cells are part of the afferent arteriole, and they contain renin granules.

4–8. The answers are: 4-C, 5-B, 6-D, 7-C, 8-A. (*III A 1, 2*) This is a scanning electron micrograph of a fractured renal glomerulus. Podocytes (A) have large numbers of primary processes (D), which are interdigitated in a complex fashion. Blood flows through the capillaries of the renal glomerulus (C), and a filtrate of blood passes through the glomerular basement membrane (B) into the urinary space (E) of Bowman's capsule.

24
The Eye

I. INTRODUCTION

A. COMPONENTS

1. The **cornea** is the transparent anterior portion of the capsule of the eye.
2. After passing through the cornea, light passes through a variable diaphragm called the **iris**.
3. The aperture of the iris, the **pupil**, changes automatically by the contraction of smooth muscle fibers. Triggering this reaction are changes in the light levels in the environment, in the subject of inspection, or in both.
4. After passing through the anterior chamber and the iris, light passes through the **lens**.
5. Light subsequently traverses the **vitreous body** (humor) and strikes the **retina**, where the light evokes an action potential in cells that synapse with the **cones** and **rods** (photoreceptors).
6. Nerve impulses then are carried to the brain via the **optic nerve**.

B. FUNCTIONS

1. The eye receives a variety of visual stimuli that can have significant effect on human life and behavior.
2. The eye is an elegantly constructed transducer that turns light into electrical impulses and transmits these electrical impulses to the brain for processing.
3. In addition, the eye has a mechanism for the formation of images.

II. SCLERA AND CORNEA

A. SCLERA

1. This tough fibrous layer of connective tissue forms in the outer layer of the eye. Tendons of the **oculomotor muscles** insert into the sclera.
2. The connective tissue of the sclera is rich in collagenous fibers which run in different directions and have fibroblasts. It also has numerous elastic fibers.

B. CORNEA

1. This anterior portion of the outer capsule of the eye is highly modified and clear. It is slightly thicker than the sclera and has a smaller radius of curvature.
2. The high refractive index (1.38) of the cornea and its small radius of curvature are two properties that make the cornea extremely important in image formation.
3. The cornea has five layers.
   - **a.** The first layer, the **corneal epithelium**, is a stratified squamous epithelium which is five to seven cell layers thick (Fig. 24-1).
     - **(1)** There are several free nerve endings in the corneal epithelium, and when stimulated they cause the blinking reflex (i.e., the eyelids close over the cornea to protect it).
     - **(2)** The corneal epithelium has a remarkable capacity to heal its own wounds and to regenerate. Minor wounds are closed by cell migration; larger wounds are healed by mitosis in the basal cell layers and by the production of new cells.
   - **b.** The corneal epithelium rests on the second layer, **Bowman's membrane**. This clear, acellular layer is 6–9 μm thick and actually is not a membrane but a composite of the basement membrane of the corneal epithelium and the outer layer of the underlying substantia propria.

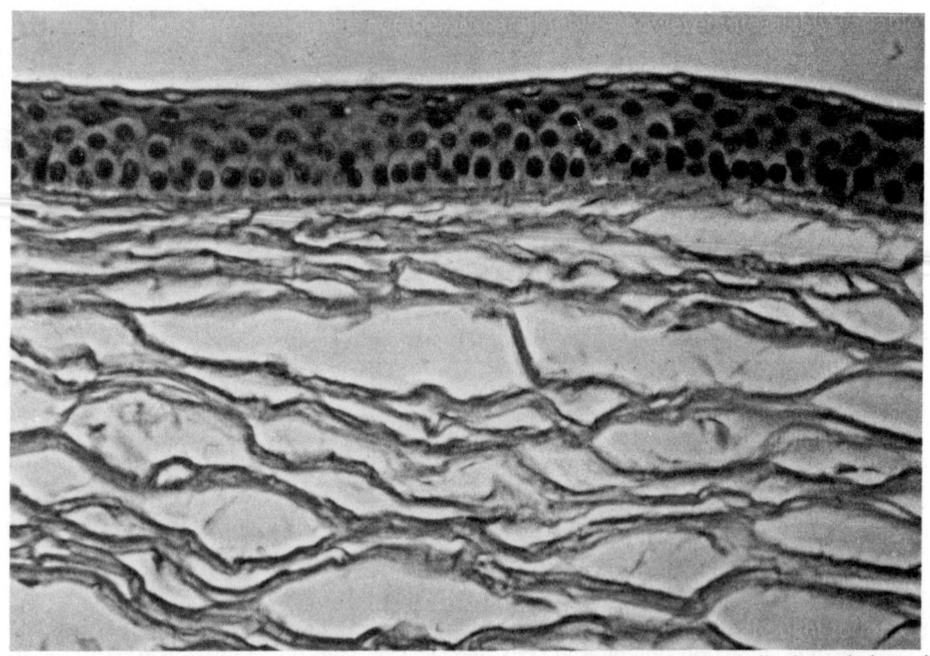

Figure 24-1. Light micrograph of the corneal epithelium. The stroma below it is abnormally distended as a fixation artifact. (Reprinted with permission from Johnson KE: *Histology: Microscopic Anatomy and Embryology*. New York, John Wiley, 1982, p 380.)

c. The **substantia propria (stroma)** is the third and thickest layer, comprising about 90 percent of the total corneal thickness.

(1) It is a mixture of fibroblastic cells, collagen fibers, and an amorphous ground substance that is rich in the glycosaminoglycans, chondroitin sulfate (the same material found in cartilage) and keratan sulfate.

(2) There are no blood vessels in this thick layer of connective tissue.

d. The fourth layer, **Descemet's membrane**, is an extremely thick basement membrane.

(1) It contains a peculiar kind of hexagonally arranged collagen, which is especially prominent in older persons.

(2) Probably, Descemet's membrane is the basement membrane of, and is secreted by, the fifth corneal layer.

e. The **corneal endothelium**, the fifth layer, is a simple squamous epithelium.

4. Corneal Transparency.

a. The transparency of the cornea is due to the regular arrangement of its fibrous elements.

b. The fibers in turn are probably kept in close register by the glycosaminoglycans in the cornea.

c. The embryonic cornea is not transparent because the stroma fibroblasts have not yet secreted and accumulated glycosaminoglycans around the collagen fibers.

d. The cornea is avascular and receives nutrition by diffusion from the aqueous humor, at the center, and from blood vessels in the limbus (corneal-scleral junction), at the periphery.

C. LIMBUS (Fig. 24-2)

1. The cornea and the sclera are continuous structures and are part of the outer layer of the eye. They join at the limbus.

2. The anatomy of this region is complex, and the area is quite significant clinically. Structures that regulate the outflow of aqueous humor from the eye are located here.

3. Failure in the regulation of outflow can cause an increase in the intraocular pressure, which is a characteristic feature of a severe and common disease called **glaucoma**.

4. At the junction of the cornea and iris is a cavity occupied by the **trabecular meshwork**. This complex network of anastomosing connective tissue cords is covered by an endothelium that is continuous with the corneal endothelium.

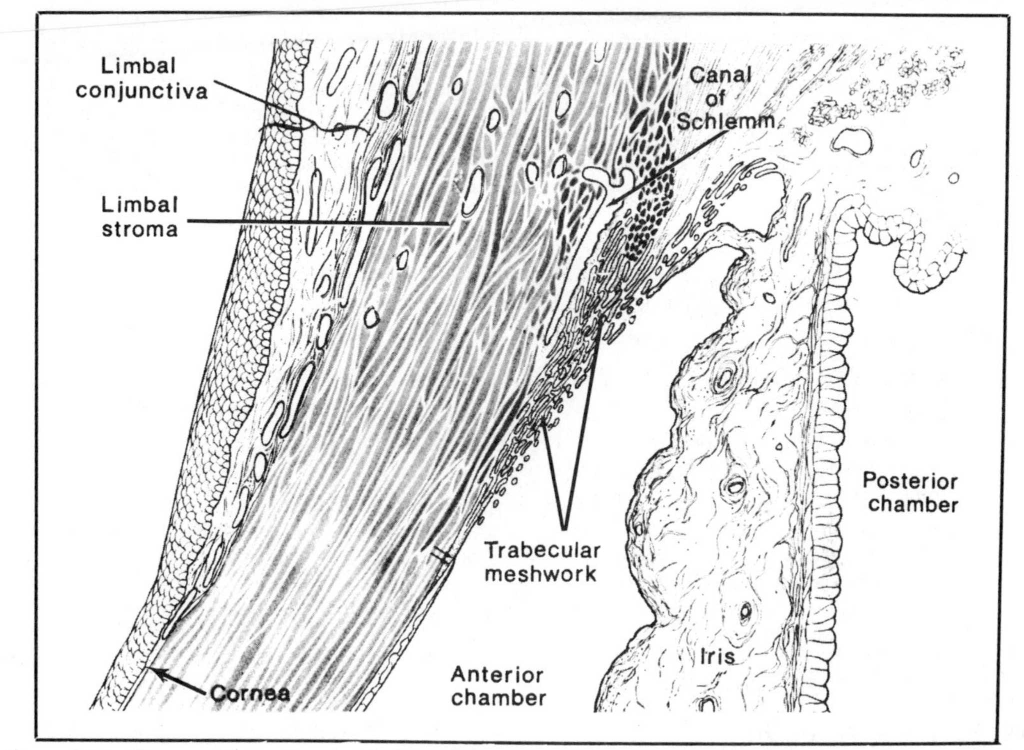

Figure 24-2. Diagram illustrating the limbus of the human eye. (Reprinted with permission from Hogan MJ, Alvarado JA, Weddell JE: *Histology of the Human Eye.* Philadelphia, WB Saunders, 1971.)

5. Deep to the trabecular meshwork is the **canal of Schlemm**; together these two structures comprise the apparatus that regulates the outflow of aqueous humor from the anterior chamber.

III. VASCULAR LAYER

A. CHOROID LAYER

1. This thin pigmented layer lies just beneath the sclera.
2. It is composed of the following layers (from outer to inner layer).
   a. A vessel layer of arterioles and venules
   b. A capillary layer
   c. Bruch's membrane

B. CILIARY BODY

1. The ciliary body occupies the space between the edge of the visual retina, called the **ora serrata** or ora terminalis, and the edge of the lens.
2. A number of long, radially arranged **ciliary processes** project from the thick ciliary body toward the lens.
3. The main mass of the ciliary body, exclusive of the ciliary processes, is composed of the muscles of accommodation, the **ciliary muscles**.
   a. **Contraction** of the ciliary muscles **reduces** tension on the ligaments that attach to the lens, and the lens become more convex.
   b. This allows the lens to focus images of **nearby** objects on the retina. (To remember this mechanism, recall that eyes become fatigued due to smooth muscle contraction after prolonged reading.)
4. The surface of the ciliary body that faces the central cavity of the eye is covered by a heavily pigmented layer of the retina. This layer lacks photoreceptors and extends anteriorly from the ora serrata as the **ciliary epithelium**.

a. The ciliary epithelium actually covers the ciliary body with two layers, an outer layer of unpigmented cells and a deep layer of pigmented cells. Both cell layers rest on a basement membrane.

b. Also, the ciliary epithelium secretes the aqueous humor.

(1) The aqueous humor produced in the posterior chamber nourishes the lens and other inner structures in the eye, after which it flows over the lens and through the **pupil**.

(2) The aqueous humor then flows into the anterior chamber, through the trabecular meshwork, and drains out through the canal of Schlemm.

C. IRIS

1. The posterior surface of the iris rests on the lens and together with the lens forms the boundary between the anterior and posterior chambers.

2. The iris has a ciliary margin at the ciliary body and a pupillary margin at the pupil.

3. The main mass of the iris consists of a loose, highly pigmented connective tissue mass with many blood vessels.

4. The iris is colored and its different colors depend on the quantity and arrangement of melanin and the thickness of the connective tissue lamellae.

5. The pupil (aperture in the iris) varies in size with the amount of light in the environment.

a. Cells derived from the anterior pigmented layer of the iris lose pigment and differentiate into contractile cells, the **myoepithelium** of the pupillary muscles.

b. There are two groups of these cells.

(1) One group is arranged concentrically around the pupil; contraction of these cells reduces the aperture of the iris.

(2) The other group is arranged radially around the pupil and upon contraction increases the aperture of the iris.

IV. REFRACTIVE MEDIA.

In addition to the previously described cornea and anterior chamber of the eye, the refractive apparatus includes the **lens** and the **vitreous humor**.

A. LENS

1. The lens is a transparent biconvex structure located immediately posterior to the pupil.

2. Its diameter is approximately 7 mm, and depending on the degree of contraction or relaxation of the ciliary muscle, it is 3.7–4.5 mm thick.

3. The lens proper is made up of **lens fibers**, which are highly modified epithelial cells.

a. Each is a hexagonal prismatic structure that is 8–12 μm wide, about 2 μm thick, and 7–10 mm long.

b. Each lens fiber contains a nucleus but is relatively devoid of organelles and intracellular inclusions.

c. The plasma membranes of adjacent lens fibers are separated by intercellular clefts 15 mm wide, and these membranes are elaborately interdigitated.

4. The lens is held in position by a series of fibers called the **ciliary zonule**.

a. These fibers arise from the epithelium of the ciliary body.

b. With the electron microscope they are seen as small hollow tubular structures that are different from microtubules.

B. THE VITREOUS BODY

(vitreous humor) fills the cavity between the lens and the retina.

1. The vitreous body is 99 percent water; the bulk of the nonaqueous portion is composed of hyaluronic acid—the glycosaminoglycan that is found also in cartilage and in the umbilical cord as Wharton's jelly.

2. Also found here is a random network of collagen fibers which lack the usual collagen periodicity.

3. **Hyalocytes** exist here in addition to higher concentrations of collagen and hyaluronate at the periphery of the vitreous body. The hyalocytes are thought to secrete the collagen and hyaluronate.

4. The remnant of the embryonic hyaloid artery passes through the middle of the vitreous body as the hyaloid canal.

V. RETINA.

The retina of the eye has two main components, a **pigmented retina** and a **neural retina**. The pigmented retina is next to the choroid layer and the neural retina is next to the vitreous body.

A. PIGMENTED RETINA

1. The retina contains a pigmented retina, which is a simple layer of cells rich in melanin granules.
2. The pigmented retinal epithelium has an extensive network of tight junctions between the individual cells.
   a. It is thought that this tightly joined layer of cells represents a barrier between the blood and the neural retina.
   b. The pigmentation prevents extraneous light from stimulating the retina and absorbs light after it has passed through the neural retina.
3. The basement membrane of the pigmented retina is part of Bruch's membrane.

B. NEURAL RETINA

B. NEURAL RETINA (Fig. 24-3). The neural retina has two types of photoreceptor neurons: rods and cones. Both types have an **outer segment** consisting of multiple layers of stacked plasma membranes; this segment is rich in the photoreceptive substances **rhodopsin** (in rods) and **iodopsin** (in cones).

1. **Rods.**
   a. Rods are long and thin and make connection with a large number of other cells.
   b. Rods function at their peak in low levels of illumination and are responsible for **night vision**.

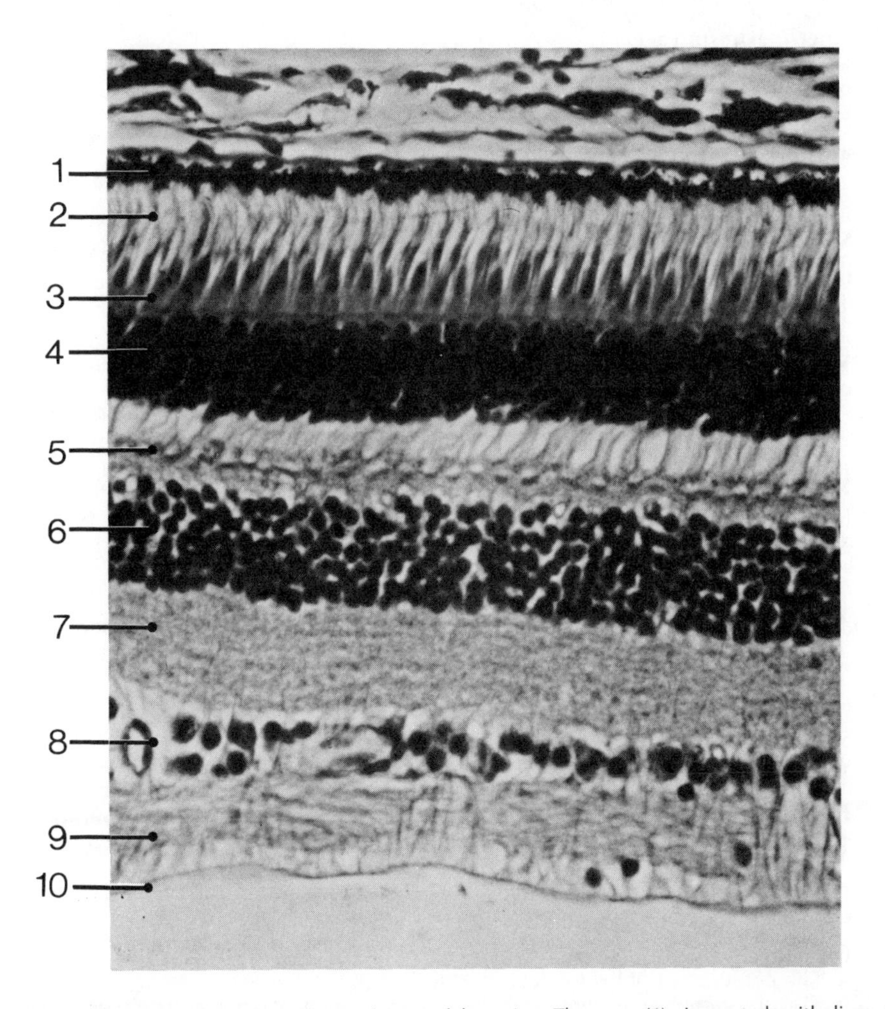

Figure 24-3. Light micrograph showing the ten layers of the retina. They are: (*1*) pigmented epithelium, (*2*) rod and cone outer segments, (*3*) outer limiting membrane, (*4*) outer nuclear layer, (*5*) outer plexiform layer, (*6*) inner nuclear layer, (*7*) inner plexiform layer, (*8*) ganglion cell layer, (*9*) layer of optic nerve fibers, and (*10*) inner limiting membrane. (Reprinted with permission from Johnson KE: *Histology: Microscopic Anatomy and Embryology*. New York, John Wiley, 1982, p 385.)

2. **Cones.**
   a. Cones are shorter and fatter than rods and make connection with a smaller number of other cells.
   b. Cones are responsible for **visual discrimination** and **color vision**.

3. **Ellipsoids.** An inner segment of each rod and cone contains a concentration of mitochondria called the ellipsoid.

4. **Outer Nuclear Layer.**
   a. The cell bodies and the nuclei of the rods and cones make up this thick layer of the retina.
   b. Radially arranged **Müller's cells** separate individual rods and cones, form contacts with them at the **outer limiting membrane** (which is not actually a membrane, but a large number of cell contacts), and then extend radially between the ellipsoids.

5. Nerve processes and synapses between rods and cones and bipolar cells make up the **outer plexiform layer** and the **synaptic layer**.

6. **Inner Nuclear Layer.**
   a. The rods and cones are connected to a **ganglion cell layer** by the inner nuclear layer of bipolar cells.
   b. This bipolar cell layer contains the nuclei of the Müller's cells as well as horizontal cells, amacrine cells (large monopolar cells), and a few ganglion cells.
   c. This layer is responsible for integration of the rods and cones, and there is some summation occurring from rod impulses.

7. **Inner Plexiform Layer.**
   a. This layer contains synapses between bipolar cells and ganglion cells.
   b. It also contains a few glial cells and small blood vessels.

8. **Ganglion Cell Layer.**
   a. This is the most anterior layer of cells in the retina.
   b. The dendrites of cells in this layer make synapses with bipolar cells, and their axons are gathered together to form the **optic nerve**.

9. **Nerve Fiber Layer.** This layer is composed of many axons of the ganglion cells.

10. **Inner Limiting Membrane.** This basement membrane represents the most anterior layer of the retina.

C. **FOVEA CENTRALIS.** The cellular composition of the retina varies considerably from place to place. There is a preponderance of rods in most regions of the light-sensitive portions of the eye. The fovea centralis **(macula lutea)**, however, has no rods but has a high concentration of tall, thin, modified cones.

1. This region of the retina is responsible for maximal visual acuity.

2. The cones in the fovea synapse one-to-one with bipolar cells, and the other layers of the retina are pushed from the fovea so that there is minimal structural interference with entering light rays.

VI. ACCESSORY ORGANS OF THE EYE.

The conjunctiva, eyelids, meibomian glands, and lacrimal glands are accessory organs of the eyes and are important for protecting them.

A. CONJUNCTIVA

1. **Structure.**
   a. The anterior surface of the eyeball, the cornea, is covered by a stratified squamous epithelium which is reflected onto the posterior surface of the eyelid as the **conjunctiva**.
   b. The conjunctiva and the corneal epithelium are continuous at the **conjunctival fornices**.
   c. The apical cells of the conjunctiva are tall and either cone-shaped or cylindrical. The basal cells are cuboidal.
   d. As the conjunctiva passes over the margin of the eyelid, the conjunctival mucous membrane gives way to epidermis.

2. **Function.**
   a. The conjunctiva helps to clean the surface of the cornea.
   b. Mucous secretions also moisten the conjunctiva and cornea and help prevent the evaporation of tears.

B. EYELIDS

1. **Structure.**
   a. The eyelids are coated on the inside by conjunctival epithelium and on the outside by thin epidermal epithelium.
   b. The eyelids contain numerous coarse hairs—the eyelashes—and highly modified sebaceous glands—the tarsal or **meibomian glands** as well as sweat glands with straight ducts **(the glands of Moll)**.

2. **Function.**
   a. The eyelids protect the cornea from damage.
   b. They also act as windshield wipers to clean the cornea.
   c. Oily secretions from the meibomian glands mix with mucus from the conjunctival glands to form a thin film on tears, helping to prevent the evaporation of tears.

C. LACRIMAL GLANDS

1. **Structure.**
   a. The lacrimal glands are situated in the superior lateral portion of the orbit.
   b. Each gland is composed of many serous acini that are drained into the orbit by 10 to 20 ducts.

2. **Function.** The lacrimal glands produce tears which cleanse and lubricate the cornea.

VII. DEVELOPMENT OF THE EYE.

The eye forms from an outpocketing of the wall of the brain and from mesenchymal elements that surround this outpocketing.

A. OPTIC VESICLE AND LENS PLACODE

1. The eye develops from a lateral evagination of the wall of the prosencephalon. This evagination (the **optic vesicle**) arises from the diencephalic subdivision of the prosencephalon prior to the closing of the neural groove. By 27 days gestation, the optic vesicle has grown laterally enough to make contact with the overlying surface ectoderm.

2. At the point of contact, the optic vesicle induces a thickening known as the optic or **lens placode**. This placode invaginates and forms a closed vesicle, which represents the precursor of the lens.

3. While the lens placode is invaginating, a second invagination forms, converting the primary optic vesicle into an invaginated **optic cup**. On the ventral side of the optic cup, an **optic fissure** forms.

4. Through the optic fissure, **hyaloid arteries** and veins grow after forming from mesenchymal cells that surround the optic cup.

5. The internal layer of the optic cup forms the neural retina, and the external layer of the optic cup forms the pigmented retina.

6. The optic cup is connected to the diencephalon by the **optic stalk**.

B. RETINAL DIFFERENTIATION

1. Melanosomes form in the external layer of the optic cup, and the optic cup becomes pigmented.

2. Neuroblasts in the neural retina differentiate into the photoreceptors and connecting cells characteristic of the adult retina. Axons from retinal neurons grow along the optic stalk into the brain. The optic stalk eventually becomes the **optic nerve**.

STUDY QUESTIONS

Directions: The question below contains five suggested answers. Choose the **one best** response to the question.

1. All of the following statements concerning corneal layers are true EXCEPT

(A) the epithelium is stratified squamous
(B) the endothelium has a basement membrane
(C) the stroma is acellular
(D) the stroma contains collagen
(E) Bowman's membrane is anterior to Descemet's membrane

Directions: The question below contains four suggested answers of which **one or more** is correct. Choose the answer

A if **1, 2, and 3** are correct
B if **1 and 3** are correct
C if **2 and 4** are correct
D if **4** is correct
E if **1, 2, 3, and 4** are correct

2. True statements concerning the development of the eye include which of the following?

(1) The optic vesicle induces thickening and invagination of the lens placode
(2) The optic vesicle gives rise to both the neural and the pigmented retina
(3) The central artery of the retina comes from the hyaloid artery
(4) The optic fissure normally remains open in the optic stalk

Directions: The group of questions below consists of lettered choices followed by several numbered items. For each numbered item select the **one** lettered choice with which it is **most** closely associated. Each lettered choice may be used once, more than once, or not at all.

Questions 3–8

For each description of layers of the retina, choose the appropriate lettered layer shown in the micrograph below.

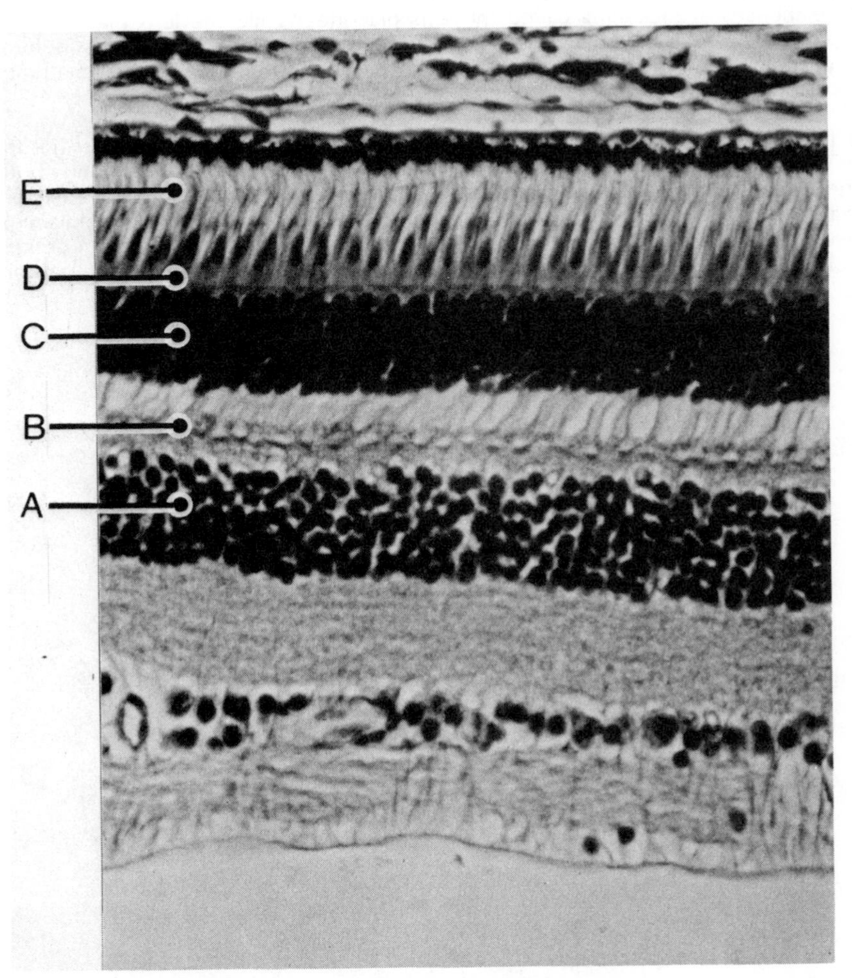

Reprinted with permission from Johnson KE: *Histology: Microscopic Anatomy and Embryology*. New York, John Wiley, 1982, p 385.

3. Contains axons of rods and cones
4. Contains nuclei of rods and cones
5. Contains outer segments of rods and cones
6. Contains bipolar cell bodies
7. Receives light before all other (lettered) layers
8. Caused by rows of contacts between Müller's cells and photoreceptors

ANSWERS AND EXPLANATIONS

1. The answer is C. *(II B 3)* The first of the five corneal layers is a stratified squamous **epithelium**. The epithelium rests on the second layer, **Bowman's membrane**. The third corneal layer, the **stroma**, has no blood supply but contains a large number of fibroblasts and collagen fibers. **Descemet's membrane**, the fourth layer, is a thick basement membrane. The final corneal layer, the **endothelium**, is a simple squamous epithelium.

2. The answer is A (1, 2, 3). *(VII A)* The optic vesicle arises from the wall of the prosencephalon. Eventually it grows laterally and makes contact with the surface ectoderm. At this point of contact the optic vesicle induces a thickening called the lens placode. While the lens placode is invaginating, the primary optic vesicle is converted into an invaginated optic cup. An optic fissure forms on the ventral side of the optic cup. Normally, the optic fissure fuses and disappears after head mesenchyme forms hyaloid blood vessels.

3–8 The answers are: 3-B, 4-C, 5-E, 6-A, 7-A, 8-D. (*V B*) In this micrograph, (A) is the layer that contains the outer segments of rods and cones. Contacts between Müller's cells and photoreceptors occur at the outer limiting membrane (B). The outer nuclear layer (C) contains the nuclei of rods and cones. The axons of rods and cones are contained in the outer plexiform layer (D). Bipolar cell bodies are contained in the inner nuclear layer (E). Light strikes the retina first at the bottom of the photograph and travels through all the retinal layers before striking the outer segments of the photoreceptors.

25
The Ear

I. INTRODUCTION

A. COMPONENTS (Fig. 25-1)

1. The **external ear** is composed of the auricle and the **external auditory meatus**.
2. The **middle ear** contains the **auditory ossicles** and connects with the **auditory tube**, which in turn empties into the nasopharynx.
3. The **tympanic membrane** separates the external ear from the middle ear.
4. The **internal ear** contains the **cochlea** with the **organ of Corti**.
   a. The organ of Corti is a specialized mechanoreceptor formed by a highly modified epithelium.
   b. In addition to the organ of Corti, the internal ear contains a vestibular apparatus with several varieties of neuroepithelia.

B. FUNCTIONS

1. The ear contains complex mechanoreceptors; motion of these mechanoreceptors is converted into electrical impulses.
2. These impulses are transmitted to the brain where they are interpreted as sound, a sense of the body's position in space, or a motion of the head.

II. EXTERNAL EAR

A. THE AURICLE (pinna) contains large amounts of elastic cartilage and is covered by skin with hairs, sebaceous glands, and a few eccrine sweat glands. The elastic cartilage extends partially down the external auditory meatus, which penetrates the temporal bone.

B. EXTERNAL AUDITORY MEATUS

1. The epithelium lining the meatus is similar to the skin of the pinna, but it has especially large sebaceous glands.
2. It also has a highly modified variety of apocrine sweat glands called **ceruminous glands**.
   a. The secretions of these glands are expressed by myoepithelial cells.
   b. The ducts of ceruminous glands are quite large and may either enter into hair follicles with sebaceous glands or empty directly onto the surface of the external auditory meatus.
   c. **Cerumen** repels insects and other vermin trying to get into the ear.

III. MIDDLE EAR

A. TYMPANIC CAVITY

1. This irregular cavity in the temporal bone has its lateral boundary at the tympanic membrane and its medial boundary at the bony wall of the inner ear. Anteriorly, it is continuous with the auditory tube.
2. The tympanic cavity is lined by a simple squamous epithelium except near the tympanic membrane and the beginning of the auditory tube, where it is lined by a cuboidal epithelium or columnar epithelium with cilia.

B. OSSICLES

1. The ossicles—**malleus**, **incus**, and **stapes**—are supported by miniature ligaments.

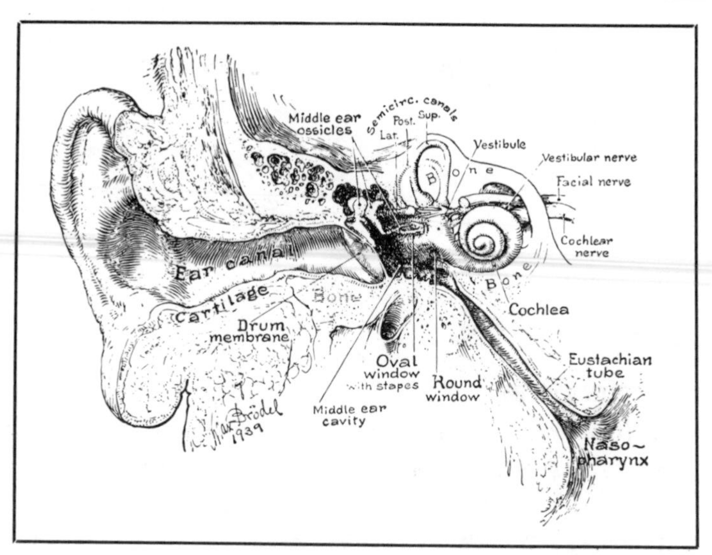

Figure 25-1. Illustration of the gross relations of the different parts of the ear. (Reprinted with permission from Brodel M: *Three Unpublished Drawings of the Anatomy of the Human Ear*. Philadelphia, WB Saunders, 1946.)

2. They are covered by reflections of the simple squamous epithelium that lines the middle ear.

C. TYMPANIC MEMBRANE

1. The tympanic membrane is covered on its lateral side by a thin layer of the stratified squamous epithelium of the external auditory meatus (but without glands and hairs) and on its medial side by a thin simple squamous epithelium.
2. Sandwiched between these two layers is a connective tissue domain with a layer of radially arranged collagenous fibers and a layer of circularly arranged fibers, elastic fibers, and fibroblasts.
3. In the anterosuperior quadrant there is no connective tissue. This thin layer is called **Shrapnell's membrane**.

D. AUDITORY TUBE

1. This tube courses anteromedially to an opening in the nasopharynx.
2. The portion near the middle ear is surrounded by the temporal bone, and the portion near the nasopharynx is partially surrounded by a spiral of elastic cartilage.
3. The mucosa has a low ciliated columnar epithelium in the temporal bone, but near the nasopharynx this gives way to a pseudostratified layer with tall columnar ciliated cells, some goblet cells, and mucous glands.
4. The lamina propria in the more medial portion also is thicker and may be extensively infiltrated with lymphocytes, which may form discrete nodules called the **tubal tonsils** (of Gerlach).
5. The auditory tube usually is closed, but the pharyngeal orifice becomes patent during yawning and swallowing, thereby equalizing pressure in the tympanic cavity and the outside world.

IV. INTERNAL EAR

A. MAJOR DIVISIONS

1. The internal ear occupies a complex cavity in the petrous portion of the temporal bone called the **osseous labyrinth**.
2. The osseous labyrinth is divided into two major cavities.
   a. The **vestibule** contains the **saccule**, **utricle**, and three **semicircular canals**.
   b. The **cochlea** contains the **organ of Corti**.

B. COCHLEA (Fig. 25-2)

1. The cochlea is a spiral-shaped cavity in which the axis of the spiral is formed by a pillar of bone called the **modiolus**.
2. The cochlea is broad at the base and tapers as a cone.
3. The broad base of the modiolus opens into the cranial cavity at the internal acoustic meatus.
   a. Here, afferent nerve processes belonging to the cochlear division of the eighth cranial nerve pass through numerous tiny openings.
   b. The cell bodies of these nerve processes are arranged together in the **spiral ganglion**.
   c. Dendrites of these cells innervate the hair cells of the internal ear, and axons project into the central nervous system.
4. The cochlea is divided into two chambers by
   a. A bony shelf called the **osseous spiral lamina**
   b. A membrane, called the **basilar membrane** or **membranous spiral lamina**
5. The cochlea is subdivided further by the **vestibular membrane (Reissner's membrane)**, which extends between the spiral lamina and the wall of the cochlea.
6. The cochlea has three compartments.
   a. The upper passage is called the **scala vestibuli**.
   b. The lower passage is called the **scala tympani**.
   c. The intermediate passage, the **scala media**, also is called the **cochlear duct**.
7. The cochlear duct is connected to the vestibular apparatus via the **ductus reuniens**.
8. The scala vestibuli and scala tympani are perilymphatic spaces—the former ending at the **fenestra ovalis** and the latter ending at the **fenestra rotunda**. The scala vestibuli and scala tympani connect at the apex of the cochlear duct through a small opening called the **helicotrema**.

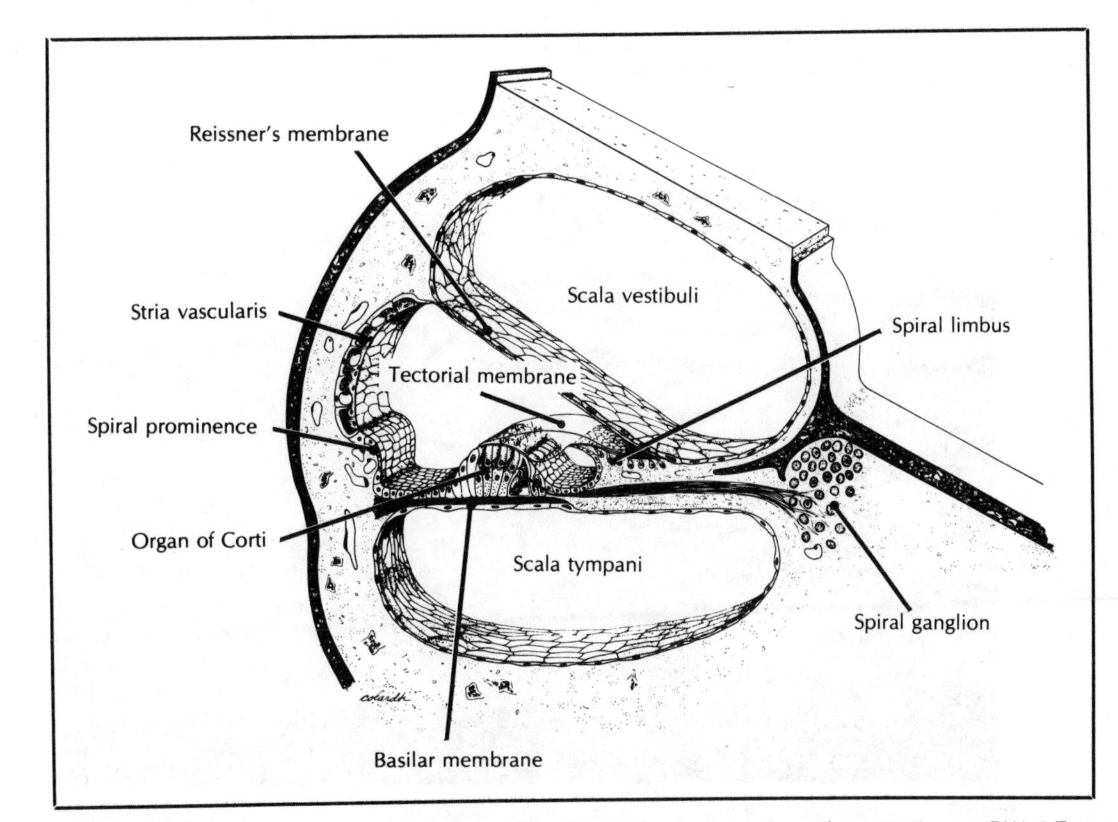

Figure 25-2. Diagram illustrating the internal ear. (Reprinted with permission from Bloom W, Fawcett DW: *A Textbook of Histology*. Philadelphia, WB Saunders, 1975.)

C. VESTIBULAR APPARATUS

1. Three **semicircular canals** connect with the utricle.
   a. These canals are filled with endolymph and contain three dilations called ampullae.
   b. In each ampulla is a patch of neuroepithelium called the crista, composed of hair cells.
   c. The cristae ampullares have an extracellular material called the cupula, which rests on hair cells.
2. **Utricle and Saccule** (Fig. 25-3).
   a. Each of these structures has a macula of neuroepithelium.
   b. Both the macula utriculi and the macula sacculi have an extracellular otolithic membrane with embedded (calcified) otoliths.
   c. Hair cells in the maculae are stimulated by movements of otoliths and otolithic membrane which, in turn, are generated by movements of the head.
3. **Endolymphatic Sac.**
   a. The endolymphatic sac has a columnar epithelium that contains some cells with microvilli and numerous apical pinocytotic vesicles. Macrophages and neutrophils also can cross the epithelium with ease.
   b. Endolymphatic fluid and cellular debris of the endolymph probably are removed at the endolymphatic sac.

D. COCHLEAR DUCT (Fig. 25-4)

1. The cochlear duct, which contains the organ of Corti, is a highly specialized diverticulum of the saccule. It is probably more complex histologically than any other area in the body.
2. The **vestibular membrane** has two layers of back-to-back flattened cells.
   a. One layer faces on and lines the scala vestibuli.
   b. The other layer faces on and lines the roof of the cochlear duct. On the outer wall of the cochlear duct, the inner layer of cells of the vestibular membrane becomes continuous with a stratified epithelium called the **stria vascularis**.
      (1) The basal cells in the stria vascularis have deep basal infoldings and numerous mitochondria.
      (2) It is believed that the stria vascularis is involved in the maintenance of the unusual ionic composition of the endolymph.

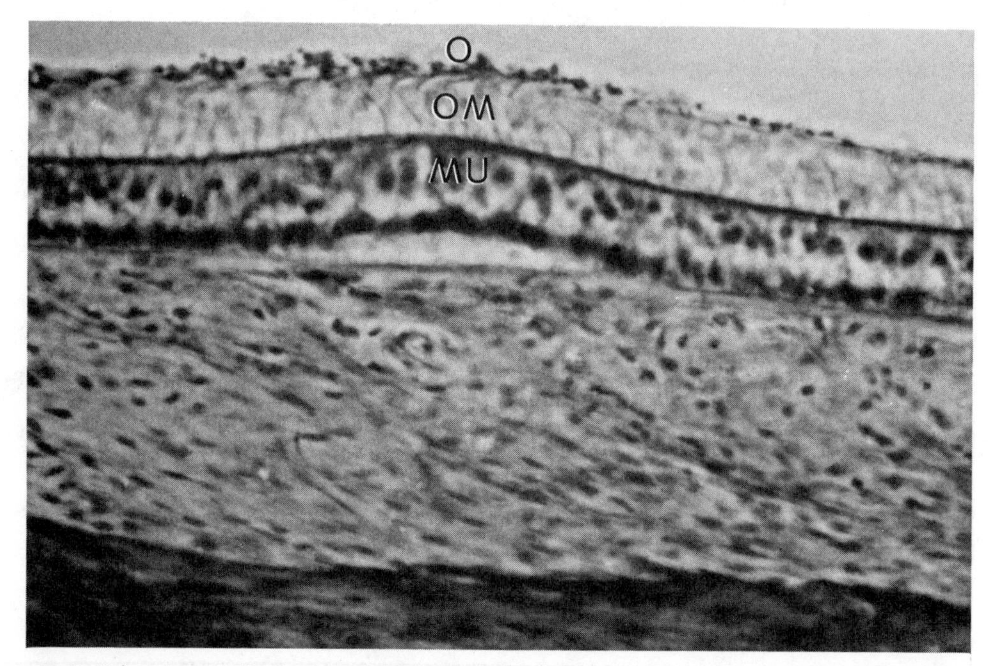

Figure 25-3. Light micrograph of macula utriculi (*MU*) with its otolithic membrane (*OM*) and otoliths (*O*). (Reprinted with permission from Johnson KE: *Histology: Microscopic Anatomy and Embryology*. New York, John Wiley, 1982, p 397.)

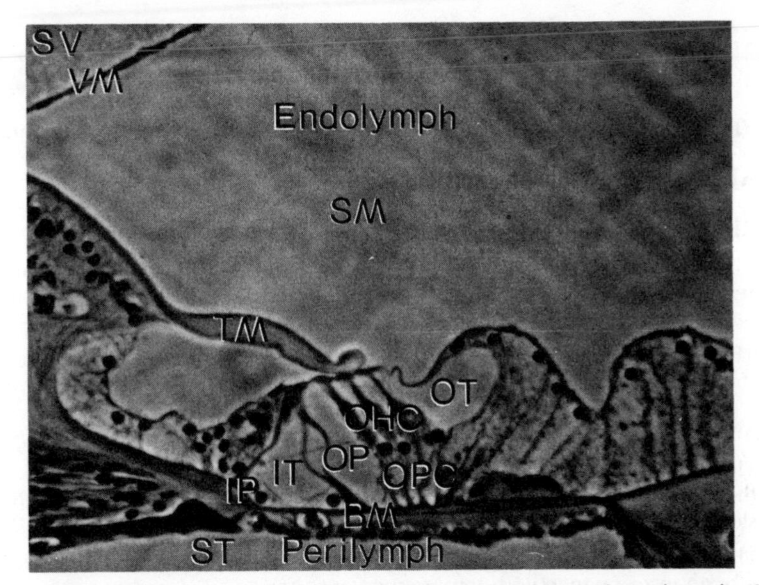

Figure 25-4. Light micrograph of the organ of Corti. This complex structure lies in the scala media (*SM*) between the scala vestibuli (*SV*) and the scala tympani (*ST*). It rests on a basilar membrane (*BM*). Outer hair cells (*OHC*) and outer phalangeal cells (*OPC*) are contacted directly by the tectorial membrane (*TM*). The outer tunnel (*OT*) and inner tunnel (*IT*) also can be seen. Inner pillar (*IP*) and outer pillar (*OP*) cells border the inner tunnel. The vestibular membrane (*VM*) is the boundary between the scala vestibuli and the scala media. (Reprinted with permission from Johnson KE: *Histology: Microscopic Anatomy and Embryology.* New York, John Wiley, 1982, p 398.)

3. As the epithelium of the cochlear duct is reflected from the stria vascularis onto the basilar membrane, the cells of Claudius and cells of Böttcher are encountered before the organ of Corti is reached.

V. ORGAN OF CORTI

A. Cells. The organ of Corti is composed of many different cell types (see Fig. 25-2). It has six kinds of **supporting cells** and two kinds of **hair cells**.

1. Supporting Cells.

a. The supporting cells are tall and slender and contain conspicuous tonofibrils.

b. Their apical surfaces are in contact with each other, with hair cells, or with both to form a continuous surface called the **reticular membrane**.

c. The **inner tunnel** is in the middle of the organ of Corti and rests on the basilar membrane.

d. Inner and **outer pillar cells** make up the floor and wall of the inner tunnel. Inner pillar cells contact inner hair cells, and outer pillar cells contact outer hair cells.

e. The inner border of the organ of Corti is made up of **border cells**, and the outer border is made up of **cells of Hensen**.

f. Phalangeal cells rest on the basilar membrane, and each has a cup-like indentation in its apical surface. The inferior third of a hair cell rests in this depression.

(1) The phalangeal cells have tunnels through them, which contain afferent and efferent nerve processes; these in turn synapse with the hair cells.

(2) The apical portions of the phalangeal cells are expanded into umbrella-like phalangeal processes, which make cell contact with the apical portions of the hair cells.

2. Hair Cells.

a. Both inner hair cells and outer hair cells are associated with phalangeal cells.

b. Each inner and outer hair cell has long apical microvilli called stereocilia and an apical centriole.

c. Outer hair cells have approximately 100 stereocilia arranged in a "W" pattern.

d. Inner hair cells have fewer stereocilia, and they are arranged in a straight line.

e. Hair cells are contacted by afferent and efferent nerve endings.

B. TECTORIAL MEMBRANE

1. In the inner angle of the scala media, the connective tissue covering the osseous spiral lamina is thrown into a crest called the **spiral limbus**.

2. The epithelium covering the spiral limbus secretes the tectorial membrane.
3. The tectorial membrane projects into the scala media, away from the spiral limbus.
4. The tips of the hair cells are embedded directly in the tectorial membrane.

VI. VIBRATIONS IN THE BASILAR MEMBRANE

A. PERILYMPH

1. The vibrations of the tympanic membrane are transmitted through the auditory ossicles to the **fenestra ovalis** and thus to the **perilymph** of the scala tympani. Vibrations in the perilymph cause vibrations in the basilar membrane.
2. The basilar membrane contains about 20,000 **basilar fibers**. These fibers project away from the bony modiolus but have free ends that can vibrate like reeds in a mouth organ.

B. BASILAR FIBERS

1. The basilar fibers near the base of the cochlea are blunt and short. Moving toward the apex of the cochlea (the helicotrema), the fibers gradually increase in length and become more slender.
2. The short fat fibers vibrate in resonance with high frequency sounds, and the long thin fibers vibrate in resonance with low frequency sounds.
3. These differences in the mechanical properties of the basilar fibers have an interesting result.
   - a. Movements in the perilymph cause a traveling wave in the basilar membrane.
   - b. The traveling wave is damped out where basilar fibers vibrate in resonance.
   - c. The maximum amplitude of the displacement of the basilar membrane varies with the frequency of the sound stimulus.
     - (1) High frequency sounds cause a maximum displacement of the basilar membrane near the base of the cochlea.
     - (2) Low frequency sounds cause a maximum displacement of the basilar membrane furthest from the base of the cochlea and nearest to the helicotrema.
   - d. The sensation of different sounds, then, results from the varied ways those sounds stimulate hair cells.
4. It should be noted that the tips of the hair cells are in contact with the tectorial membrane, which is fixed with respect to the moving hair cells.

VII. EMBRYOLOGY OF THE INTERNAL EAR

A. OTOCYSTS

1. The epithelium lining the membranous labyrinth begins as a simple invagination of the surface ectoderm. These invaginations eventually become separated from the surface of the embryos as two otocysts.
2. Soon after formation, a hollow diverticulum forms from the otocyst along its medial surface. This later becomes the **endolymphatic sac**.
3. By differential growth, one otocyst forms a **vestibular pouch** and the other forms a **cochlear pouch**.

B. THE VESTIBULAR POUCH gives rise to the semicircular canals and the utricle.

C. COCHLEAR POUCH

1. The saccule forms from this ventral pouch.
2. The saccule also gives rise to a growing diverticulum, which eventually spirals two and one-half times to form the cochlear duct.

STUDY QUESTIONS

Directions: The group of questions below consists of lettered choices followed by several numbered items. For each numbered item select the **one** lettered choice with which it is **most** closely associated. Each lettered choice may be used once, more than once, or not at all.

Questions 1-8

For each description of characteristics of membranes found in the ear, choose the appropriate membrane.

(A) Reticular membrane
(B) Tectorial membrane
(C) Basilar membrane
(D) Vestibular membrane
(E) Tympanic membrane

1. Secreted from spiral limbus and makes contact with tips of hair cells
2. Boundary between scala media and scala tympani
3. Boundary between scala media and scala vestibuli
4. Boundary between external ear and middle ear
5. Conducts traveling wave of deformation
6. Formed from phalangeal cell processes and apices of hair cells
7. Covered by stratified squamous epithelium on lateral surface
8. Remains nearly stationary when stapes moves

ANSWERS AND EXPLANATIONS

1–8. The answers are: 1-B, 2-C, 3-D, 4-E, 5-C, 6-A, 7-E, 8-B. (*IV B, D*) The reticular membrane is a planar structure formed by the phalangeal cell processes and the apices of hair cells. The tectorial membrane contacts the tips of hair cells and remains essentially motionless because it is fixed to the skull. All the other membranes mentioned here move when the stapes moves. The basilar membrane is part of the boundary between the scala media and the scala tympani. It contains the basilar fibers and conducts a traveling wave of deformation when the stapes moves. The vestibular membrane is the boundary between the scala media and the scala vestibuli. The tympanic membrane is the boundary between the external auditory meatus and the middle ear. On its lateral surface, the tympanic membrane is lined by a thin stratified squamous epithelium.

Post-test

QUESTIONS

Directions: Each question below contains five suggested answers. Choose the **one best** response to each question.

1. The immune system is composed of several cell types. Which of the following components of this complex system is derived from the third pharyngeal pouch (i.e., endoderm)?

(A) Cells lining splenic sinuses
(B) Plasma cells
(C) Thymic epithelial cells
(D) Lymphocytes
(E) Macrophages

2. Which property best identifies a metachromatic structure?

(A) A positive periodic acid-Schiff (PAS) reaction
(B) A net positive charge
(C) Staining with eosin
(D) Staining with toluidine blue
(E) A high concentration of DNA

3. A blood film from a normal adult female reveals all of the following characteristics of neutrophils EXCEPT

(A) histamine-containing granules
(B) azurophilic granules
(C) nuclei with three to five lobes
(D) chemotactic ability
(E) phagocytotic activity

4. When examined with an electron microscope, the cells of the adrenal medulla reveal all of the following characteristics EXCEPT

(A) large amounts of rough endoplasmic reticulum
(B) large mitochondria with peculiar tubular cristae
(C) prominent perinuclear Golgi apparatus
(D) prominent granules that are 200 nm in diameter
(E) epinephrine-containing granules

Questions 5 and 6

The diagram below is of a developing heart. Letters A through E designate components of this fetal organ.

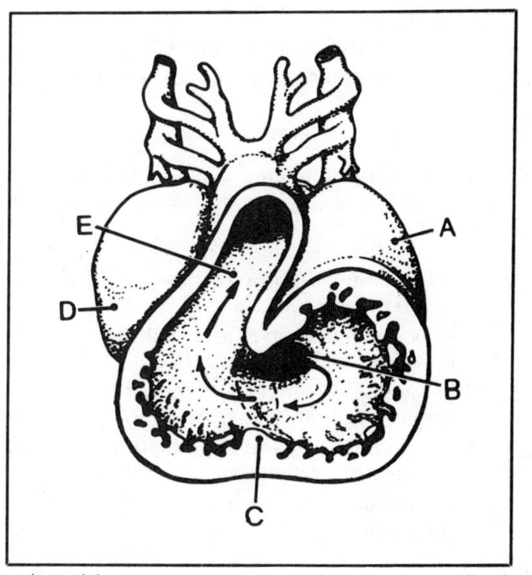

Adapted from Langman J: *Medical Embryology,* 4th ed. Baltimore, Williams and Wilkins, 1981, p 161.

5. The structure that receives oxygenated blood from the inferior vena cava in an unborn fetus is designated by

(A) A
(B) B
(C) C
(D) D
(E) E

6. The structure that eventually separates the primitive ventricles is designated by

(A) A
(B) B
(C) C
(D) D
(E) E

7. Cells that show incorporation of ^{3}H-thymidine are found in which of the following epidermal strata?

(A) Corneum
(B) Lucidum
(C) Granulosum
(D) Spinosum
(E) Basale

8. The cornea is composed of five layers; these layers are characterized by all of the following statements EXCEPT

(A) the stroma is rich in glycosaminoglycans
(B) the stroma contains fibroblasts and orthogonal layers of collagen
(C) Descemet's membrane rests on the endothelium
(D) the endothelium is a stratified epithelium
(E) the epithelium is continuous with the skin

9. All of the following statements concerning thyroglobulin synthesis and secretion are true EXCEPT

(A) it is secreted with peroxidase into the lumen of the thyroid follicles
(B) it is iodinated in the rough endoplasmic reticulum and Golgi apparatus
(C) its synthesis can be promoted by thyroid-stimulating hormone (TSH)
(D) its polypeptides are assembled in the rough endoplasmic reticulum
(E) glycosylation occurs in part in the rough endoplasmic reticulum

10. The release of hydrochloric acid is stimulated by which of the following types of enterochromaffin cells?

(A) G cells
(B) I cells
(C) S cells
(D) EC cells
(E) A cells

11. The epithelial cells of the small intestine are characterized by all of the following statements EXCEPT

(A) they are joined together by prominent apical zonulae occludentes
(B) the basement membrane usually is absent
(C) goblet cells have a basal nucleus
(D) Paneth's cells contain apical granules of lysosome
(E) sucrase is a component of the glycocalyx

12. Which of the following statements does not characterize erythropoiesis?

(A) Polychromatic erythroblasts have more cytoplasmic RNA than do basophilic erythroblasts
(B) Polychromatic erythroblasts have less hemoglobin than do orthochromatic erythroblasts
(C) Reticulocytes lack a nucleus and can occur in normal peripheral blood
(D) Orthochromatic erythroblasts have more hemoglobin than does peripheral blood
(E) Erythroblasts are 15 μm in diameter

13. Normally, erythrocytes are best described as being

(A) more than 20 percent hemoglobin by weight
(B) less than 20 percent of the total volume of the blood
(C) loaded with mitochondria
(D) 3–6 μm in diameter
(E) in the blood circulation for less than 40 days

Questions 14–16

Pictured below is a rat epithelium resting on a field of connective tissue.

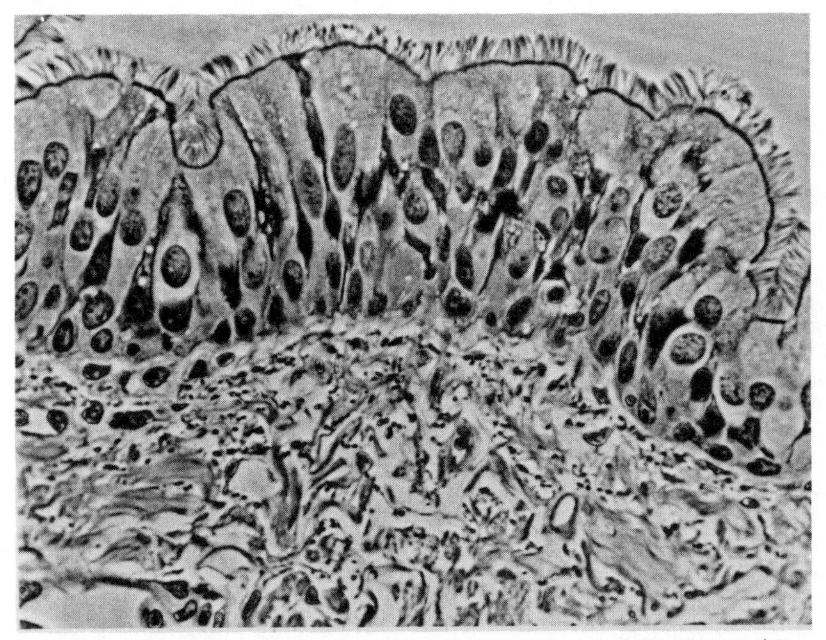

Reprinted with permission from Johnson KE: *Histology: Microscopic Anatomy and Embryology*. New York, John Wiley, 1982, p 33.

14. What type of connective tissue is pictured above?

(A) Dense regular
(B) Dense irregular
(C) Loose (areolar)
(D) Adipose
(E) Reticular

15. What type of epithelium is pictured above?

(A) Simple columnar
(B) Simple cuboidal
(C) Stratified cuboidal
(D) Stratified squamous
(E) Pseudostratified ciliated columnar

16. This type of epithelium lines which of the following organs?

(A) Trachea
(B) Stomach
(C) Gallbladder
(D) Epididymis
(E) Duodenum

Directions: Each question below contains four suggested answers of which **one or more** is correct. Choose the answer

- **A** if **1, 2, and 3** are correct
- **B** if **1 and 3** are correct
- **C** if **2 and 4** are correct
- **D** if **4** is correct
- **E** if **1, 2, 3, and 4** are correct

17. The injection of large molecular weight (MW) tracer molecules into the blood supply of a renal glomerulus reveals which of the following facts concerning glomerular filtration mechanisms?

(1) Proteins with a MW greater than 160,000 cannot cross the podocyte slit diaphragms
(2) Proteins with a MW of 40,000 enter the glomerular filtrate
(3) Proteins with a MW greater than 400,000 cannot cross the basement membrane
(4) Proteins with a MW greater than 400,000 cannot cross the fenestrated capillary

18. The endosteum can be described as being a

(1) population of osteoblasts
(2) participant in bone repair
(3) mesoderm derivative
(4) layer that lies adjacent to the marrow compartment

19. The adrenal cortex exhibits a striking zonation. Components of one zone, the zona glomerulosa, include

(1) parenchymal cells with less lipid than parenchymal cells in the zona fasciculata
(2) mitochondria with tubular cristae
(3) abundant smooth endoplasmic reticulum
(4) cells with an elongated nucleus

20. True statements concerning dentin include which of the following?

(1) It contains hydroxyapatite crystals
(2) It contains collagen
(3) It is a calcified extracellular matrix
(4) It is secreted by odontoblasts

21. Sections of the brain and spinal cord that are derived from the prosencephalon include the

(1) telencephalon
(2) metencephalon
(3) diencephalon
(4) myelencephalon

22. True statements concerning the histology of the ureters include which of the following?

(1) They have a sparse lamina propria
(2) The proximal two-thirds have only an inner longitudinal layer of smooth muscle
(3) They are lined by a stratified epithelium
(4) The distal third has only an outer circular layer of smooth muscle

23. The microscopic anatomy of the thyroid gland is characterized by

(1) apical microvilli, which disappear following thyroid-stimulating hormone (TSH) stimulation
(2) TSH binding to follicular epithelial cells, which causes an increase in cyclic adenosine monophosphate levels in these cells
(3) a lack of fenestrated capillaries
(4) thyroglobulin phagocytosis promoted by TSH

24. True statements concerning the structure and function of the colonic epithelium include which of the following?

(1) Goblet cells are more prominent here than in the duodenum
(2) Hydrolysis of foodstuff occurs in the ascending colon
(3) Epithelial surface area is increased by deep invaginations
(4) Villi line the epithelial pits

25. Thin skin on the forearm of a black person has which of the following components?

(1) Prominent stratum lucidum
(2) Thick dermis
(3) Apocrine sweat glands
(4) Hemidesmosomes in the stratum basale

26. Functions of the secretion product from bulbourethral glands include

(1) antibacterial activity
(2) protection from urine acidity
(3) nourishment of spermatozoa
(4) sexual lubricant

27. True statements concerning autoradiography include which of the following?

(1) The technique can be used to elucidate the synthesis site of proteins
(2) The exposure period needed ranges from minutes to days
(3) The emulsion used contains silver halide
(4) The emulsion used is sensitive only to β-particles

28. T cells are more numerous than B cells in which of the following locations?

(1) Periarterial lymphatic sheaths
(2) Peripheral blood
(3) Tertiary cortex of a lymph node
(4) Secondary nodule of a lymph node

29. True statements concerning the development of the pituitary gland include which of the following?

(1) Rathke's pouch is endodermally derived
(2) The infundibulum forms as an evagination of the floor of the diencephalon
(3) The pars nervosa forms from the anterior wall of Rathke's pouch
(4) The cleft in the adult pituitary is anterior to the pars intermedia

30. Parts of the body that are lined by transitional epithelium include the

(1) esophagus
(2) cervical os
(3) lumen of a blood vessel
(4) urinary bladder

31. Features that characterize steroid-secreting endocrine cells include

(1) mitochondria with tubular cristae
(2) extensive smooth endoplasmic reticulum
(3) lipid droplets
(4) surface microvilli

32. Myelin can be described as being

(1) produced by Schwann cells
(2) produced by oligodendroglia
(3) an insulator for axons
(4) an accelerator of nerve conduction velocity

33. Gastrin cells (G cells) are characterized by

(1) parietal cell secretion of hydrochloric acid
(2) stimulation of antral motor activity
(3) microvilli that are stimulated by food
(4) acidity that contributes to secretion activity

34. The bands of a sarcomere that shorten during muscle contraction include

(1) A bands
(2) I bands
(3) M bands
(4) H bands

35. Components of saliva include

(1) lactoperoxidase
(2) secretory immunoglobulins
(3) ions
(4) ptyalin

SUMMARY OF DIRECTIONS

A	B	C	D	E
1,2,3 only	1,3 only	2,4 only	4 only	All are correct

36. The mesonephric ducts form which of the following components of the male reproductive system?

(1) Vas deferens
(2) Ductus epididymis
(3) Seminal vesicles
(4) Seminiferous tubules

37. True statements concerning the layers of the retina include which of the following?

(1) The inner plexiform layer has synapses between cones and bipolar cells
(2) The inner limiting membrane is the basement membrane for the neural retina
(3) The outer nuclear layer contains ganglion cell nuclei
(4) The outer limiting membrane is an area of contact between Müller's cells and the photoreceptors

38. Superior hypophyseal arteries carry blood to which of the following hypophyseal components?

(1) Pars tuberalis
(2) Median eminence
(3) Pars distalis
(4) Pars nervosa

Directions: The groups of questions below consist of lettered choices followed by several numbered items. For each numbered item select the **one** lettered choice with which it is **most** closely associated. Each lettered choice may be used once, more than once, or not at all.

Questions 39–42

For each description of cells involved in the process of oogenesis, choose the appropriate cell type.

(A) Oogonia
(B) Polar bodies
(C) Primary oocytes
(D) Secondary oocytes
(E) Primordial germ cells

39. Cells that migrate to the genital ridges from the yolk sac

40. Cells that are present in primary follicles in sexually mature females

41. Cells that are ovulated and capable of being fertilized

42. Cells that often remain in the same stage of development for more than 30 years

Questions 43–46

For each of the following descriptions of components of the trachea, choose the appropriate lettered structure shown in the micrograph below.

43. Basement membrane for the epithelium

44. Multicellular gland

45. Cell that secretes mucus

46. Flexible piece of avascular connective tissue that helps to maintain the tracheal lumen

Questions 47–50

For each description of a component of a chondrocyte, choose the appropriate lettered structure shown in the micrograph below.

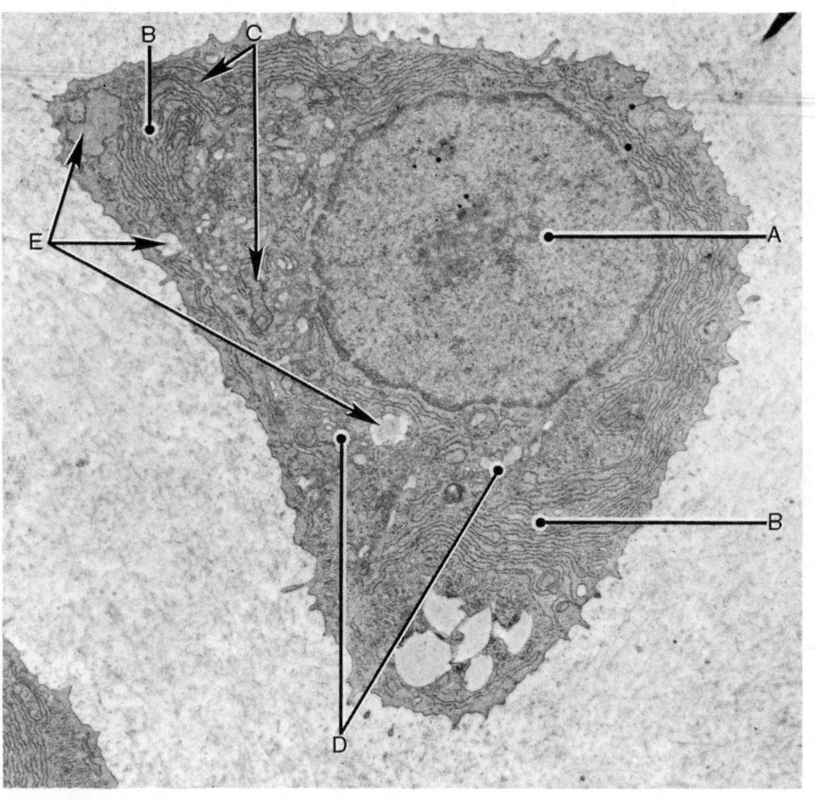

Courtesy of Dr. Daniel P. DeSimone, George Washington University School of Medicine.

47. Site of initial protein synthesis for proteoglycan aggregates

48. Produces ATP for synthetic activities of the cell

49. Contains hyaluronic acid, proteoglycan aggregates, and collagen molecules

50. Rich in glucuronyltransferases

Questions 51–55

For each description of a component of the inner ear, choose the appropriate lettered structure shown in the micrograph below.

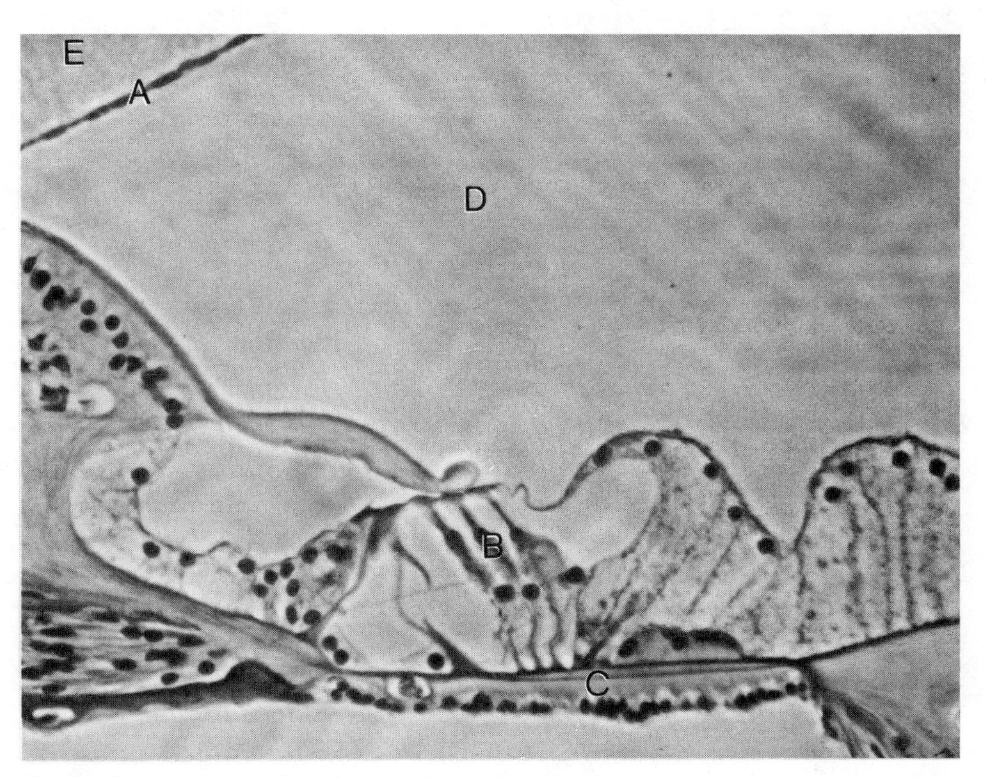

Reprinted with permission from Johnson KE: *Histology: Microscopic Anatomy and Embryology*. New York, John Wiley, 1982, p 403.

51. Contains basilar fibers

52. Contains endolymph

53. Contains perilymph

54. Has apical stereocilia

55. Boundary between scala media and scala vestibuli

Questions 56–60

For each description of the fine structure of a pancreatic acinar cell, choose the appropriate lettered structure shown in the micrograph below.

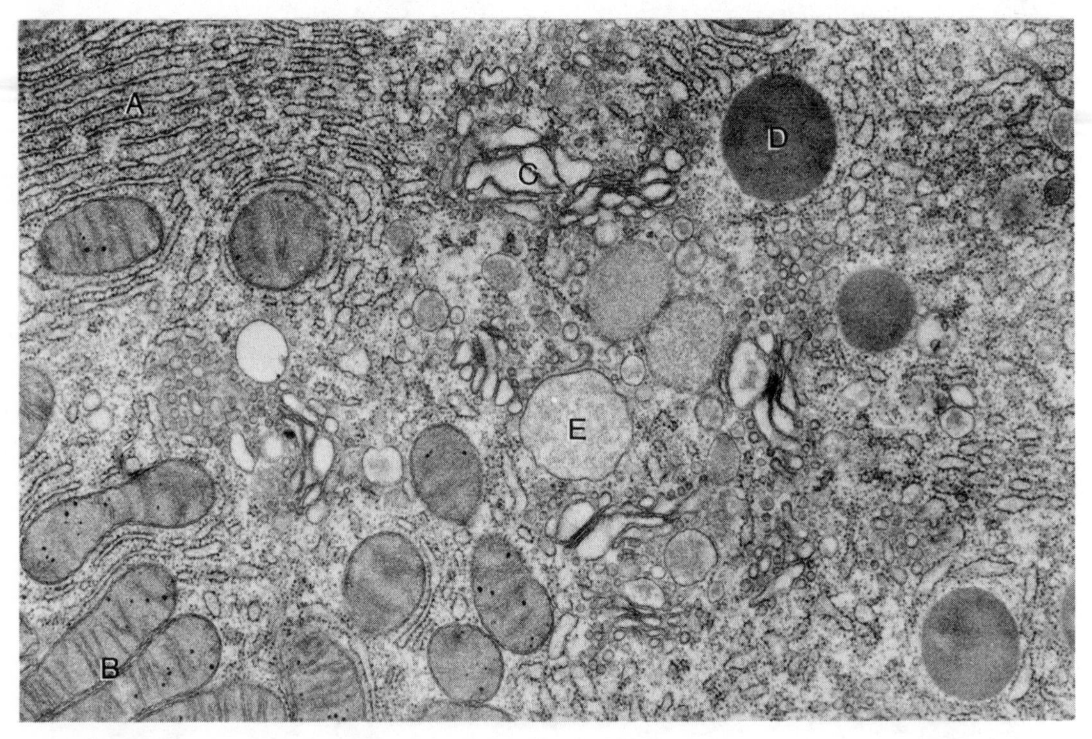

Courtesy of Dr. Frank J. Slaby, Department of Anatomy, George Washington University.

56. Synthesis of nascent chains of hydrolytic enzymes begins here

57. Membrane-bound vesicle of enzymes, ready for secretion

58. Involved in glycosylation and packaging of secretion product

59. Contains unit membranes with attached ribonucleoprotein particles

60. Supplies ATP used in protein synthesis and contains enzymes for oxidative phosphorylation

ANSWERS AND EXPLANATIONS

1. The answer is C. *(Chapter 10 III A)* Thymic epithelial cells are derived from the third pharyngeal pouch (endoderm). Monocytes leave the closed circulation and become macrophages. Lymphocytes have a very small amount of cytoplasm and thus a very high nuclear to cytoplasmic ratio. The cells lining splenic sinusoids are contractile, allowing cells to leave the closed splenic circulation for the open spaces in the red pulp of the spleen.

2. The answer is D. *(Chapter 1 III D 2–4)* If a specimen shows toluidine blue metachromasia (i.e., stains purple), it indicates that the specimen has a high net negative charge. Such a specimen would be expected to bind other basic dyes such as hematoxylin and methylene blue. It is unusual for a specimen to stain metachromatically with acidic dyes such as eosin. DNA molecules, which are not as highly negatively charged as structures that stain metachromatically, usually stain orthochromatically (blue). There is no direct connection between metachromasia and the PAS status of a specimen.

3. The answer is A. *(Chapter 8 III B 1 a, b; 3 b)* Neutrophils and basophils are two types of granule-containing leukocytes or **granulocytes**; neutrophils are the most common leukocyte in human blood, and basophils are the rarest leukocyte. The vasoconstrictor, histamine, is contained in basophilic granules. Neutrophilic granules do not contain histamine but can be azurophilic. These azurophilic granules play a major role in the phagocytotic activity of neutrophils. Phagocytized bacteria release peptides and other substances, to which neutrophils are chemotactically attracted. Neutrophils reveal nuclei with three to five lobes.

4. The answer is B. *(Chapter 19 III A 2; V B 1, 2)* The cells of the adrenal cortex contain numerous large mitochondria with tubular cristae. The cells of the adrenal medulla, however, have unremarkable mitochondria. Other ultrastructural characteristics of a cell in the adrenal medulla include an abundance of rough endoplasmic reticulum and prominent Giolgi apparatus located all around the nucleus. Prominent membrane-bound granules are synthesized in the Golgi. These granules are 200 nm in diameter and are thought to store epinephrine.

5 and 6. The answers are: 5-D, 6-C. *(Chapter 11 I I 3)* In the early stages of the development of the heart, both the atria and the ventricles are unseptated and in direct communication. The primitive left atrium (A) and the primitive right atrium (D) remain patent right to the end of fetal life; they are not divided until after birth. The primitive right ventricle is separated from primitive left ventricle by the interventricular septum (C). This division occurs in two steps. The truncus arteriosus and the conus cordis (E) also become subdivided, eventually leading to a separation of the systemic and pulmonary output from the heart.

7. The answer is E. *(Chapter 17 II B)* Cells in the stratum basale are engaged in DNA synthesis and, therefore, would show ^{3}H-thymidine incorporation. Lamellated granules are synthesized early during keratin formation and are near the Golgi apparatus in cells in the stratum spinosum. Amorphous granules of basophilic material are prominent in the stratum granulosum. In the stratum corneum, lamellated granules have been discharged into the spaces between cells.

8. The answer is D. *(Chapter 24 II B 3)* The corneal **epithelium**, on the anterior corneal surface, is a stratified squamous epithelium; it rests on the second corneal layer, **Bowman's membrane**. The **stroma**, the third layer, represents about 90 percent of the total corneal thickness and is a mixture of fibroblasts, collagen fibers, and amorphous ground substance rich in glycosaminoglycans. The fourth corneal layer, **Descemet's membrane**, is a thick basement membrane. The posterior corneal layer, the **endothelium**, is a single layer of flattened cells (i.e., a simple squamous epithelium).

9. The answer is B. *(Chapter 18 II D)* Thyroglobulin is a glycoprotein. Thyroglobulin polypeptide chains are synthesized in the rough endoplasmic reticulum (RER). Sugar residues then are added in the RER and Golgi apparatus. Thyroglobulin is iodinated by extracellular peroxidases that are secreted into the follicular lumen along with thyroglobulin.

10. The answer is A. *(Chapter 14 V B)* G cells are found in the pyloric antrum and secrete gastrin, a polypeptide hormone that regulates hydrochloric acid secretion. I cells are found in most parts of the stomach and secrete a polypeptide called cholecystokinin (pancreozymin). S cells, found in the distal stomach, secrete a polypeptide called secretin, which stimulates alkaline secretions from the pancreas. EC cells are common throughout the gastrointestinal mucosa and secrete serotonin. A cells are scattered throughout the duodenum and jejunum and secrete glucagon.

11. The answer is B. *(Chapter 15 II B)* The small intestinal epithelium is a columnar epithelium with goblet cells. The cells all are joined together by robust apical junctional complexes. Like all other epithelial layers, the small intestinal epithelium rests on a basement membrane. The Paneth's cells are thought to contain apical granules of lysozyme, and the glycocalyx is peculiar in that it appears to contain extracellular carbohydrate-hydrolyzing enzymes.

12. The answer is A. *(Chapter 9 IV A, B)* A basophilic erythroblast has much more cytoplasmic RNA than does a polychromatic erythroblast. The former uses cytoplasmic ribosomal RNA to synthesize hemogloblin. As the hemoglobin accumulates, the ribosomal RNA is replaced, and the tinctorial quality of the cytoplasm changes from basophilic, to polychromatic, to orthochromatic.

13. The answer is A. *(Chapter 8 II A, B)* Erythrocytes normally make up about 40 percent of the total volume of the blood. They lack mitochondria and are larger than 3–6 μm in diameter. They circulate for approximately 120 days.

14–16. The answers are: 14-C, 15-E, 16-A. *(Chapter 3 II B; Chapter 4 II B)* This is an example of loose (areolar) connective tissue (CT). This CT type is common in the subcutaneous fascia and the mesenteries; here it is pictured in the lamina propria, which is the CT domain underlying moist epithelia in the gastrointestinal tract and elsewhere. In this sample, the loose CT is underlying the pseudostratified ciliated columnar epithelium of the trachea.

17. The answer is A (1, 2, 3). *[Chapter 23 III A 1 e (2)]* There are two barriers to prevent plasma proteins from entering the urinary filtrate. One is the glomerular basement membrane, which prevents passage of proteins with a molecular weight (MW) of 400,000 and above. The other is the podocyte slit diaphragm, which prevents passage of proteins with a MW of 160,000 and above. Low MW substances (MW = 40,000) pass both barriers and enter the urinary filtrate.

18. The answer is E (all). *(Chapter 9 II B)* The endosteum is a layer of osteoblasts. This layer lines the marrow cavity of bone and like the rest of the bone and bone marrow, it is mesodermally derived. The osteoblasts of the endosteum are involved in bone remodeling and repair.

19. The answer is A (1, 2, 3). *(Chapter 19 II A 1; III A, B)* All cells found in the adrenal cortex have a nucleus that is quite round; however, these cells sometimes are arranged in long columns. The cells of the zona fasciculata have abundant lipid droplets, far more than the cells of the zona glomerulosa. The cells of the zona glomerulosa have mitochondria with tubular cristae and an abundance of smooth endoplasmic reticulum.

20. The answer is E (all). *(Chapter 13 III B)* Dentin is a bone-like secretion product of odontoblasts. As such, it contains collagen fibers and is calcified by the accumulation of hydroxyapatite crystals, a complex salt containing calcium, phosphate, and other ions. Dentin is secreted as uncalcified, collagenous predentin which becomes converted to dentin by the accumulation of hydroxyapatite crystals.

21. The answer is B (1, 3). *(Chapter 7 IV D)* The prosencephalon forms the diencephalon, which forms the optic cup and the hypothalamus. The telencephalon also is formed by the prosencephalon, and in turn forms the cerebral hemispheres. The rhombencephalon forms the metencephalon, which gives rise to the cerebellum and pons, and the myelencephalon, which forms the medulla oblongata.

22. The answer is B (1, 3). *(Chapter 23 VI B)* The ureters have transitional epithelium that is multilayered (i.e., stratified). This transitional epithelium rests on a basement membrane and is associated with a sparse underlying lamina propria. The entire ureter has at least an inner longitudinal and an outer circular layer of smooth muscle. In addition, the distal portion has a middle circular layer.

23. The answer is C (2, 4). *(Chapter 18 II C 1; D 2)* Thyroid-stimulating hormone (TSH) stimulation would increase cellular phagocytosis of thyroglobulin and cause an increase in the number and size of apical microvilli—surface modifications involved in phagocytosis. Fenestrated capillaries are a common feature in many endocrine organs, including the thyroid gland.

24. The answer is A (1, 2, 3). *(Chapter 15 III B)* The gross surface of the lumen of the colon is smooth (lacks villi) but does contain deep surface invaginations called pits. These serve to increase surface area. There are many more goblet cells here than in the small intestine. The mucus from the goblet cells becomes mixed with the undigested food materials to produce feces. Some continued digestion occurs in the colon.

25. The answer is C (2, 4). *(Chapter 17 II A; V C)* In thin skin, the stratum corneum and stratum lucidum are not prominent. In all epidermal types, the cells of the stratum basale are anchored to the basement membrane by hemidesmosomes. Skin on the back of the arm lacks apocrine sweat glands. Also, skin found here is thin, and so has a thin epidermis but a relatively thick dermis.

26. The answer is D (4). *(Chapter 22 VIII C)* The two paired bulbourethral glands secrete a clear viscous material when a male becomes sexually aroused. This mucoprotein secretion acts as a sexual lubricant but has no other known functions.

27. The answer is B (1, 3). *(VIII A 1–2; VII B 1–2)* The silver halide emulsion used in autoradiography is sensitive to natural light, photons, gamma rays, and other electromagnetic radiations, as well as to the β-particles released by radioactive decay. Typically, the emulsion is exposed to β-particles anywhere from several days to several months. Autoradiography can be used to study almost any metabolic event within cells, including protein synthesis.

28. The answer is A (1, 2, 3). (*Chapter 10 IV C*) T cells outnumber B cells in the periarterial lymphatic sheaths of the spleen and in the tertiary cortex of a lymph node. B cells outnumber T cells in the secondary nodules of lymph nodes but not in the peripheral circulation, where approximately 65 percent of the lymphocytes are T cells and 35 percent are B cells.

29. The answer is C (2, 4). *(Chapter 20 III B, C; V B 3)* Rathke's pouch is formed from the ectoderm. The anterior wall of Rathke's pouch forms the pars distalis. The infundibulum is a small evagination of the floor of the diencephalon. In the adult, when a remnant of the cleft of Rathke's pouch persists, the pars intermedia lies posterior to the cleft.

30. The answer is D (4). *(Chapter 3 III B)* Several components of the urinary system are lined by transitional epithelium, including the urinary bladder, ureters, and renal pelvis. The esophagus is lined by a stratified squamous epithelium, and the lumen of a blood vessel is lined by simple squamous epithelium. The cervical os is lined by either the columnar epithelium of the uterus or the stratified squamous epithelium of the vagina. Although it is a transitional zone, it does not contain transitional epithelium.

31. The answer is A (1, 2, 3). *(Chapter 2 VII B; XII B)* Mitochondria with tubular cristae are peculiar to steroid-secreting endocrine cells. Their significance is unknown. The prominent smooth endoplasmic reticulum and lipid droplets are invovled in the biosynthesis of steroids. Surface microvilli are not a feature of steroid-secreting cells.

32. The answer is E (all). *(Chapter 7 II E)* Myelin is produced by Schwann cells in the peripheral nervous system and by oligodendroglia in the central nervous system. In both locations, myelin insulates axons and speeds conduction velocity.

33. The answer is A (1, 2, 3). *(Chapter 14 V B 1)* Gastrin cells (G cells) are part of the enterochromaffin system. When their apical microvilli are stimulated, G cells produce gastrin, a polypeptide that promotes antral motility and hydrochloric acid secretion by parietal cells. Gastric acidity inhibits further secretion of gastrin, completing a feedback loop for the control of stomach acidity.

34. The answer is C (2, 4). *(Chapter 6 IV B)* When a muscle contracts, filaments slide past one another. The H band shortens and the I band shortens, but the A band remains constant. The M band also does not move during muscle contraction.

35. The answer is E (all). *(Chapter 13 V A)* Saliva, the secretory product of the salivary glands, has many different functions in the oral cavity. The two most important functions are digestion of carbohydrates (via ptyalin) and antibacterial activity (via lactoperoxidase and secretory immunoglobulin A). Saliva also contains ions, although it is hypotonic with respect to blood.

36. The answer is A (1, 2, 3). *(Chapter 22 X A, B)* The mesonephric ducts give rise to the vas deferens, the ductus epididymis, and the seminal vesicles. The seminiferous tubules have a different origin. They are derived from primordial germ cells, which give rise to spermatogonia and spermatocytes as well as spermatids and spermatozoa. The Sertoli cells of the seminiferous epithelium arise from the epithelial cells that originally covered the gonad (i.e., coelomic epithelial cells that migrated into the developing gonad and became associated with the primordial germ cells and their descendants).

37. The answer is C (2, 4). *(Chapter 24 V B)* The outer plexiform layer, not the inner plexiform layer, is the location of synapses between cones and bipolar cells. The outer nuclear layer contains nuclei of rods and cones, not ganglion cells. The inner limiting membrane, a basement membrane, represents the most anterior layer of the neural retina. The outer limiting membrane is an area of contact between Müller's cells and individual rods and cones.

38. The answer is A (1, 2, 3). *(Chapter 20 IV A 1, 2)* The superior hypophyseal arteries are the main afferent blood supply for the pars tuberalis, median eminence, and pars distalis. The inferior hypophyseal artery supplies the pars nervosa.

39–42. The answers are: 39-E, 40-C, 41-D, 42-C. *(IV A–C)* Primordial germ cells originate in the wall of the embryonic yolk sac and migrate to the genital ridges. In the primitive gonads, they then differentiate into diploid oogonia. The oogonia are the only highly proliferative cells in the ovary. Some of the oogonia then begin their first meiotic division and become primary oocytes contained within primordial follicles. By birth, all follicles in the ovary are primordial ones containing primary oocytes. Not until puberty do the primordial follicles begin to mature, changing into primary, secondary, and eventually preovulatory (mature) follicles. Primary oocytes do not complete their first meiotic prophase until immediately prior to ovulation (i.e., at the end of the maturation period of the secondary [antral] follicle). With the long-delayed completion of the first meiotic division, a secondary oocyte and a first polar body are formed. The ovulated cell capable of being fertilized is a secondary oocyte.

43–46. The answers are: 43-E, 44-B, 45-C, 46-D. *(Chapter 12 III A 3)* The tracheal epithelium consists mainly of ciliated cells with cilia (A) and goblet cells (C). These ciliated cells move mucus and entrapped debris. The epithelium rests on a basement membrane (E). Under the basement membrane is found a lamina propria with multicellular glands (B), and beneath the lamina propria is cartilage (D).

47–50. The answers are: 47-B, 48-C, 49-E, 50-D. *(Chapter 5 III E 1)* Protein synthesis begins in the rough endoplasmic reticulum (B). Mitochondria (C) produce ATP. The secretion product vacuoles (E) contain those materials destined for the extracellular matrix. Glucuronyltransferases are involved in glycosaminoglycan synthesis and are localized on Golgi membranes (D).

51–55. The answers are: 51-C, 52-D, 53-E, 54-B, 55-A. *(Chapter 25 IV B, D)* This is a light micrograph of the organ of Corti. It is contained within the scala media (D), which is also known as the cochlear duct. The vestibular membrane (A), also known as Reissner's membrane, is the boundary between the scala media and the scala vestibuli (E). The hair cells of the organ of Corti (B) are connected to the basilar membrane (C). When the basilar membrane vibrates, the tips of the hair cells are deformed and an action potential is initiated.

56–60. The answers are: 56-A, 57-D, 58-C, 59-A, 60-B. *(Chapter 16 V A 2)* The pancreatic acinar cells manufacture protein for export. Protein synthesis begins in the rough endoplasmic reticulum (A) and continues in the Golgi apparatus (C), where proteins are glycosylated. Next, condensing vacuoles (E) are converted into zymogen granules (D), just prior to release from the cell. Protein synthesis is an energy-dependent process, driven by ATP from mitochondria (B).

Index

Page numbers in *italics* refer to illustrations; those followed by a (t) denote tables.

A

A bands, 69, 277
 of a sarcomere, *63*, 64
A cells, 144, 275
 See also Alpha cells
Acetone, 11
Acetylcholine, 67, 74, 75
Acidophilia, 7, 12
Acidophilic structure, 2, 7
Acidophils, 18, 206
 of adenohypophysis, 200, *200*
Acinar cells, pancreatic, *5*, 10, 163, 164, *164*, *165*, *274*, 278
Acinus(i), 123
 hepatic, 161
 pancreatic, 163, *164*
Acromegaly, 202
ACTH, *see* Adrenocorticotropic hormone
Actin, 25, 61, *65*
Actomyosin complex, 64
Adenohypophysis, 10, 197, 199–202
 cells of, 200, *200*
 histology of, 199–202, *200*
 hormones of, 199, 200–202
 pars distalis of, 197, *198*, 199–200
 pars intermedia of, 198, *198*, 199
 pars tuberalis of, 198, *198*, 199
Adenosine triphosphate (ATP), 20, 24, 28, 233, 245
ADH, *see* Antidiuretic hormone
Adipocytes, 44–45
Adipose tissue, 3, 8, 9, 44–45
Adrenal glands, 189–196
 blood supply to, 189
 cortex of, 1, 189, 268, 276
  embryology of, 190, 194
  histology of, 189–190, *190*
  histophysiology of, 192–193
  ultrastructure of, 190–192
  zona fasciculata of, 189, *190*, 191, *191*, 194, 195, *195*
  zona glomerulosa of, 189, *190*, 191
  zona reticularis of, 190, *190*, 191
  zonal variation of, 191
 function of, 189
 gross anatomy of, 189
 medulla of, 189, 193, 196, 265
Adrenocortical function, 194
 control of, 192
Adrenocortical hormones, 192
Adrenocorticotropic hormone (ACTH), 7, 192–193, 200–201, *201*, 206
Adventitia, 113, 118
 of esophagus, 141
 of large intestine, 153
 of small intestine, *151*, 152
 tracheal, 9
Afferent arterioles, 240, 245
Agranulocytes, 10, 84, 85–86
Agranulopoiesis, 92
Airways, conducting, 119
Alar plate, 77
Aldosterone, 7, 192, 196, 240
Alkaline phosphatase, 152
Alpha cells, of islets of Langerhans, 165
Alpha chains, 42
Alveolar cells, great, 124
Alveolar ducts, 122, *123*
Alveolar epithelium, 127, 128
Alveolar sacs, *123*
Alveoli, 123–124
Ameloblasts, 9, 130, *130*, 138
 in teeth, 130–131
Aminopeptidases, 152
Amorphous ground substance, 49
 in bone, 51
 in cartilage, 51
 glycoproteins in, 39, 44
 glycosaminoglycans in, 44
 in hyaline cartilage, 52
 minerals in, 39
 protein components in, 44
 proteoglycans in, 39, 44
Ampulla, of male reproductive system, *224*
 of oviducts, 213
Androgens, 10, 192
Angiogenic cell clusters, 90, 115
Angiotensin, 192, 196, 240, *240*
Angiotensinogen, 192, 196, 240
Annuli fibrosi, 111
ANS, *see* Nervous system, autonomic
Antidiuretic hormone (ADH), 8, 204
Antrum, of ovarian follicles, 210
 of stomach, 141
Aorta, *162*
Apocrine mechanism, 179
Appendices epiploicae, 154, 157
Appositional growth, in endochondral ossification, 56
Arachnoid, 76
Arachnoid villi, 76
Arches, aortic, 115
Arcuate artery, of kidney, *236*, 240
Area cribrosa, 236, 245
Argyrophilic fibers, 46

Arrectores pilorum, of hair follicles, 179
Arteriovenous shunt, 175
Artery(ies), arcuate, *236,* 240
central, 103, *103*
coiled, 215
elastic, 109, 111–112, *112*
follicular, 103, *103*
hepatic, 8, 160, 162, *162,* 168
hyaloid, of the eye, 253
hypophyseal, inferior, 199
superior, 270, 277
interlobar, *236,* 240
muscular, 109, 112–113, *113*
renal, *236,* 240
trabecular, 103, *103*
Arterioles, 109, 117, 118
afferent, 240, 245
efferent, 240
function of, 113
microscopic anatomy of, 113
Artifacts, of fixation, 1
Astrocytes, 74, 80, 81
ATP, *see* Adenosine triphosphate
Atresia, of ovarian follicles, 208
Atrioventricular bundle, 111
Atrioventricular node, 111
Atrium, primitve, *265,* 275
Auditory meatus, external, 257, 264
Auditory ossicles, 257–258
Auditory tube, 3, 8, 135, 257, 258
Auerbach's plexus, of small intestine, 151–152, *151*
Auricle, 257
Autofluorescence, 192
Autonomic nervous system, *see* Nervous system, autonomic
Autoradiography, 16, 269, 277
Axon, 71, *80,* 81
unmyelinated, *203*

B

Banding pattern, of a sarcomere, 64, *65*
Basal body, 121
Basal cells, 120
of sebaceous glands, 173, *173,* 179
Basal lamina, 8, 118
See also Basement membrane
Basal plate, 77
Basement membrane, 29, 33, 37
of esophageal mucosa, *146,* 147
fine structure of, 34
glomerular, 276
ileal, *150*
and periodic acid-Schiff reaction, 13, 18
of Schwann cells, *80,* 81
of skin, *178,* 179
of stratified squamous epithelium, *171*
of tracheal epithelium, *271,* 278
Basilar fibers, 262
Basilar membrane, 263, 264
of cochlea, 259, *259*
of organ of Corti, *261, 273,* 278
vibrations of, 262
Basophilia, 12
Basophilic structure, 7
Basophils, 10, 84, 85, 88
of adenohypophysis, 200, *200,* 201
B cells, of immune system, 92, 99
of islets of Langerhans, *see* Beta cells
Benign prostatic hyperplasia, 229
Beta cells, of islets of Langerhans, 165
Bilateral adrenalectomy, 205, 206
Bile, 159, 162–163
canaliculi, 163
composition, 163
of intestines, 149
Bile ducts, 160, 163
Birth control pills, and luteinizing hormone releasing hormone (LH-RH), 203
Blood, 83–88
erythrocytes in, 83–84, *85, 86*
fibers in, 83
function of, 83
leukocytes in, 84–86, *86*
platelets in, 86
Blood-brain barrier, 118
Blood cells, 83
Blood circulation, through cardiovascular system, 190–110
Blood clotting, 83, 86
B lymphocytes (B cells), 99, 106, 277
characteristics of, 99–100
function of, 99–100
in lymphopoiesis, 92
memory, 100, 108
Bodies, multilamellar, 124
multivesicular, 124
Bolus, 139
Bone, 35
articular surfaces of, 53
canaliculi of, 53
cancellous, 54
components of, 51
cytologic structure of, 54–55
developing, regions of, 58, *58*
endochondral formation of, 59,
endosteum of, 53
formation of, 55–56
See also Ossification
function of, 51
histophysiology of, 57
lacunae of, 53
lamellae of, 53–54
macroscopic structure of, 53
matrix of, chemical composition of, 55
membrane of, 56
microscopic structure of, 53–54
periosteum of, 53, 54
regions of, 53
vascular channels in, 54
Bone marrow, 4, 89–95, *90*
agranulopoiesis in, 92
erythropoiesis in, 91–92, *91*

granulopoiesis, in, 92
hematopoiesis in, 90–91
microscopic anatomy of, 89
stem cells of, 97, 108
thrombocytopoiesis in, 92–93
types of, 89
Bony collar, 56
Border cells, of organ of Corti, 261
Boutons, 73, 75
Bowman's capsule, 237–238, *238,* 244, *244,* 245
Bowman's glands, 128
Bowman's membrane, 247, 256, 275
Brain, 71, 77–78, 118, 268
development of, 77–78
Brain parenchyma, 118
Brightfield microscopy, 14
Bronchi, 122
Bronchial tree, 122–123, *123*
Bronchiolar cells, 122
Bronchioles, 7, 122, *123*
Brown fat, 44
Brunner's glands, 157, 158
of duodenum, 152
of intestines, 149
Brush border, 121, 150
Brush cells, in tracheal epithelium, 121
Buccopharyngeal membrane, 154
Bulbourethral glands, 229, 269, 276–277
Bulbus cordis, 115
Bundle, atrioventricular, 111
Bursa of Fabricius, 99

C

Calcitonin, 7, 57, 181, 183
Calcium in bone, 57
Call-Exner bodies, 10, 210
Calsequestrin, 66
Calyces, major, 235, 236, 245
minor, 235, 245
Canaliculi, 53
of bile, 163
of parietal cell, *143*
Canal(s), pericardioperitoneal, 124
portal, 160, *161, 167*
of Schlemm, 249, *249*
semicircular, 258, 260
Capillaries, 109, *162*
continuous, 114, 117, 118
discontinuous, 114, 117, 118
distribution of, 114
fenestrated, 114, 117, 118
glomerular, renal, 118
lymphatic, 109, 117
microscopic anatomy of, 113–114
varieties of, 114
Capsules, of kidney, 236
of liver, 159–160
of lymph nodes, 99, *102*
Cardiovascular system, (CVS), 109–118
arterioles of, 113
basic histology of, 109–118
blood circulation through, 109–110
capillaries of, 113–114
early development of, 115
elastic arteries of, 111–112, *112*
embryology of, 115–116
endothelium in, 109
layers of, *5,* 110
lymphatic system and, 109
muscular arteries of, 109, 112–113, *113*
veins of, 115
venules of, 115
Cartilage, 35, 51–59
components of, 51
distribution of, 51
elastic, 52–53
function of, 51
growth of, 51
hyaline, 8, 52
structure of, 51
of tracheal epithelium, *271,* 278
varieties of, 51
Castration cells, 227
Catecholamines, 189
Cavernous bodies, 224
Cavity, pericardial, 111
tympanic, 257
C cells, of thyroid, 182, 183
Cecum, 154
Cells(s), acinar, pancreatic, *5,* 10, 163, *274,* 278
alpha, of islets of Langerhans, 165
alveolar, great, 124
apical, of esophageal mucosa, 146, 147
of Auerbach's plexus, *140*
basal, 120
of esophageal mucosa, 146, 147
of sebaceous glands, 173, *173,* 179
beta, of islets of Langerhans, 165
in bone, 51
border, of organ of Corti, 261
bronchiolar, 122
brush, in tracheal epithelium, 121
in cartilage, 51
castration, 227
centroacinar, 164, *164*
chief, of gastric glands, 3, 9, 142
of parathyroid, 184, *184*
chromaffin, 193
ciliated, 2
in tracheal epithelium, 121, *121*
clear, of eccrine sweat glands, 172
dark, of eccrine sweat glands, 172
differentiated, adult human, 27, 28
embryonic, 26
endrocrine, steroid-secreting, 269, 277
endothelial, capillary, 118, 128
enterochromaffin, 9, 142–144, 147, 158, 266
ependymal, 74
epithelium, *see* Epithelial cells
fat, of parathyroid, 184
follicular, ovarian, 210–211
gastrin, 269, 277

germ, primor
glial, 71, 81
goblet, 22, 2
in tracheal
granule, in tr
granulosa, 10
of ovarian f
granulosa-lut
hair, of organ
of Hensen, 2
inner pillar, o
interstitial, of
juxtaglomerul
keratinized, 3
Kupffer, 97, 1
lacis, 239
Leydig, 9, 22
of Meissner's
mesangial, 23
mesenchymal
mucous, 9
in tracheal e
Müller's, 252
muscle, smoo
myoepithelial,
in eccrine sw
in salivary gl
neural crest, 7
olfactory, 119
osteoblast, 54,
osteoclast, 55
osteocyte, 54-
osteoprogenit
outer hair, of
outer phalang
outer pillar, of
oxyphil, *184,*
Paneth, 149, 2
parafollicular,
parathyroid, 1
parenchymal,
hepatic, 22,
parietal, 142,
perichondrial,
phagocytic, 97
phalangeal, of
plasma, 100, 1
Purkinje, 73, 7
pyramidal, of c
reticular, 89
satellite, 75
Schwann, 4, 9,
and myelinati
78, *80,* 81, 2
Sertoli, 224, 22
short, in trache
spermatogenic,
stem, erythrocy
steroidogenic,
sustentacular, 1
synovial, 57
theca interna, o

interlobular, of pancreas, 164
proximal, *137,* 138
mesonephric, 270, 277
pancreatic, 159
papillary, 236, 245
striated, 133
testicular, 230, 232
thoracic, 109
thyroglossal, 181
vitelline, 154
Ductus arteriosus, 116
Ductus deferens, 227–228, 231, 233
gross anatomy of, 227
microscopic anatomy of, 227–228, *228*
Ductus reuniens, 259
Duodenum, 149, 152, 158
Dura mater, 75–76
Dwarf, achondroplast, 58, 59
Dwarfism, 202
Dynein, 32

E

Ear, 257–264
basilar membrane of, 262
components of, 257, *258*
external, 257
functions of, 257
internal, 257, 258–261, *259,* 273, *273*
cochlea of, 259
cochlear duct of, 260–261, *261*
embryology of, 262
major divisions of, 258
vestibular apparatus of, 260
middle, 257–258, 264
organ of Corti of, 261
EC cells, 144, 275
Eccrine sweat glands, 4, 9
Ectoderm, 206, 277
Efferent ductules, of testes, 227
Ejaculatory ducts, *224,* 227, 228
Elastic arteries, 111–112, *112*
Elastic fibers, *113*
of dermis, 175
distribution of, 44
structure of, 42–43
synthesis of, 42
Elastic lamina, *113*
Elastica, externa, 112
interna, 111, 112
Elastin, 42, 52
Electron microscopy, 11, 14–15
ferritin in, 13
scanning (SEM), 11–12, 15
transmission (TEM), 11, 15
Embedding, epoxy in, 11
paraffin in, 11
Embryogenesis, 26
Emphysema, 126
Enamel, 4, 9, 130, *130,* 131
Enamel organ, 130
Endocardial cushions, 116
Endocardium, 9, 110
Endochondrial bone formation, 3, 8
Endocrine cells, steroid-secreting, 269, 277
Endocrine glands, of parathyroid, 181
of thyroid, 181
Endoderm, 265, 275
Endolymphatic sac, 260, 262
Endometrium, 215–216
Endomysium, 62
Endoneurium, 75
Endoplasmic reticulum (ER), 19
function of, 22
functional relationship to nuclear envelope, 24
of parenchymal cells, 160
rough (RER), 9, 22, 28, 275
of chondrocytes, *272,* 278
of pancreatic acinar cells, *165, 274,* 278
of parietal cells, *143*
of thyroid cells, 188
of zona fasciculata, *191, 195,* 196
smooth (SER), 9, 22, 196, 277
of granulosa-lutein cells, *220,* 221
of zona fasciculata, *191, 195,* 196
structure of, 22
Endorphins, 144
Endosteum, 53, 54, 268, 276
osteogenic, 89
Endothelium, in cardiovascular system, 109
corneal, 248, 256, 275
Enterochromaffin cells, 9, 142–144, 158, 266
functional considerations of, 142
Enterochromaffin system, 4, 9, 277
Enzyme histochemistry, 13
Enzymes, surface, 86
Eosin, 7, 12
Eosinophilia, 12
Eosinophils, 84, 87, 88
Ependymal cells, of central nervous system glia, 74
Epicardium, 9, 110, 111
Epidermis, 31, 169–172
epithelial, 178–179
layers of, 169–171, *170, 171*
Epididymis, 37, *224,* 227, 233
Epimysium, 62
Epinephrine, 193, 275
Epineurium, 75
Epiphysis, 53
Epithelial apices, 32
Epithelial cells, *30*
ciliated, 27
coelomic, 218, 230
follicular, 182, *182,* 186, 188
reticular, *105,* 108
of small intestine, 266, 275
squamous alveolar, 123–124
thymic, 275
Epithelium, 1, 29–39, *30,* 118
alveolar, 7, 123–124, 127, 128
apical specialization of, 30
apical surface of, 29
apices of, 31–32

avascularity of, 30
basal surface of, 29
basement membrane of, 29, 33
blood supply of, 30
ciliary, 7, 249–250
columnar, 7, 31, 31(t)
simple, 36, 37
colonic, 268
corneal, 247, *248,* 256, 275
cuboidal, 31, 31(t), *32,* 168
double-layered, 7
for gas exchange, 119
germinal, 208
junctional complex of, 30, 33–34
keratinization of, 171–172
of large intestine, 153
neural crest of, 34–35
nonkeratinized stratified squamous, 138
olfactory, 119–120, 127
polarization of, 29
primary germ layers of, 34
of the proximal convoluted tubules of the kidney, 31, *32*
pseudostratified, 7, 8, 31, 37
ciliated columnar (PCC), 7, 9, 119, 138, *267,* 276
of rat, 267, *267*
seminiferous, 31, 224–226, *225,* 233
simple, 31
of small intestine, 149–151, *150*
squamous, 31
stratified, 31, 36, 37
squamous, 147, *171*
of oral cavity, 129
in tongue, 131
syncytial, 31
of the trachea, 31(t), *120, 121,* 125, *267,* 276, 278
transitional, of urinary bladder, 241, *241,* 269, 277
types of, 30–31, 31(t)
of the vagina, 31, 31(t), 216
Epitrychium, 176
Epoxy, 11
ER, *see* Endoplasmic reticulum
Erythroblasts, 95
basophilic, *90,* 91, 276
in erythropoiesis, 91
orthochromatic, 91
polychromatic, *90,* 91, 276
Erythrocytes, *85, 86,* 88, 89, 266, 276
function of, 83–84
hemoglobin in, 83
nucleated, *91*
primitive, 90
structure of, 83
Erythropoiesis, 91–92, 95, 266
Erythropoietin, in erythropoiesis, 91
Esophagus, 139, 154, 277
function of, 141
microscopic anatomy of, 140–141
Estrogens, 10, 211
Ethanol, 11
External genitalia, of female, 217, 218
of male, 230
External root sheath, of hair follicles, 174
Eye, 247–256
accessory organs of, 252–253
cornea of, 247–249
development of, 253, 254
function of, 247
iris of, 250
limbus of, 248–249, *249*
pupil of, 250
refractive media of, 250
retina of, 250–252
retinal differentiation of, 253
sclera of, 247–249
vascular layer of, 249–250
vitreous body of, 250
Eyelids, 253

F

Face, 136, 138
development of, 133, *134*
F-actin, 61, *63*
Fasciae adherentes, 66
Fascicles, 62, 75
Fat cells, of parathyroids, 184
Female reproductive system, 207–221
components of, 207
external genitalia, 217, 218
ovaries, 208
oviducts, 213–215
uterus, 215–216
vagina, 216–217
cyclic changes of, 207
hormonal control of, 207, 217
development of, 217–218
follicular development in, 208–212
functions of, 207
oogenesis in, 212–213
Fenestra ovalis, 259, 262
Fenestra rotunda, 259
Ferritin, 13
Feulgen reaction, 7, 13, 18
FFF, *see* Freeze-fracture etching
Fiber(s), basilar, 262
in bone, 51
in cartilage, 51
collagen, 83, *112*
elastic, *112,* 175
of the lens, 250
Purkinje, 110, 111
reticular, 89
Fibrin, 86
Fibrinogen, 83, 86
Fibrinolysin, 229
Fibroblasts, 8, 40
of dermis, 175
Fibrocartilage, 52–53
Fila olfactoria, 120, 128
Filopodia, *99*

Fimbriae, of oviducts, 213, *214*
Fixation, 11
Flagella, 32
Fluorescent probes, 13
Follicles, ovarian, antrum of, 210, *211*
atresia of, 208
Call-Exner bodies of, 210
corpus luteum of, 211–212
cumulus oophorus of, 210
granulosa cells of, *210*
liquor folliculi of, 210, *211*
oocytes of, *210*
ovulation of, 211
preovulatory, 210
primary, 209, *209, 210*
primordial, 208, 209, *209*
secondary, 210, *212*
theca externa of, 209, *211*
theca interna of, 209, *210, 211*
zona pellucida of, 209, *210, 211,* 219, 221
thyroid, 181, *182*
Follicle-stimulating hormone (FSH), 201–202, 217
Follicular cells, of ovarian follicles, 210–211
Foramen, ovale, 116
of Magendie, 77
Foramina of Luschka, 77
Foregut, 154
Formaldehyde, and fixation, 11
Fovea centralis, of the retina, 252
Freeze-fracture etching (FFE), 3, 8, 11
methodology of, 15–16
use of, with plasma membranes, 21, *21*
Frontal prominence, 133, *134,* 138
FSH, *see* Follicle-stimulating hormone
Fundic glands, 3
Fundus, of stomach, 141
of uterus, 215
Fusiform vesicles, of transitional epithelium, 241

G

G-actin, 61, 63
Gallbladder, 159, 163
Gamete, 211
Ganglia, 35
Gap junctions, in cardiac muscle, 66
in osteocytes, 55
Gastric glands, 142
Gastric pits, *141,* 142
Gastrin, 143
Gastrin cells (G cells), 142–143, 269, 275, 277
Gastrointestinal tract, 9, 162
upper, 129–133
oral cavity in, 129
pharynx in, 133
salivary glands in, 132
teeth in, 129–131, *130*
tongue in, 131–132
tonsils in, 133
G cells, *see* Gastrin cells
Genes, immune response, 100
Genital ridge mesenchyme, 218, 230
Germ cells, primordial, 278
of testes, 230
Germ layers, primary, skin, 176
Germinal centers, in lymph nodes, 101, *101*
Germinal matrix, of hair follicles, 174
GH, *see* Growth hormone
Gigantism, 202
Glands, adrenal, *see* Adrenal glands
Bowman's, 128
bronchial, 122
Brunner's, 149, 152, 157
bulbourethral, 229, 269, 276–277
ceruminous, of external auditory meatus, 8, 257
conjunctival, 9
Cowper's, *224*
endocrine, 181
gastric, 142
lacrimal, 9, 253
meibomian, 9, 253
mixed-type, of oral cavity, 129
of Moll, 253
mucous, 129, *140*
multicellular, of tracheal epithelium, *271,* 278
parathyroid, 181–188
pituitary, 197–206, 269
See also Pituitary gland
prostate, 2, 7, 229
salivary, 132–133
See also Salivary glands
sebaceous, 173–174, *173,* 177, 179
development of, 176
and fetal development, 176
suprarenal, *see* Adrenal glands
sweat, 172–173
apocrine, 172, 179
eccrine, 172–173, *178,* 179
development of, 176
thyroid, 181–188
vestibular, 217
Glans penis, *224*
Glaucoma, 248
Glia, 74–75
Glial cells, 71, 81
Glioblasts, 77, 81
Glomerulus, 237, 238, *239*
basement membrane of, *239*
renal, 244, *244,* 245, 268
Glucagon, 9, 144, 159, 165
Glucocorticoids, 192, 206
Glutaraldehyde, 8
use of, in fixation, 11
Glyceride synthetase, 152
Glycerol, 8
Glycocalyx, 13, 18, 31–32
of small intestine, 149
Glycoconjugates, 13
in bone matrix, 55
Glycogen, 13, 18, 20
of parenchymal cells, 160
vaginal, 216–217
Glycolipids, 20

Glycoproteins, 8, 9, 10
Glycosaminoglycans, 8, 44
Glycosyltransferases, 10
Gn-RH, *See* Gonadotropin releasing hormone
Goblet cells, 22, 27, 28, 138
in tracheal epithelium, *271,* 278
see also Mucous cells
Golgi apparatus, 19, 22–23
forming face of, 23
function of, 23
of granulosa-lutein cell, *220,* 221
maturation face of, 23
of pancreatic acinar cell, *165, 274,* 278
of liver parenchymal cell, 160
of parietal cell, *143*
of structure of, 22–23
of thyroid cell, *187,* 188
Golgi membranes, of chondrocyte, *272,* 278
Golgi zone, *150*
Gonadal steroids, in histophysiology of bone, 57
Gonadotropin releasing hormone (Gn-RH), 203, 217
Gonadotropins, 217
Gonadotrops, 200, 201
Granule cells, in tracheal epithelium, 122
Granules, azurophilic, 7, 92, 275
catecholamine, 275
eosinophilic, 84
keratohyalin, 170
lamellated, 169
lipofuscin, 20
melanin, 174
trichohyalin, of hair follicles, 174
zymogen, 163, *164, 274,* 278
Granulocytes, 7, 10, 84–85, 92, 95
Granulomere, in platelets, 86
Granulopoiesis, 2, 92, 95
Granulosa cells, 10, 22
of ovarian follicles, 209, 210, *210, 211*
Granulosa-lutein cells, 10, 211–212
of corpus luteum, 220, *220,* 221
Gray matter, of central nervous system, 72
Greater omentum, 155
Growth hormone (GH), 201, 202
histophysiology of bone and, 57
Gut, developmental anatomy of, 154–155
endodermal derivatives of, 154
later development of, 154–155
primitive, 154

H

Hair, 169, 174
development of, 176
follicles of, 174, 176, 177, 179
lanugo, 176
shaft of, *173*
vellus, 176
Hair cells, of organ of Corti, 261, *273,* 278
Hassall's corpuscles, in thymus, 105, *105,* 108
Haustra, 153, 157, 158
Haversian canals, 54
Haversian systems, 54, 114
H band, 10, *63,* 64, *65,* 277
HCG, *see* Human chorionic gonadotropin
Heart, 109
conducting system of, 111
development of, 115, 265, *265,* 275
microscopic anatomy of, 110–111
Heavy meromyosin (HMM), 25, *63*
Helicotrema, 259
Hematocrit, 83
Hematoxylin, 12
Hematopoiesis, 90, 95, 159
theories of, 90–91
Hematopoietic compartment, 89
Heme, 83
Hemidesmosomes, 33, 34, 276
Hemocytoblasts, 115
Hemoglobin, 276
in erythrocytes, 83
Heparan sulfate, 44
Heparin, 44
in basophils, 85
Hepatic parenchymal cells, 22
Hilus of kidney, 235, *236*
of lymph node, 101
of ovary, 208
Hindgut, 154
Histamine, in basophils, 85
Histiocytes, 41
and monocytes, 86
HMM, *see* Heavy meromyosin
Holocrine secretion, 173, 179
Homeostasis, 83
Hormone, antidiuretic, 204
follicle-stimulating, 201–202, 217
gastrointestinal, 142–144
gonadotropin-releasing, 203, 217
hypothalamic releasing, 202
interstitial cell stimulating, 201
luteinizing, 201–202, 217
thyrotropin releasing, 203
^{3}H-thymidine, *see* Tritiated-thymidine
Human chorionic gonadotropin (HCG), 212
Hyaline cartilage, 52
in trachea, 120
Hyalocytes, 250
Hyaloid arteries, 253
Hyalomere, in platelets, 86
Hyaluronate, 8
Hyaluronic acid, 44, 175
Hypophyseal arteries, inferior, 199
superior, 199, 270, 277
Hydroxyapatite, 9
17-α-Hydroxylase, 7, 192
Hydroxylysine, 33
Hyperplasia, 45
Hypertrophy, 45
Hypophyseoportal system, 199, 202–203
Hypophysis, *see* Pituitary gland
Hypothalamicohypophyseal tract, 203
Hypothalamic releasing hormone, 202
Hypothalamus, 9

I
I band, *10, 63,* 64, *64, 65,* 277
I cells, 143, 275
Ileal villus, 6, *6,* 10
Ileum, 149, 152, 156, 158
Immune system, 83, 97–108, 265
 components of, 97
 function of, 97
 lymphocytes in, 99–100
 lymph nodes in, 100–102
 macrophages in, 98, *99*
 mononuclear phagocyte system in, 97–98
 spleen in, 102–104, *103*
 thymus in, 104–105, *105*
Immunity, cellular, 100
Immunoglobulins, 152
Immunohistochemistry, 13
Incus, 257
Inferior vena cava, *162*
Infundibular process, *see* Pars nervosa
Infundibular stem, of neurohypophysis, 198
Infundibulum, 205, 206, 213, 277
Inner pillar, of organ of Corti, 261, *261*
Inner tunnel, of organ of Corti, 261, *261*
Inorganic salts, in bone matrix, 55
Insulin, 159
Intercalated ducts, of pancreas, 164
Interlobar arteries, *236,* 240
Interlobular ducts, of pancreas, 164
Intermediate filaments, 25
Internal root sheath, of hair follicle, 74
Interstitial cell-stimulating hormone (ICSH), 201
Interstitial cells, 224
Interstitial growth, in endochondrial ossification, 56
Interstitial systems, 54
Interstitium, blood vessels of, 227
 connective tissue of, 227
 constituents of, 226–227
 Leydig cells of, 226–227
Intestines, 149–158
 absorption by, 149
 Brunner's glands of, 149
 digestion in, 149
 large, 153–155
  adventitia of, 153
  epithelium of, 153
  function of, 153
  lamina propria of, 153
  microscopic anatomy of, 153–154
  muscularis externa of, 153
  rectum of, 154
 small, 149–153
  anatomic variations of, 152
  duodenum of, 152
  epithelium of, 149–151
  external layers of, 151–152, *151*
  extracellular digestion in, 152
  gross anatomy of, 149
  histologic features of, 149–152, *150, 151*
  ileum of, 152
  jejunum of, 152
  lipid absorption of, 152–153
  muscularis mucosa of, 151
  lamina propria of, 151
  surface elaboration of, 149
Intrinsic factor, gastric, 146
 of parietal cells, 142, *143*
Iris, 250
Islets of Langerhans, 159, 163, 165
Isodesmosine, 42
Isotopes, radioactive, 16
Isthmus, of oviducts, 213
 of thyroid, 81

J
Jejunum, 149, 152, 156, 158
JGA, *see* Juxtaglomerular apparatus
Joints, 57
J polypeptide, 152
Junctional complex, 33–34, 37
 apical, 150–151, *150*
 function of, 33
 gap junction of, 34
 macula adherens of, 34
 structure of, 33
 zonula adherens of, 33–34
 zonula occludens of, 33
Junctional folds, 66
Junctions, myoendothelial, 111
 tight, luminal, 118
Juxtaglomerular apparatus (JGA), 239–240, *240,* 243
Juxtaglomerular cells, 239, 245

K
Keratan sulfate, 8, 44
Keratin, 169
 biochemistry of, 172
Keratinocytes, 169
Keratinization, 171–172
Kidneys, 235–240, 243
 blood supply of, 240
 juxtaglomerular apparatus of, 239–240
 lymph supply of, 240
 macroscopic anatomy of, 235–236, *236*
 nerve supply of, 240
 renal cortex of, 236–237
 renal medulla of, 236–237
Kupffer cells, 41, 97, 114, 161, 168

L
Labia majora, 217
Labia minora, 217
Labyrinth, osseous, 258
Lacis cells, 239
Lacrimal glands, 9, 253
Lacteals, 151
Lactotrops, 10, 200, 202
Lacunae, of bone, 53–54
 of cartilage, 52
Lamellar patterns, 54

Lamellipodia, *99*
Lamina propria, 10, 158
 of digestive system, 41
 of esophageal mucosa, 147
 of gastrointestinal tract, 39, 88, *140*
 of respiratory system, 41
 of small intestine, 151
 in trachea, *120*
 of vagina, 217
Laminae, elastic, 110
Lanugo hair, 176
Laryngopharynx, 133
Larynx, 120
Lateral palatine processes, *135*
Lecithin, dipalmitoyl, 124
Lens, of the eye, 247, 250
Lens placode, 253, 256
Lesser omentum, 155
Leukocytes, 5, 10, 83, 84–86, *86*
 classification of, 84
 polymorphonuclear, *87,* 88
 vaginal, 217
Leydig cells, 9, 224, *225,* 226–227
LH, *see* Luteinizing hormone
LH-RH, *see* Luteinizing hormone releasing hormone
Light meromyosin (LMM), 61, *63*
Light microscope, components of, 13–14
Limbus, of the eye, 248–249, *249*
Lipid droplets, 277
 of granulosa-lutein cell, *220,* 221
 of zona fasciculata, *191, 195,* 196
Lipids, 152–153, 156, 158
Lipofuscin, in zona fasciculata, *195,* 196
 in zona reticularis, 191, *191*
Lipoproteins, very, low-density (VLDL), 160
Lipotropin, 200, 201
Liquor folliculi, 10, 210, *211*
Liver, 118, 155, 159–166
 blood flow through, 162, *162*
 capsule of, 159–160
 central veins of, 3, 8
 development of, 165–166
 function of, 159
 hematopoiesis in, 90
 histologic organization of, 159–162
 lobules of, 160–161
 parenchymal cells of, 160
 porta hepatis of, 160
LMM, *see* Light meromyosin
Lobes, in bronchial tree, 122–123
Lobules, 138
 of liver, 160–161, 168
 primary, *123*
 renal, 236
 secondary, *123*
Loop of Henle, *236,* 238, 239, 245
Loops, midgut, 15
L-thyroxine (T_4), 183
Lumen, 118
 of blood vessel, 277
 of gastrointestinal tract, 158
 of oviduct, *214*
 of stomach, *143*
Lungs, 118
 and cigarettes, 125–126
 defenses of, 125
 deleterious materials to, 125
 early development in, 124
Luteinization, 211, *212*
Luteinizing hormone (LH), 201–202, 217
Luteinizing hormone releasing hormone (LH-RH), 9, 203
Lymph, circulation of, 100
Lymph nodes, 97, 100–108
 distribution of, 100
 function of, 102
 lymphatic channels in, 101–102, *101*
 lymphocyte distribution in, 101
 structure of, 100–102, *101, 102*
Lymphatic channels, 101–102, *101*
Lymphatic systems, 109
Lymphatic vessels, 109
Lymphoblasts, 90
Lymphocytes, 85, *86,* 88, 275
 function of, 85
 heterogeneity in, 99
 in immune system, 97
 in pharynx, 138
 in respiratory system, 125
 structure of, 85
Lymphoid organs, 97
Lymphokines, 100, 105
Lymphopoiesis, in thymus, 90
Lysosomes, 8, 10, 20, 24–25, 95
 distribution of, 24
 function of, 25
 of Paneth cell, 150
 of parenchymal cell, 160
 of parietal cell, *143*
 structure of, 24
 of thyroid cell, *187,* 188
Lysozyme, 98
Lysyl oxidases, 42

M

Macrophage(s), 108, 123–124, 128, 275
 alveolar, 88, 124
 development of, 98
 distribution of, 41
 features of, 2, 8
 function of, 40–41, 98
 lysosomes in, 24
 monocytes and, 86
 in mononuclear phagocyte system, 97
 in respiratory system, 125
 structure of, 40, 98, 99
Macula adherens, 34, 124
Macula densa, 239, 245
Macula lutea, 252
Male reproductive system, 223–233
 components of, 223, *224*
  glands, 228–229

interstitium, 226–227
penis, 229–230
seminiferous epithelium, 224
testes, 223–224
testicular drainage ducts, 227–228
development of, 223, 230
function of, 223
spermatogenmesis in, 226
Mandibular swellings, in face, 133
Mantle layers, 77
Marginal layers, 77
Mast cells, 41, 85
Mastication, muscles of, 134
Maxillary processes, 138
Maxillary swellings, in face, 133
M band, 64, *65,* 277
Meatus, external auditory, 8, 135, 257, 264
Meckel's cartilage, 134, *135*
Median eminence, of neurohypophysis, 198, *198*
Mediastinum testis, 223
Medulla, adrenal, 265
Medullary pyramid, of kidney, *236*
Medullary rays, 236, *236*
Medullary sinuses, *101,* 101–102
Megakaryocytes, 94, 95
in platelets, 86
in thrombocytopoiesis, 92–93
Meibomian glands, 9, 253
Meissner's plexus, of small intestine, 152
Melanin, 174–175
Melanocytes, 35, 169, 174
of hair follicles, 179
Melanosomes, 20
of hair follicles, 179
Membrane(s), basilar, 263, 264
of organ of Corti, *273,* 278
vibrations of, 262
buccopharyngeal, 154
cloacal, 154
Descemet's, 275
inner limiting, 277
outer limiting, 277
otolithic, *260*
Reissner's 259, *259, 273,* 278
reticular, 263, 264
of organ of Corti, 261
Shrapnell's, 258
synovial, 57
tectorial, *259,* 263, 264
of organ of Corti, 261–262
tympanic, 8, 257, 258, 263, 264
vestibular, 263, 264
of cochlea, 259, *259*
of cochlear duct, 260, *261*
Meninges, 75–76
Menstrual cycle, 207–208
Merocrine mechanism, 172
Meromyosin, heavy (HMM), 61, *63*
light, (LMM), 61, *63*
Mesangial cells, 239
Mesaxon, 78
Mesencephalon, 77
Mesenchyme, 46, *46,* 95
Mesentery, *140*
splanchnic, 127, 128
Mesonephric ducts, 270, 277
Mesonephros, 235, 242
Mesosalpinx, of oviducts, 213
Mesothelial cells, of adrenal cortex, 190
of adrenal glands, 189
Mesothelium, 140
Metochromasia, 12–13, 265, 275
Metamyelocytes, 7
in granulopoiesis, 92
neutrophilic, *90*
Metaphysis, of long bone, 53
Metanephros, 235, 242
Metencephalon, 77, 276
Methylene blue, 12
Microfilaments, 8, 20, 25
Microglia, 75, 81
in mononuclear phagocyte system, 97
Microscopy, electron, 11, 14–15
light 11, 13–14, 17–18
Microtubules, 20, 25, 28
Microvilli, 32, 118
apical, of thyroid, *187,* 188
of small intestine, 149
of tall columnar epithelial cells, 150, *150*
Midgut, 154
Mineralocorticoid, 7, 191
Mitochondria, 20, 28, *64, 65,* 276
of chondrocyte, *272,* 278
cristae of, 24
function of, 24
of granulosa-lutein cell, *220,* 221
of pancreatic acinar cell, *165, 274,* 278
in production of adenosine triphosphate, 24
structure of, 24
with tubular cristae, 277
of zona fasciculata, 191
Mitosis, 150
Modiolus, 259
Monoblasts, in monopoiesis, 92
Monocytes, 10, *87,* 88, 275
in development of macrophages, 98
function of, 86
in mononuclear phagocyte system, 97
in monopoiesis, 92
structure of, 85–86
Monoglycerides, 153, 158
Mononuclear phagocyte system (MPS), 41
97–98, 106, 108
cells of, 97
in immune system, 97–98
Morphogenesis, 46
MPS, *see* Mononuclear phagocyte system
Mucosa, of the esophagus, 140, 146, *146,* 147
olfactory, 7, 119–120
oviductal, 214–215, *214*
respiratory, of the nasal cavity, 119
of stomach, 141
of tubular viscera, 139, *140*
Mucous acini, *137,* 138

Mucous cells, in tracheal epithelium, 121, *121*
Müller's cells, 252, 256
Multilocular adipocytes, 44
Muscle fasciculus, *63*
Muscle fiber, *63*
See also Muscular tissue
Muscles, ciliary, of the eye, 249
oculomotor, 247
skeletal, 118
Muscular arteries, 112–113, *113*
Muscular coats, smooth, of ureter, 241
of urinary bladder, 241, *241*
Muscular tissue, 29, 61–69
cardiac, 61, 66, *67*
neuromuscular junction of, 66–67
sarcomere, 64–66
skeletal, 61, 62–66, *63, 64, 65*
smooth, 61, 62
types of, 61
Muscularis externa, of the esophagus, 141, 147
of large intestine, 153
microanatomy of, 145
of small intestine, 151–152, *151*
of stomach, *141*
of tubular viscera, 139–140, *140*
Muscularis mucosae, of gastrointestinal tract, 140, *140*
of small intestine, 151
of stomach, *141*
Myelencephalon, 77, 276
Myelin, 72, *80,* 81, 269, 277
Myelin sheaths, 9, *80,* 81
Myelination, 72, 78–79
Myeloblasts, 92
Myelocytes, in granulopoiesis, 92
Myeloperoxidase, 84
Myoblasts, 26
Myocardium, 110, 111
Myoepithelial cells, 122
of eccrine sweat glands, 172
Myoepithelium, 250
Myofibrils, 62, *63*
Myofilaments, *63*
Myometrium, 216, 219, 221
Myosin, 61–62, *63, 65*

N

Nails, 169
Nasal cavity, 138
Nasal conchae, 134
Nasal placodes, 133, *134,* 138
Nasal swellings, 138
Nasopharynx, 120, 133
Nephrons, 245
of uriniferous tubules, 237–239, *237*
Nerve endings, free, 75
Nerve fibers, 75
myelinated, 80, *80*
Nerves, peripheral, 75
Nervous system, autonomic, 71
central, 71
gray matter of, 72
neuronal variations in, 72–74
white matter of, 72
development of, 76–79
mantle layer of, 77
marginal layer of, 77
myelination of, 78–79
neuroepithelium of, 76–77
peripheral, 71, 80, 81
Nervous tissue, 29, 71–81
components of, 71
development of, 72
myelination of, 72
neuron structure of, 71
Neural crest, 34–35, 72, 189, 196
derivatives of, 35
neural tube formation of, 34
Neural crest cells, 76
Neural groove, 76
Neural plate, 72, 76
Neural tube, 72, 76
formation of, 34
neural plate of, 34
Neuroblasts, 77
Neuroectoderm, 206
Neuroepithelium, 76–77
Neurofilaments, 73
Neuroglia, 74–75
See also Glia
Neurohypophysis, 197, 203–204, *203,* 206, 217
function of, 204
infundibular stem of, 198
median eminence of, 198, *198*
microscopic anatomy of, 203, *203*
pars nervosa of, 198, *198*
Neurolemma, 75
Neuron, 71
Golgi type I, 73
Golgi type II, 73
motor, 73–74
preganglionic sympathetic, 193
pseudounipolar, 72
pyramidal, 72, *73*
structure of, 71
Neurophysins, *203,* 204
Neuropil, 72
Neurotransmitters, 75
Neurotubules, 73
Neutrophil, 7, 10, 84, *85, 87,* 88, 108, 265, 275
Night vision, 251
Nissl bodies, 71
Node, atrioventricular, 111
sinoatrial, 111
Nodes of Ranvier, 9, 78, *78*
Nodules, lymphoid, 8
primary, *102*
in lymph nodes, 101
secondary, in lymph nodes, 101
in spleen, 103
Norepinephrine, 75, 193, 275
Nuclear apparatus, cell cycle of, 24
components of, 23–24

Nuclear envelope, 19, 24
Nuclear layer, outer, 277
Nuclear pores, 23
Nucleolus, 23
Nucleus, 23

O

Oculomotor muscles, 247
Odontoblasts, 130, *130,* 131, 138, 276
Olfactory cells, 119–120, 128
Olfactory epithelium, 119–120, 127
Olfactory mucosa, 119–120
Oligodendroglia, 74, 277
Oocyte, 10, 207, 221
  of ovarian cortex, *209*
  of ovarian follicles, *210, 211*
  primary, 212–213, 278
  secondary, 213, 278
Oogenesis, 212–213, 270
Oogonia, 212, 278
Oophorus, cumulus, *211*
Optic cup, 253, 256
Optic fissure, 253
Optic nerve, 247, *251,* 252, 253
Optic stalk, 253
Optic vesicle, 253, 256
Oral cavity, 129, *135,* 138
Orange G, 7, 12
Ora serrata, of the eye, 249
Organ of Corti, 257, 258, *259,* 260, 261–262, *261, 273,* 278
Organelles, 19, 22, 27, 28, 138
Oropharynx, 133
Os, cervical, 277
Osseous labyrinth, 258
Osseous spiral lamina, 259
Ossicles, auditory, 257–258
Ossification, 55–56
Osteoblasts, 8, 53, 54, 276
  in bone marrow, 89
Osteoclasts, 55
Osteocytes, 53, 59
  processes of, 58
Osteoid, 56
Osteolysis, 7
Osteoprogenitor cells, 8
Ostium primum, 116
Ostium secundum, 116
Otocysts, 262
Otolithic membrane, *260*
Otoliths, *260*
Outer hair cells, of organ of Corti, *261*
Outer phalangeal cells, of organ of Corti, *261*
Outer pillar cells, of organ of Corti, 261, *261*
Outer tunnel, of organ of Corti, *261*
Ovaries, cortex of, *209*
  oocyte in, *209*
  development of, 217–218
  follicles of, 208, 212
  development of, 5
  See also Follicles, ovarian
  gross anatomy of, 208
  microscopic anatomy of, 208
Oviductal muscularis, *214,* 215
Oviducts, 213–215, *214*
  gross anatomy of, 213
  mucosa of, 214–215
  and ovum transport, 213–214
  serosa of, 215
Ovulation, 211
Oxyphil cells, of parathyroid, 184–185, *184*
Oxyphils, 188
Oxytocin, 204

P

Palate, 136, 138
  anterior hard, 138
  development of, 133–134
  posterior, 138
  primary, 134, 138
  secondary, 134
Palatine, lateral, 138
PALS, *see* Periarterial lymphathic sheath
Pancreas, 159, 163–165
  development of, 166
  endocrine, 165
  exocrine, 163–164, *164*
  islets of Langerhans, 159
Pancreatic ducts, 159
Pancreozymin, 147
Paneth cells, 149, 275
Papilla(e), circumvallate, 1, 7, 132
  dermal, 175, *178,* 179
  renal pyramid, 236
  types, 7, 132
Papillary ducts, 236, 245
Papillary layer, of dermis, 175
Paraffin, 11
Parafollicular cells, 182–183
Parathormone (PTH), 2, 7, 57, 181, 183, 184, 185
Parathyroid cells, 184–185, *184*
Parathyroid glands, 181–188
  distribution of, 183–184
  embryology of, 184
  histology of, 184
  inferior, 135
  superior, 135
Parenchymal cells, of liver, 160
Pars distalis, 205, 206, 277
  of adenohypophysis, 197, *198,* 199–200
  and releasing hormones, 202–203
Pars intermedia, 206, 277
  of adenohypophysis, 198, *198,* 199
Pars membranacea, of male urethra, 242
Pars nervosa, of neurohypophysis, 198, *198,* 203, *203*
Pars spongiosa, of male urethra, 242
Pars tuberalis, 206
  of adenohypophysis, 198, *198,* 199
PAS reaction, *see* Perdiodic acid-Schiff reaction
PCC epithelium, *see* Epithelium, pseudostratified ciliated columnar
PCT, *see* Tubule, proximal convoluted

Penis, 229–230
Pepsin, of chief cells, 142
Pepsinogen, 9
 of chief cells, 142
Periarterial lymphatic sheath (PALS), 102, 108
Perichondrial cells, in tracheal cartilage, 122
Perichondrium, 52
Pericytes, 113
Periderm, 176
Perikaryon, 71
Perilymph, 262
Perimetrium, 216
Perimysium, 62
Perineurium, 75
Periodic acid-Schiff (PAS) reaction, 7, 13, 17, 18
Periosteum, 53, 54
Peripheral cortex, of ovaries, 208
Peripheral nervous system, *see* Nervous system, peripheral
Peritonium, of male reproductive system, *224*
Peroxidase, 183
Peroxisomes, of liver parenchymal cells, 160
Peyer's patches, of ileum, 152, 158
Phagocyte, 8, 10, 88, 128
Phagocytic cells, in mononuclear phagocyte system, 97
Phagocytosis, 84, 88
 of bacteria, 275
 in macrophages, 98
Phagolysosome, 84
 of thyroid cell, *187*, 188
Phagosome, 25, 84, 188
Phalangeal cells, of organ of Corti, 261
Pharyngeal arches, *135*
 development of, 134–135
Pharyngeal pouches, fourth, 184, 188
 third, 184, 188, 265, 275
Pharynx, 133, 136, 138
Phase-contrast microscopy, 7, 14
Phospholipids, 20, *21*
Photoreceptors, 7
Pia-arachnoid layer, 76
Pia mater, 76
Pinocytosis, 8
 in macrophages, 98
Pituicytes, of the pars nervosa, 203, *203*
Pituitary gland, 5, 299
 circulation of, 199
 development of, 198–199
 embryology of, 198–199
 endocrine regulation of, 197
 gross anatomy of, 197–198
 lobes of, 197
 major divisions of, 197
 origin of, 198
 subdivisions of, 197–198, *198*
Placenta, chorionic villus in, 31
Plasma cells, 100, 152
Plasma membrane, 19, 20–21, *21*, 28
 freeze-fracture etching technique with, 21, *21*
Platelets, 83, 86, 87, 88, 89, 95
 in thrombocytopoiesis, 93
Pleura, parietal, 125
 visceral, 125
Plexiform layer, outer, 277
Plicae circulares, 157, 158
 of gastrointestinal tract, *140*
 of small intestine, 149
PNS, *see* Nervous system, peripheral
Podocyte slit diaphragms, 276
Podocytes, 237, *238, 244,* 245
Polar body, 278
Polysomes, 19
Pores, transcellular, 89
Porta hepatis, 160
Pouches, pharyngeal, 134, 135
Predentin, 130, *130*, 268, 276
Pregnenolone synthetase, 7, 192
Primary germ layers, 34
Primordial germ cells, 212, 217
Processus vaginalis, 223
Progesterone, 212
Prolactin, 10
Promonocytes, in monopoiesis, 92
Promyelocytes, in granulopoiesis, 92
Pronephros, 235, 242
Prosencephalon, 77, 268, 276
Prostate gland, *224*, 229
Prostatic concretions, 229
Prostatic utricle, *224*
Proteoglycan aggregate, 44
Proteoglycans, 8, 44
Protofilaments, 25
Proximal tubule, 238
PTH, *see* Parathormone
Pubis, 224
Pulp cavity, *130*
Pupil, 247, 250
Purkinje cells, 73, *74*, 78
Purkinje fibers, 110, 111
Pyloric portion, of stomach, 141
Pyramidal cells, of cerebral cortex, 72
Pyramids, renal, 236, 245

R

Rathke's pouch, 198–199, 205, 206, 277
RBC, *see* Red blood cell
Rectum, 154, *224*
Red blood cells (RBC), 83, *87*, 88
 production of, 94
 See also Erythrocytes
Red marrow, 89, 94, 95
Red pulp, in spleen, 102
Reichert's cartilage, 134
Reissner's membrane, 259, *259, 273,* 278
Renal column, 236
Renal artery, *236*, 240
Renal corpuscle, 238
Renal cortex, 236–237, *245*
Renal medulla, 236–237, *245*
Renal pelvis, 235
Renal sinus, 235

Renin, 235, 239
Rennin, of chief cells, 142
RER, *see* Endoplasmic reticulum, rough
Residual body, of Sertoli cell, *225*
Respiratory mucosa, 119
Respiratory distress syndrome, 125
Respiratory tract, upper, 119–120
Respiratory system, 119–128
 alveoli in, 123–124
 bronchial tree in, 122–123, *123*
 lung defenses of, 119
 lung development in, 124–125
 olfactory epithelium in, 119–120
 trachea in, 120–122, *120, 121*
 upper respiratory tract in, 119–120
Rete cutaneum, of dermis, 175
Rete subpapillare, of dermis, 175
Rete testis, 227
Reticular cells, 89, 101
Reticular connective tissue, 45–46
Reticular fibers, in lymph nodes, 101
Reticular lamina, 33
Reticular layer, of dermis, 175
Reticular membrane, 263, 264
 of organ of Corti, 261
Reticulocytes, 91
Retina, 250–252, 270
 fovea centralis of, 252
 layers of, 255, *255*
 neural, 250–252, *251*
  cones of, 251, 252
  ellipsoids of, 252
  ganglion cell layer of, *251,* 252
  inner limiting membrane of, *251,* 252
  inner nuclear layer of, *251,* 252, *255,* 256
  inner plexiform layer of, *251,* 252
  nerve fiber layer of, *251,* 252
  outer limiting membrane of, *251,* 252, *255,* 256
  outer nuclear layer of, *251,* 252, *255,* 256
  outer plexiform layer of, 251, 252, *255,* 256
  rods of, 251, *251*
 pigmented, 250–251
  epithelium of, 251, *251*
Rhodopsin, 251
Rhombencephalon, 77, 276
Ribonucleic acid (RNA), 23, 276
Ribosomes, 19, 22, 28
RNA, *see* Ribonucleic acid
Rods, of the eye, 247
Roots, aortic, 115
Rugae, of stomach, 141

S

Sac, endolymphatic, 260
Saccule, 258, 260
Saliva, 132, 269, 277
Salivary glands, major, 132–133
 minor, 129
 submaxillary, 137, *137,* 138
Sarcolemma, 62
Sarcomere, 6, 10, 28, *63,* 64–66, *65,* 69, 269
 regulation of contractions of, 64–65
  regulation of, 64–65
Sarcoplasmic reticulum (SR), 26, 62, 65
Satellite cells, 75
Scala media, 259, *261,* 264, 273, 278
Scala tympani, 259, *259, 261,* 264, 273, 278
Scanning electron microscopy (SEM), 11, 15
 preparation for, 11–12
S cells, 144, 275
Schistosomiasis, 84, 88
Schmidt-Lantermann clefts, 9, 79
Schwann cells, 4, 9, 35
 and myelinization of peripheral nervous system, 75, 78, *80,* 81, 277
Sclera, 247–249
Sclerotome, 46
Scrotum, *224*
Sebaceous glands, 173–174, *173,* 177, 179
 and fetal development, 176
Sebum, 169, 173, 179
Secretion, 9, 144
 aprocrine, 121
 holocrine, 173
 prostatic, 9
Sectioning, 12
Sella turcica, of pituitary gland, 197
SEM, *see* Scanning electron microscopy
Semicircular canals, 258, 260
Seminal vesicle, *224,* 227
 function of, 229
 structure of, 222–229
Seminiferous epithelium, 224–226, *229,* 233
 Sertoli cells of, 224, *225,* 226
 spermatogenic cells of, 224
Seminiferous tubules, 277
Septum, esophagotracheal, 124
 interatrial, 111
 interventricular, *265,* 275
 nasal, 134, *135,* 138
Septum membranaceum, 111
Septum primum, 116
Septum secundum, 116
Septum transversum, 155, 166
SER, *see* Endoplasmic reticulum, smooth
Serosa, 140, *140*
 of large intestine, 154
 of the oviducts, 215
Serotonin, 9, 75, 86, 144
Serous demilunes, *137,* 138
Sertoli cells, 233, 277
 compartmentalization of, 226
 functions of, 226
 structure of, 224, *225*
Sharpey's fibers, 54
Short cells, in tracheal epithelium, 121
Shrapnell's membrane, 258
Sinoatrial node, 111
Sinus(es), cervical, 135
 cortical, 19
 medullary, 101–102, *101*
 subcapsular lymph, *102*

vascular, 89
Sinus horns, 115
Sinus venosus, 116
Sinusoids, 167, *167*, 168
of bone marrow, 9
hepatic, 8, 161–162, *161*, *162*
Skeletal muscle, in tongue, 131
Skeleton, cardiac, 111
Skin, 169–179, *178*
dermis of, 169, 175–176, *178*
embryology of, 176
epidermis of, 169–172, *178*
function of, 169
melanin of, 174–175
melanocytes of, 174–175
thin, 269, 276
Slow-reacting substance (SRS), in basophils, 85
Somatostatin, 165, 203
Somatotrops, 200, 202
Somites, 46
Space of Disse, 161
Specimen preparation, 11–12
Spermatids, 224, *225*
Spermatocytes, 224, *225*
Spermatogenesis, 9, 226, 233
Spermatogenic cells, of seminiferous epithelium, 224
Spermatogonia, 224, *225*
Spermatozoa, 7, 224, *228*, 233
Spermatozoon, *225*, 231, 233
Spermiogenesis, 226
Spinal cord, 268
Spiral ganglion, *259*
Spiral limbus, *259*, 261
Spiral prominence, of cochlea, *259*
Spleen, 8, 95, 108, 118, 275
blood circulation in, 103, *103*
cords of, 104
functions of, 104
hematopoiesis in, 90
microscopic anatomy of, 102–103, *103*
reticular elements of, 104
sinuses of, 104
SR, *see* Sarcoplasmic reticulum
SRS, *see* Slow-reacting substance
Staining methods, 12–13
Stapes, 257
Stereocilia, 32
Steroid hormones, 189
Steroidogenic cells, 221
Stomach, 139, 154–155
microscopic anatomy of, 141–142, *141*
Stratum basale, 275, 276
of epidermis, 169
Stratum corneum, 170–171, *171*, 275, 276
Stratum granulosum, 170, *171*, 275
Stratum lucidum, 170, *171*, 276
Stratum malpighii, 170
Stratum spinosum, 169, *170*, *171*, 275
Stria vascularis, 259, 260
Striated muscle differentiation, 26
Stroma, 275
of the cornea, 248, 256
Subarachnoid space, 76
Subdural space, 76
Submucosa, of the esophagus, 140
of stomach, *141*
of tubular viscera, 139, *140*
Substantia propria, 248
Sucrase, 152
Sulcus limitans, 77
Surfactant, pulmonary, 124
Suspensory ligament, of ovaries, 208
Sustentacular cells, 120
Sweat, 169
Sweat glands, 172–173
apocrine, 172–179
eccrine, 172–173, *178*, 179
development of, 176
Synapse, morphology of, 75
Synarthroses, 57
Syncytiotrophoblast, 32
Systole, 109

T

Tanycytes, 204, 206
Taste buds, 132
TCA, *see* Tricarboxylic acid
T cells, 2, 8, 99, 108, 269, 277
characteristics of, 100
effector, 100
function of, 100
helper, 100, 108
killer, 100
in lymphopoiesis, 92
suppressor, 100
Tears, 4, 9
Tectorial membrane, *259*, *261*, 263, 264
of organ of Corti, 261–262
Teeth, 129–130, 136
ameloblasts in, 130–131
dentin in, 130, *130*
development of, 130
enamel in, 130
odontoblasts in, 131
primary, 129
secondary, 129
Telencephalon, 276
TEM, *see* Transmission electron microscopy
Teniae coli, 153, 157, 158
Terminal bars, *150*
Tertiary cortex, of lymph nodes, 101
Testes, 223–224, *224*, 231, 233
descent of, 223
development of, 230
microscopic anatomy of, 223–224
Testicular drainage ducts, 227–228, 232
Testosterone, 3, 9, 224, 233
Theca interna, of ovarian follicles, 209, *210*, 211
Theca interna cells, 10, 210–211
Theca-lutein cells, 10, 212
Thermoregulation, 8
Thrombocytopoiesis, 92–93
Thymosin, 8, 105

Thymus, 9, 106, 107, *107,* 188
development of, 105
function of, 105
microscopic anatomy of, 104, *105*
Thyroglobulin, 181, *182,* 188
synthesis of, 266, 275
Thyroid cell, 187, *187*
Thyroid gland, 181–188
embryology of, 181
histology of, 181–182, *182*
histophysiology of, 183
ultrastructure of, 182–183, 268, 276
Thyroid-stimulating hormone (TSH), 9, 10, 183, *187,* 188, 276
Thyrotropin, 202
Thyrotropin releasing hormone (TRH), 9, 183, 202, 203
Thyrotrops, 10, 200, 202
Thyroxine, 181, 188
blood levels of, 4, 9
T lymphocytes, 99
See also T cells
Toluidine blue, 7, 12
Tonofilaments, 25, 34
Tongue, in developing palate, 134, *135*
microscopic anatomy of, 131
papillae in, 132
Tonsils, 133, 138
lingual, 133
palatine, 133, 135
tubal, 258
Trabecula(e), of liver, 160
of lymph node, *102*
Trabecular meshwork, of the eye, 248, *249*
Trachea, 120–122, *120, 121, 271*
cartilage in, 122
microscopic anatomy of, 4, 120–122, *120*
Transferrin, 92
Transitional cells, of parathyroid, 184, 185
Transitional epithelium, 269, 277
of urinary bladder, 241, *241*
Transmission electron microscopy (TEM), 11, 15
TRH, *see* Thyrotropin releasing hormone
Tricarboxylic acid (TCA), 24
Triglycerides, 153, 158
Trigona fibrosa, 111
3, 5, 3,-Triiodo-L-thyronine (T_3), 183
Tritiated-thymidine (^{3}H-thymidine), 266, 275
Tropocollagen, 41–42
Tropomyosin, 61, 65, 68, 69
Troponin, 61, 65, 68, 69
Truncus arteriosus, 115
TSH, *see* Thyroid-stimulating hormone
T tubules, 65–66
Tubular viscera, anatomy of, 139–140, *140*
Tuber cinereum, 198
Tubules, *65*
dentinal, 131
distal convoluted (DCT), 8, 245
of renal cortex, *236*
of parietal cells, *143*
proximal convoluted (PCT), 3, 8, *236,* 242, 245
seminiferous, 277
uriniferous, 237–239
Tubulin, 25
Tubuli recti, of testicular drainage ducts, 227
Tunica adventitia, 9, 110
Tunica albuginea, of ovaries, 208
of penis, 229
of testes, 223
Tunica intima, 9, 110
Tunica media, 110, *113*
Tunica vaginalis, of male reproductive system, *224*
Twins, dizygotic, 208
Tympanic cavity, 257
Tympanic membrane, 257, 258, 263, 264
Tyrosinase, 174

U

Unilocular adipocytes, 44
Ureter, 236, 268, 276
Ureteric bud, 242
Urethra, *224,* 241–242
Urinary bladder, *224,* 241–242, *241*
Urinary pole, 238
Urinary system, 235–245
components of, 235
juxtaglomerular apparatus, 239–240
kidneys, 235–240
ureters, 241–242
urethra, 241–242
urinary bladder, 241–242, *241*
uriniferous tubules, 237–239
development of, 235, 242
function of, 235
Urogenital sinus, 242
Uterus, 215–216
Utricle, 258, 260, *260*

V

Vacuoles of chondrocytes, *272,* 278
of pancreatic acinar cells, *274,* 278
Vagina, 216–217
Valves, cardiac, 110
Vasa recta, *236,* 240
Vasa vasorum, 9, 111
Vascular bud, 56
Vascular channels, of bone, 54
Vascular layer, of the eye, 249–250
Vascular pole, of renal corpuscles, 238
Vein(s), arcuate, 240
central, 160, *161, 162,* 168
of liver, 3, 8
cardinal, 116
cortical, 240
function of, 115
hepatic, *162,* 168
interlobular, 240
microscopic anatomy of, 115
portal, 160, 162, *162,* 168
of the pulp, 103
renal, 240

splenic, 103
sublobular, 8
trabecular, 103, *103*
umbilical, 116
vitelline, 116
Vellus hair, 176
Vena cava, inferior, *265,* 265
Ventral mesogastrium, 154
Ventricle, primitive, 115, *265,* 265
Venules, 109, 115
Vermiform appendix, 154
Vernix caseosa, 176
Vesicles, pinocytotic, 118, 124
synaptic, 67, 75
Vessels, lymphatic, 109
Vestibular apparatus, of internal ear, 260
Vestibular glands, 217
Vestibular membrane, 263, 264, *273,* 278
of cochlea, 259, *259*
of cochlear duct, 260, *260*
Vestibular pouch, 262
Vestibule, of osseous labyrinth, 258
Villi, intestinal, 140, 149
Vitelline duct, 154
Vitreous body, of the eye, 250
VLDL, *see* Lipoproteins, very low-density
Volkmann's canals, 54

W

White fat, unilocular adipocytes in, 44
White matter, of central nervous system, 72
White pulp, in spleen, 102

Y

Yellow marrow, 89, 95
Yolk sac, 212

Z

Z line, 10, *63, 65,* 69
Zona fasciculata, 7, 276
of adrenal cortex, 189, *190,* 191, *191,* 194, 195, *195*
Zona glomerulosa, 7, 268, 276
of adrenal cortex, 189, *190,* 191
of adrenal gland, 240
Zona pellucida, of ovarian follicles, 209, *210, 211,* 219, 221
Zona reticularis, of adrenal cortex, 1, 7, 190, *190,* 191
Zonula adherens, 33–34, 36, 37, 124
Zonula occludens, 33, 36, 37, 124
Zymogen granules, 163, *164*
of pancreatic acinar cells, *165, 274,* 278